Applied Hydraulic Transients for Hydropower Plants and Pumping Stations

This book tries to be a modest homage to the civil engineers who realized, over time, those human works which are so simply named: "Hydrotechnical Constructions and Hydroelectric Plants"

Applied Hydraulic Transients

for Hydropower Plants and Pumping Stations

Mihail Popescu
Faculty of Civil Engineering
University of Constantza, Romania

Dumitru Arsenie
Faculty of Civil Engineering
University of Constantza, Romania

Paul Vlase
Technical Expert – H.P.2E. Company, France

CRC Press
Taylor & Francis Group
Boca Raton London New York

CRC Press is an imprint of the
Taylor & Francis Group, an **informa** business
A BALKEMA BOOK

Translation: Professor Adelina Georgescu, translation edited by Hubert Turner

Typesetting: Charon Tec Pvt. Ltd, Chennai, India

Library of Congress Cataloging-in-Publication Data

Popescu, Mihail, 1936-
Applied hydraulic transients for hydropower plants and pumping stations / Mihail Popescu, Dumitru Arsenie, Paul Vlase.
p. cm.
Includes bibliographical references and index.
ISBN 90-5809-395-6
1. Fluid dynamics–Mathematical models. 2. Hydraulic transients–Mathematical models. 3. Hydroelectric power plants–Romania. 4. Pumping stations–Romania. I. Arsenie, Dumitru, 1934- II. Vlase, Paul, 1941- III. Title.

TA357.P69 2003
621.2–dc21

2003042758

Published by:
CRC Press/Balkema
P.O. Box 447, 2300 AK Leiden, The Netherlands
e-mail: Pub.NL@taylorandfrancis.com
www.crcpress.com – www.taylorandfrancis.com

ISBN 13: 978-90-5809-395-0 (hbk)

Typeset by MPS Limited, Chennai, India

Contents

Preface

This book is not a standard monograph based on previously published material. In this respect, it differs from similar books in the field, which deal mainly with the methods of hydraulic computation applied to design and check the operation of hydroelectric power plants and pumping stations.

The book is primarily based on the theoretical and applied results obtained by the authors during more than 20 years of practice devoted to problems of hydraulic computation in the design of a significant number of hydroelectric power plants and pumping stations in Romania. That is why the authors have endeavoured to offer a large variety of carefully chosen case studies covering a wide range of computational methods applied and especially of the modern engineering issues on which selection of the best technical and operational hydraulic schemes is based when designing hydropower plants and pumping stations.

In the course of time the authors had the opportunity to present these scientific and technical results in more than 100 articles and papers published in Romania and abroad (United Kingdom, France, Hungary, Italy, Spain, the former Yugoslavia, U.S.A., the former Soviet Union, and so on) and to compare their performance and practical efficiency with similar papers in the literature. In fact, we paid particular attention to this in selecting the references for this book.

On the other hand, it is worth mentioning that one of the authors (M. Popescu) was awarded the "Traian Vuia" prize for Technical Sciences of the Romanian Academy in the year 1982 for a series of papers on "The theoretical and experimental study of the unsteady motions in hydraulic surge systems with application to hydroelectric power plants and pumping stations", which, in fact, are included in the present book.

This book is addressed equally to engineers and practitioners in the design, construction and operation of hydraulic structures as well as to research workers and Professors in universities. At the same time, it may be very useful to undergraduate and graduate students in engineering (hydraulic engineering, hydroenergetics and land reclamation).

Most of the elementary theory or the theory that may be found in well-known textbooks and monographs in the field has deliberately been omitted from this book. The theory has been developed only where it was considered to help the reader understand the methods of hydraulic computation and the scientific and technical information concerning practical engineering aspects of the questions studied, which represent—at least they were intended to by the authors—the main purpose of this book.

Paris, April 18, 2002 — The Authors

Preface

Introduction

Generalities

The analysis of the operation of hydraulic surge systems (hydroelectric power plants, pumping stations, industrial hydraulic installations and others) in transient regimes kindled extensive theoretical and practical interest, which explains the large number of papers devoted to this topic in our century.

The authors have been aware of the difficult task they assumed when trying a unitary approach to the problems of hydraulic computation for hydroelectric power plants and pumping stations. Whether they succeeded or not to reach their goal can be judged by the reader only. On the other hand, the unitary treatment of the computational methods and of the technical solutions for the design and operation of hydraulic schemes for hydroelectric power plants and pumping stations, which are usually presented separately, eventually proved to be an advantage. From it, the authors tried to get the highest profit, both in dealing with the specific theoretical and practical aspects and promoting modern solutions and hydraulic schemes applicable to both types of hydraulic structures.

As far as the choice of the methods of hydraulic computation is concerned, preference has generally been given to numerical methods—with their well-known computational possibilities and performances. At the same time, whenever analytical methods were found that were able to provide complimentary information on the hydraulic functioning of complex systems or to offer the advantage of quick preliminary computations for the design, these methods have been described and developed accordingly. As one may see from the contents and the structure of the book, the authors have deliberately treated the *methods of hydraulic computation* in close connection both with the *experimental results* obtained in laboratory installations and with the *practical solutions* adopted in the design of several hydroelectric power plants and pumping stations in Romania.

On the other hand, when estimating the accuracy of computational methods, the usual procedure to plot under the same graph, the results of computations and experiments in order to check the degree of their "overlapping" has not been applied. Instead, the large number of computational examples and experimental results concerning the same experimental installation enable the reader to check directly both the accuracy attained at present with the hydraulic computation methods and to apply these methods in current engineering practice.

Usually, in hydraulic surge systems the fluid flow is steady but there exist many other situations when this flow is unsteady due to perturbations induced by the change in the consumed discharges, the manoeuvres at the valves or some accidents. Hence, the unsteady motions in hydraulic surge systems are produced by the passage from a steady regime to another and are referred to as *hydraulic transient regimes*.

The hydraulic computation of the steady operation of hydraulic power plants and pumping stations is avoided since, in the authors' opinion, the textbooks and handbooks on hydraulic computation ensure a satisfactory coverage of the present engineering practice requirements regarding the design and the surveillance of these hydraulic projects in operation.

The transformation of the kinetic energy into potential energy (of pressure, of position) and conversely, occurring in a transient regime, is influenced in various ways by the elastic oscillations of the system and by the mass transfer of the fluid. In a transient regime, the liquid motion may vary slowly or rapidly according to: the nature of the perturbation and the initial conditions, the elastic characteristics of the fluid and of the respective hydraulic systems, the protection devices applied.

The rapidly varied motion is characterized by large amplitudes and frequencies of the pressure oscillations, the development of which is strongly influenced by the fluid compressibility and the elasticity of the material the pipe is made of. In the rapidly varied motion the mass transfer is negligible. Usually, the rapidly varied motion in surge pipes is called a *waterhammer* or an *undulatory motion*.

The slowly varied flow in hydraulic surge systems is a motion with small amplitudes and frequencies of the pressure oscillations but with an important mass transfer in the system, the elastic effects being negligible. The slowly varied motion in hydraulic surge systems is also called an *oscillatory motion* or a *mass oscillation*.

The book is divided into nine chapters. *Chapters 1, 2* and *3* are devoted to problems of waterhammer computation and protection systems of pumping stations against waterhammer. The introduction of the concept of *waterhammer sensitivity*, the description of a laboratory set-up enabling the experimental study of all types of unsteady motions in hydraulic surge systems met in hydraulic engineering practice and, finally, the presentation of a highly efficient protection system for pumping stations are the major topics of these chapters. In a synthetic form, the main characteristics of the protection systems used for seventy pumping stations are given.

Chapter 4 describes several computation methods of hydraulic resonance power plants and pumping stations and introduces the concept of *resonance sensitivity*. The main idea is to advocate a strategy based on resonance sensitivity computations associated with practical actions of surveillance of the resonance sensitive hydraulic systems during operation in order to avoid risks and prevent accidents.

In *Chapters 5, 6* and *7* questions related to computation and experimental study of mass oscillation in hydraulic surge systems as well as several analytical and numerical methods for analysing the hydraulic stability of hydroelectric power plants provided with surge tanks are discussed. Computational applications and experimental studies are reported for hydroelectric power plants that are already functioning for some time.

In *Chapter 8* the operation in transient hydraulic regimes of hydroelectric power plants designed on complex schemes is dealt with analytical and numerical methods, illustrated with several important hydroelectric power plants built on the rivers in Romania. The case of a pumped storage hydroelectric power plant is considered separately.

In *Chapter 9* a special class of unsteady motions in the intermediate domain between the fast and the slow motions in hydraulic surge systems is studied by analytical and experimental methods.

All the computational methods described in the book are illustrated by applications and practical solutions that refer to projects built in Romania.

A few examples of hydroelectric power plants and pumping stations in Romania

The Iron Gates I hydropower and navigation system on the Danube

The Danube is the second largest river in Europe after the Volga. With a drainage area of 817,000 km^2, the Danube runs across the territories of several countries and is suitable for navigation along 2,285 km, thus providing a beneficial link between economically rich regions. Forming a 224 km long natural border between Romania and ex-Yugoslavia, the Danube's energetic potential is considerable: it amounts to an average of 8,050 kW/km, being the largest in Europe. When passing the Iron Gates strait, the Danube's linear theoretical energetic potential is of 11,250 kW/km, due to a significant discharge (on average 5,500 m^3/s) and to a mean slope of 0.22‰. The difficult navigation conditions in this zone (high water velocity, small depth and dangerous underwater rock steps) entailed four times higher costs for transportation and a three times longer duration for the passage through the strait, as compared to other Danube reaches.

The outstanding energetic potential of this reach associated with the need to improve navigation have been the main reasons for the two countries sharing this zone—Romania and ex-Yugoslavia—to build together the Iron Gates I hydropower and navigation system. The adopted constructive solution consisted in a system of one navigation lock and one hydroelectric power plant on each of the Danube's banks, connected by an overflow dam with a 1,278 m long total impoundment front (Fig. 1).

The construction of the project "Iron Gates I" began in September 1964 and has been completed in December 1971. The overflow dam is a concrete gravity type dam with the following characteristics: maximum height 60.60 m, crest length 441 m, volume of concrete 0.7 million cubic meters, creating a reservoir with an approximate volume of 2,400 million cubic meters. The dam has been provided with 14 outlets, each 25 m

Figure 1. Iron Gates I hydroenergetic and navigation system.

long, equipped with 25 × 15 m^2 gates to ensure the water level control in the reservoir and act as flood and sludge overflow weirs; the maximum discharge capacity is of 15,400 m^3/s.

Each of the two hydroelectric power plants, located between the overflow dam and the locks, on both Danube banks (Romanian and ex-Yugoslav), cover 214 m of the Danube's impounding front and have the following characteristics: maximum head 34.50 m, installed discharge 8,700 m^3/s, installed capacity (both plants) 2,100 MW, mean annual power supplied (both plants) 10,500 GWh. Each hydroelectric power plant is provided with six groups of 178 MW vertical Kaplan turbines. The double-chamber locks are symmetrically placed against the overflow dam, their axes being perpendicular to the dam. Their characteristics are: total length of each lock 780 m; chamber length 310 m; chamber width 34 m; minimum sill depth 4.50–5.00 m; water level difference between the headrace and the tailwater 20–35 m; annual transport capacity approx. 53 million tonnes. One lock chamber can accommodate convoys of one large plus nine 1,200 tonnes tugboats or two engined two boats of 5,000 tonnes each. The passage through the lock takes 62–74 minutes. The hydroelectric power plants and the locks on the two banks operate independently of one another, according to the needs and agreements between the two countries.

The total cost of this complex project was estimated to 395 million US dollars, at the rate of 1970.

The hydroelectric power plant on the Arges River

The Arges River, flowing through a geographical zone of various relief configurations, has made possible the construction of a multiple-purpose hydraulic project. The general scheme of this complex hydraulic system on the Arges River included: the catchment of the discharges within the Arges River watershed and of the neighbouring tributaries, directing them to Vidraru reservoir; the total head was concentrated in a single underground station—the Arges HPP; the construction of 14 additional hydroelectric power plants of smaller capacity, located downstream the channel, up to the station of the city of Pitesti. The discharges of the neighbouring watersheds were concentrated in Vidraru reservoir by means of 10 secondary intake systems of conduits with a total length of 29,000 m and a cross-section area of 6–8 m^2. These inflows supply the Vidraru reservoir with a mean discharge of 19.7 m^3/s.

The Vidraru dam, belonging to the Arges HPP, is located on the upper course of the river, under special geological conditions and has the following characteristics (Fig. 2): double arch dam, 166 m high, crest length 307 m, thickness at the dam crest level 6 m, thickness at base level 25 m, total volume of concrete 480,000 m^3.

From the Vidraru reservoir, with a capacity of 465 million m^3, the water is fed through a pressure tunnel 2,130 m long and 5.15 m diameter, to an underground power station, connected to a 185 m deep vertical shaft and to a surge tank with a height of 140 m and a diameter of 7.20 m. The underground power station is located on the upper course of the Arges River, 104 m below the channel level and uses a gross head of 324 m. It has an installed discharge of 90 m^3/s and is provided with four groups of 54.5 MW Francis turbines. The total installed power is 220 MW and the annual mean power supply is 406 GWh. The turbined discharge is evacuated to the Arges River channel—under a free flow regime—through a 10,500 m long and 5.5 m diameter tailrace, continued by a free surface canal.

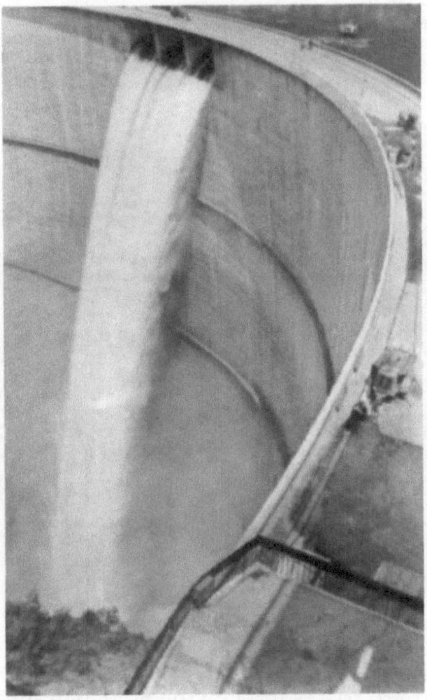

Figure 2. Vidraru dam of Arges hydroelectric power plant.

The Arges HPP is the first of a power plant chain built along the Arges, 14 other plants being located downstream the river. Out of these, five plants are of the station-dam type and nine plants are of the diversion dam type. The total installed power of the 14 stations is 179 MW and the annual mean power supply is of 347 GWh. Hence, the total installed power for the HPPs chain of the Arges River is 399 MW and the total power supply is of 753 GWh/year.

The Vidraru reservoir and the HPP chain built on the Arges River, including the seven downstream reservoirs, allow a suitable control of the hydrograph discharges as well as a flexible operation of the downstream plants and of the main plant, meeting the needs for municipal water supply in the region.

The Hydroelectric power plant 'Strimtori'

The Strimtori dam on the Firiza River has been designed and built over the period 1970–1974. With a height of 51.50 m and a crest length of 260 m (Fig. 3), this is a concrete dam 60,000 m^3 in volume, made of 15 plots of downstream open mushroom head buttresses and continued on the right-hand slope with a rockfill levee. The buttresses are provided with finished joints parallel to the downstream side of the dam and with a contraction joint parallel to the upstream side.

Three of the central plots of the dam are of overflow type; a sky jump spillway ensures the water passage over the building of the power plant and the energy dissipation, the downstream side of the dam. The hydroelectric power plant located below the sky jump spillway is equipped with a vertical shaft Kaplan turbine producing 4.2 MW.

Figure 3. Strimtori dam on the Firiza River.

The penstock ($D = 220$ mm) passes through the ninth plot mushroom and is concrete covered over its entire length except for a portion where two butterfly valves are fitted.

The Strimtori water storage belongs to a complex hydraulic project intended to serve multiple purposes. One of these: to ensure the water supply for the city of Baia Mare.

The Poarta Alba Pumping Station

The Poarta Alba pumping station belongs to the Carasu land reclamation system and is one of the largest pumping stations in Romania, with an installed discharge $Q_P = 80\,m^3/s$, geodetic pumping height $H_G = 17.5$ m and provided with five vertical axis electric pumps, each having the following characteristics: $Q_o = 16\,m^3/s$, $H_o = 18.5$ m, $n = 300$ rot/min, $N = 4{,}000$ kW (Fig. 4). The pumps discharge into five separate metallic conduits of length $L = 100$ m and diameter $D = 2{,}800$ mm. The conduits are provided with reinforcing rings and sliding supports over their length (Fig. 5).

In the upper part, in area of the pressure basin, the conduits are equipped with hydromechanical devices that operate automatically at the station's stopping manoeuvre, thus ensuring the water column breaking up whenever a reverse flow occurs from the pressure basin into the conduits. The station is not provided with electromechanical devices to close down the conduits, hence, when the pumping station is stopped, the reverse water flow that occurs until the conduits are emptied produces a reverse rotation at the pumps, though within permissible limits.

The main constructive feature of the scheme, including the water suction basin, the pumping station, the pressure basin, and the free surface channel, is that the water suction basin and the pressure basin are built in a 90° bend. In order to ensure a uniform discharge at the pump suction, underwater walls were built in the suction basin to guide the water flow. The suction basin was made after an original design as far as the sidewalls are concerned: they can take over high values of the lateral pressure of the land.

Figure 4. Poarta Alba pumping station.

Figure 5. Metallic pressure pipe ($D = 2{,}800$ mm) equipped with stiffening signs and slipping supports.

The Basarabi—Padure pumping station

This pumping station also belongs to the Carasu land reclamation system and is in fact a repumping station located downstream of the Poarta Alba pumping station. Its installed discharge is $Q_P = 76\,\mathrm{m^3/s}$; the geodetic pumping height is $H_G = 62\,\mathrm{m}$. The station is provided with eight vertical shaft centrifugal electropumps, each having the following characteristics: $Q_o = 8.2\,\mathrm{m^3/s}$, $H_o = 65\,\mathrm{m}$, $n = 375$ rot/min, $N = 7{,}500\,\mathrm{kW}$

Figure 6. Basarabi—Padure pumping station.

(Fig. 6). The metallic pressure pipes are successively reduced from eight at the station ($D = 140$ mm) to four ($D = 2{,}400$ mm) and then to two ($D = 3{,}200$ mm) over a total length $L = 380$ m.

The station is provided with hydraulic actuated valves with automatic operation at the shut-down manoeuvre. In the upper part, in area of the pressure basin, electromagnetic valves are fitted at the pipes ($D = 1{,}400$ mm), in order to break the water column and prevent the reverse flow at the station's stopping manoeuvre.

Chapter 1

The Waterhammer Phenomenon in Hydraulic Systems Under Pressure

1.1. The physics of waterhammer and its equations

In the motion of liquids, the waterhammer phenomenon appears in pressure pipes as a result of a sudden variation in the boundary conditions (for example, the rapid opening/closing of a valve, the sudden start-up/shut-down of a pump or turbine, the break of a pipe, and so on). The waterhammer calculation, namely the assessment of the highest and the lowest pressures which appear in the various sections of the pipeline, is compulsory when designing or checking the operation of a hydraulic pressure system involving such a dangerous phenomenon.

Until the development of electronic computers, the calculation of waterhammer was made by approximation, by means of simple methods or schemes. With the advent of automatic computation, the means for solving the problems related to waterhammer in hydraulic pressure systems are practically unlimited. Now, our limits lay rather in the understanding of the physical phenomenon (e.g. the waterhammer produced by the instantaneous separation and gathering of a water column, as a result of cavitation in a pipe [13, 5]), than in the lack of means for its computer simulation.

To get inside the waterhammer phenomenon, let us suppose a basic hydraulic layout consisting of one reservoir fed-pipe provided with a valve at its end (Fig. 1.1), [2, 4, 7, 9, 16, 29, 30]. The overpressure calculation for a sudden closing of the valve can be performed with Jukowski's formula:

$$\Delta p = \rho c v_o \quad \text{or} \quad \Delta p/\gamma = c v_o/g \tag{1.1}$$

The value $c = \Delta x/\Delta t$, called the celerity, is the propagation speed of the overpressure Δp, namely the propagation speed of the pressure wave. The time for the pressure wave to arrive at section A close to the reservoir is obviously ℓ/c. At that time, the whole mass of the liquid inside the pipe is in equilibrium, and the pressure is $p_o + \Delta p$. In section A, at the wall of the reservoir, the pressure is not affected by the phenomena inside the pipe and assumes the value $p_o = p_A$; the pressure inside the pipe is higher and due to this fact the fluid starts to flow with velocity v_o from the pipe to the reservoir. In the pipe a new wave starts to propagate from section A to section 0, called the reflected or back wave, which annuls the overpressure Δp.

Within the time $\tau = 2\ell/c$ the back wave reaches the valve section. For the following instant, due to inertia, the liquid inside the pipe still flows to the reservoir. In the pipe,

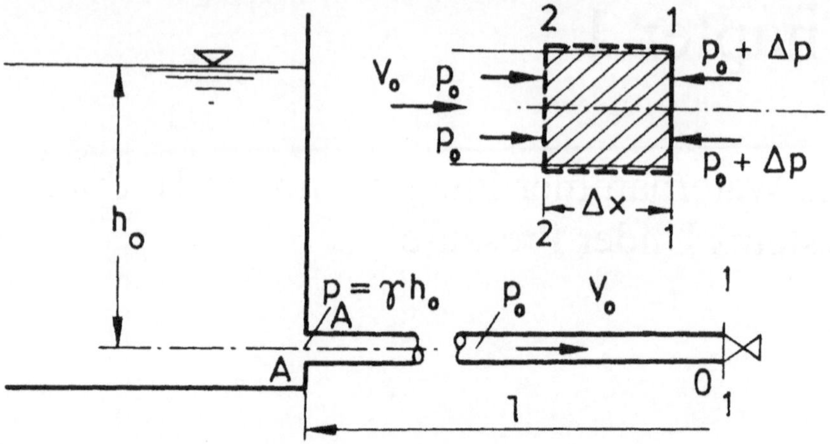

Figure 1.1. Scheme of chamber with pipe and end valve.

near the valve, a pressure drop with the value $p_o - \Delta p$ appears and propagates towards the reservoir with the same velocity c.

In the pipe, wherever the pressure is $p_o - \Delta p$, the liquid velocity is zero. At the time $\tau + \tau/2$ the back wave reaches the reservoir at section A and the whole liquid mass in the pipe has zero velocity. Due to the pressure difference between p_o and $p_o - \Delta p$, at section A, the liquid enters from the reservoir into the pipe with the velocity v_o. A new back wave of pressure p_o is born, and liquid velocity behind the wave front is v_o. At the moment 2τ, the whole liquid mass inside the pipe has the velocity v_o and the pressure p_o. This situation corresponds to the initial moment. The valve being closed, the phenomenon repeats periodically. The time τ is called the phase (reflection time). If the energy did not dissipate due to internal frictions in the liquid and due to the deformations of the pipe walls, the phenomenon would indefinitely continue. In practice, the phenomenon is soon damped due to frictions.

If the valve is closed instantaneously from the velocity v to the velocity v_o the resulting overpressure would be

$$\Delta p = \rho c(v_o - v) \quad \text{or} \quad \Delta p/\gamma = c(v_o - v)/g \tag{1.2}$$

Generally, the first wave, which propagates along the pipe due to the perturbation produced by the perturbative agent (the sudden closing/opening of the valve—either partially or completely) is called a direct wave, while its reflection is called the *reverse* or *indirect* wave.

At any pipe section there is an *incident wave*—the wave originating from the perturbative agent and coming towards the respective section, and the *reflected wave,* which propagates in the opposite direction as compared to the incident wave, and which is its transform.

The direct phase of the waterhammer represents the time interval between the moment when the first direct wave reaches a given section and the moment when its transform reaches the same section. *The indirect phase* of the waterhammer at the same section is the time interval between the moment when the reflected wave passes that section and the moment when the direct wave reaches it.

The propagation speed (celerity) of the waterhammer in the pipe liquid is expressed as

$$c = \frac{\sqrt{\frac{\varepsilon}{\rho}}}{\sqrt{1 + \frac{D\varepsilon}{eE}}} = \frac{c_o}{\sqrt{1 + \frac{D\varepsilon}{eE}}} \tag{1.3}$$

where D is the diameter of the pipe, e the thickness of the pipe walls, ε the cubic Young modulus of the liquid, E the Young modulus of the material the pipe is made of, and $c_o = \sqrt{\varepsilon/\rho}$ is the propagation velocity of the waterhammer in a pipe with rigid walls ($E \to \infty$), respectively the sound propagation velocity in liquids (e.g. $\varepsilon = 2.1 \times 10^4$ kgf/cm^2 for water, $E = 2.1 \times 10^6$ kgf/cm^2 for steel, so that in steel pipes $c \cong 1{,}000$ m/s).

Actually, the waterhammer phenomenon is much more complex, not only because of the frictions which cause the damping of pressure oscillations, but also due to secondary effects which appear and superimpose on the primary ones, as the reflections and refractions of the waves at the pipe elbows or at the variations in the magnitude of sections, or the oscillations of the valve or pipe, or the valve leakage etc. [10, 25, 26].

The waterhammer equations consist of the continuity and momentum equations. The basic hypotheses adopted when writing these equations are:

- the flow is one-dimensional;
- the head losses are expressed by the same formulas as in case of the steady flow.

Usually, the waterhammer equations are written as (Figs. 1.2 and 1.3 [24])

$$\begin{aligned} \frac{\partial H}{\partial x} &= \frac{1}{g}\frac{\partial V}{\partial t} + \frac{\lambda}{2gD} V|V| = 0 \\ \frac{\partial H}{\partial t} &+ \frac{c^2}{g}\frac{\partial V}{\partial x} = 0 \end{aligned} \tag{1.4}$$

Note that the term $(\lambda/2gD)V|V|$ is the head loss along the unit length as given by Darcy's formula.

Actually, the pipelines are not perfectly elastic, thus apart of the friction losses, the elasto-plastic behaviour of the pipe also contributes to the damping of the waterhammer effect. Recent studies concerning this problem lead either to finding lower celerity values and a more accurate method of waterhammer computation [8], or necessarily introduced a dissipation term into the flow equations due to the structural head losses [29]. The cavitation phenomenon itself, together with the water column separation and gathering in a very short time interval, is an additional cause for waterhammer damping [25, 27].

Recently, attention was focussed on the waterhammer in two-phase bubbly air–water mixtures [6, 11, 12, 14]. For example, theoretical and experimental studies concerning the variation of the celerity as a function of α (the percentage of free air) and as a function of the pressure inside the pipe p/γ, in two-phase bubbly air–water mixtures, reveal significant results as plotted in Fig. 1.4 [6, 26, 29, 32]. In this case, the equations themselves need some specific changes. In the following, we shall briefly analyse the case when the head losses in the pipe are neglected, which is practically admissible for

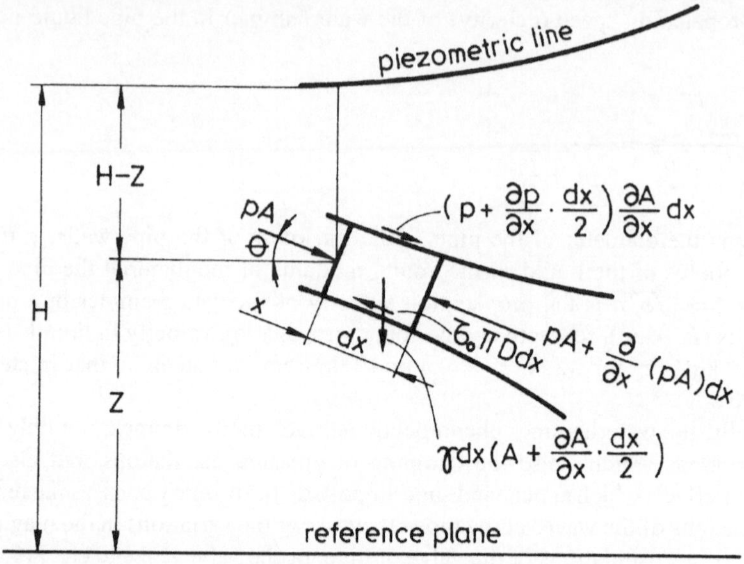

Figure 1.2. Notation for the derivation of the dynamics equation.

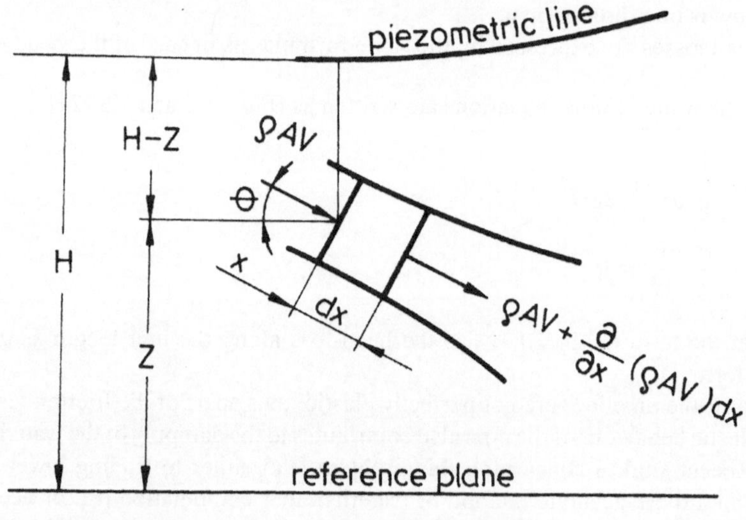

Figure 1.3. Notation for the derivation of the continuity equation.

the first pressure oscillation. The flow equations take the simplified form

$$\frac{\partial H}{\partial x} + \frac{1}{g}\frac{\partial V}{\partial t} = 0$$

$$\frac{\partial H}{\partial t} + \frac{c^2}{g}\frac{\partial V}{\partial x} = 0 \tag{1.5}$$

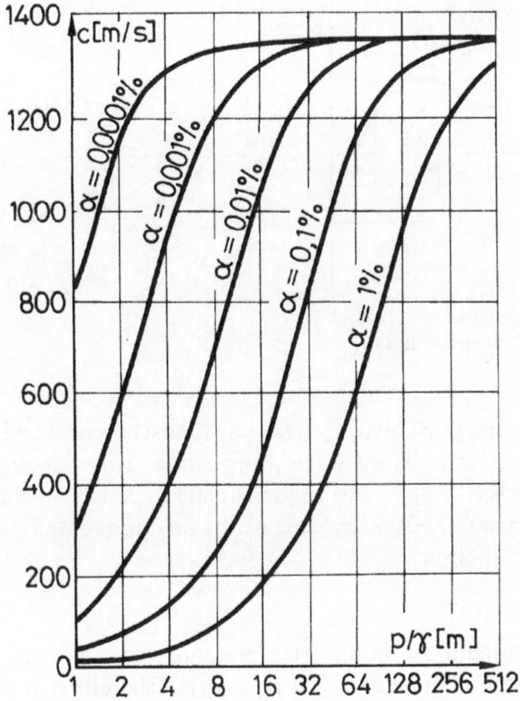

Figure 1.4. Influence of the free air and of the pressure in the pipe on the celerity magnitude.

As it is well known [16], by eliminating in turn the variables V and H between these equations, we obtain

$$\frac{\partial^2 H}{\partial t^2} = c^2 \frac{\partial^2 H}{\partial x^2} \tag{1.6}$$

and respectively

$$\frac{\partial^2 V}{\partial t^2} = c^2 \frac{\partial^2 V}{\partial x^2} \tag{1.7}$$

which are linear second-order hyperbolic equations [4, 8].

The general solution is obtained by means of two arbitrary functions F and f under the form of an associated system of waves

$$\begin{aligned} H - H_o &= F(x - ct) + f(x + ct) \\ V_o - V &= \frac{g}{c}[F(x - ct) - f(x + ct)] \end{aligned} \tag{1.8}$$

where H_o and V_o are the values of H and V at the initial time $t = 0$. The physical meaning of function $F(x - ct)$ is obtained by taking $f = 0$ and by considering an observer moving with a constant speed given by the law

$$x - ct = x_1 - ct_1 = \text{const} \tag{1.9}$$

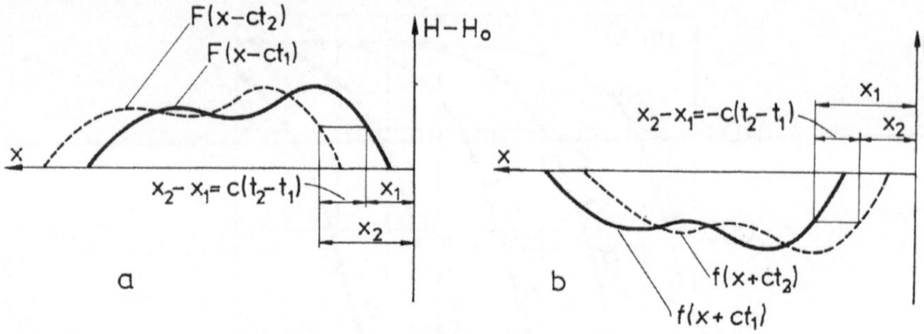

Figure 1.5. Plots of curves F and f.

The first relation in eqn. (1.8) gives $H - H_o = F(\text{const}) = \text{const}$ and, as a consequence, function F is the direct pressure wave propagating with velocity c in the positive direction of the x-axis ($t < t_1, x > x_1$) and maintains a constant value (of the velocity).

Similarly, by taking $F = 0$ and by considering an observer uniformly moving under the following law of motion

$$x + ct = x_1 + ct_1 = \text{const} \tag{1.10}$$

it follows that the function f is the back pressure wave moving with velocity c in the negative direction of the x-axis ($t > t_1, x < x_1$). Therefore, if at an instant $t = t_1$, the pressure wave has a pressure distribution $F(x - ct_1)$ along the x-axis, this wave propagates without distortions in the positive direction of the x-axis with the velocity c (Fig. 1.5a). Also, the pressure wave, which at the moment t_1 has a distribution $f(x + ct_1)$, propagates along the x-axis, but in the negative direction, with velocity c, without distortions (Fig. 1.5b).

The pressure waves are undeformed due to the neglect of head losses along the pipe.

1.2. Waterhammer computation methods

As revealed by the papers devoted to hydraulic pressure systems there are many valuable theoretical and methodological approaches to compute the unsteady motion of fluids. The use of one method or another is at the choice of the person who performs the specific computations, the only requirement being to ensure the accuracy of the results. The waterhammer computation methods fall into many categories as a function of their complexity. The complexity of the chosen computation method is related to the importance of the hydraulic installation and to the way in which waterhammer acts. Generally, the methods are of two kinds, namely the exact ones and approximate ones. The approximate methods consist in the use of graphical curves or computation tables obtained by the synthesis of some exact and experimental results, of some approximate formulas or analogical simulation. In the following, a brief discussion of the known exact methods is presented, though many of them are already well known [3, 8, 24].

The *arithmetic method* was used until 1930, when the graphical methods started to be widely used. This method neglects the effect of friction and takes into account only the horizontal pipe. Actually the equation of the pressure rise due to a variation of velocity is solved for specific boundary conditions, which are expressed under simple forms.

The equation $\Delta H = c\Delta V/g$ is solved taking into account the boundary conditions; the results can be obtained in a table for each time step, underlining the process of the wave reflection in the hydraulic system under study. This method requires a very simple theoretical approach, and a slide-rule or a mechanical computing machine is quite enough to perform the computation. The arithmetic method has the disadvantage of requiring a couple of tests concerning the boundary conditions (the law of the valve closing).

The *Schnyder–Bergeron graphical method*, though known under this name, was actually founded by Loewy and thereafter, during 1930–1935, Schnyder and Bergeron developed it. This graphical method neglects the friction during the theoretical treatment, but still uses the necessary means to take it into account by a correction. This aspect is underlined in reference [3], where it is shown that by applying the Schnyder–Bergeron graphical method to the class of periodic movements, it is necessary to take into account, even in a simplified form, the energy dissipation as a dissipative term which ensures the correction of the Schnyder–Bergeron solution. The Schnyder–Bergeron method allows one to consider the longitudinal and local head losses at some points (namely, the concentration of the losses at some sections) and the mutual transformations between the kinetic and potential energies. The practical way of taking into account these losses is also a common part of other computation methods such as the method of computation waves or the method of characteristics.

The Schnyder–Bergeron graphical method is the easiest and the most practical method to analyse the unsteady flow in hydraulic pressure systems, after the 1930s. This method has the advantage of offering a visualisation of the waterhammer phenomenon during the solution of the problem itself. Also, it allows various kinds of boundary conditions and can be easily used for single pipes. The main disadvantage of the method is the fact that great errors may be involved resulted from the graphical arrangement of the computations themselves.

The *method of physical waves* is, from the mathematical point of view, the same as the method of effect superposition. With this method, in order to compute the flow at a given section of the hydraulic system and at a given moment, one has to add the effects of all the direct and back waves travelling the given section to the initial flow and pressure values. The method might be applied to inviscid fluids, by using simple computation equations following the hypothesis of constant wave propagation velocity. The method of physical waves has the advantage of being simple and of requiring a small amount of computations, when only the first stages of the transient flow are envisaged.

The *method of computation waves* might be considered as a variant of the method of characteristics. According to this method, the propagation of the physical waves of pressure $\zeta_0(H)$ and speed $\zeta_0(V)$ is replaced by the propagation of some invariant mathematical expressions containing the global values of parameters H and V, not the variation of these parameters. Riemann graphically solved this problem in the physical plane and Schnyder and Bergeron synthesised it for the waterhammer in the hodograph plane. It is worth saying that this method lies at the basis of some numerical methods for waterhammer computation.

The *algebraic method* is another variant of the method of characteristics. It uses the two equations of the boundary conditions along the entire pipe. That is why two non-linear algebraic equations are obtained for each pipe in the pressure hydraulic system under study. The two equations for each pipe together with the boundary conditions allow the solutions to be determined by using equal time intervals. If the term standing for the head losses is neglected, from a mathematical point of view, the algebraic method is equivalent to the graphical method. The use the algebraic method on a computer requires some skill for taking into account simultaneously all the non-linear equations of the system.

The *method of impedance* is less used because it is suitable only for time-periodic unsteady flow. This method is based on the theory of electrical networks and uses the friction term in its non-linear development for the mean velocity value and the linearised expression of the friction for the oscillatory component. This method has the advantage of being easily used in the analysis of pipe systems, allowing the computation of the fundamental period and its harmonics. A short computational time is required to obtain the solution, and the complex boundary conditions can be treated by means of Fourier series.

The *Lax–Wendorff finite-difference method* is mainly applied to compute the transient flow in open channel hydraulic systems, but good results are also obtained when applied to the unsteady flow in hydraulic pressure systems, to two-phase water–air mixtures, or to cavitation phenomena with water column separation [12]. Practically, when using this method, the flow equations are written under their conservation (divergence) form and a two-stage explicit scheme is applied with second order accuracy (Fig. 1.6). The first stage is an intermediate one which uses Lax' scheme with a first-order approximation for the time interval $(t, t + \Delta t)$; the second stage is applied during the time interval $(t, t + 2\Delta t)$ with a second-order approximation. The method has the advantage of enabling us to compute shockwaves (with discontinuities), but it still needs some adjustments with correction coefficients yielded by experiments.

The *method of characteristics* will be separately presented in the next section.

Analytical methods. Besides the computation methods briefly presented above, which mainly deal with numerical simulation, there exist many analytical methods in the literature still used in the theoretical study of waterhammer [15]. As a rule, analytical methods do not offer suitable solutions to practical problems in engineering,

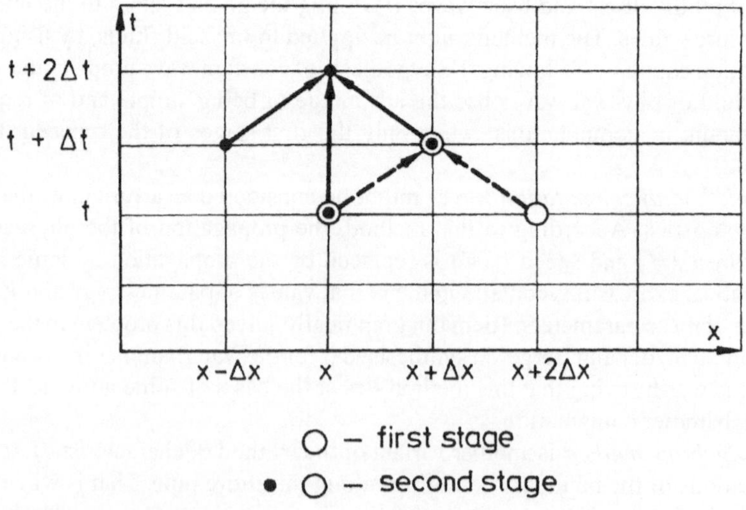

Figure 1.6. Computation scheme of the Lax–Wendorff method in two steps.

but they represent a subtle and very effective means for discovering new qualitative aspects of waterhammer and providing test cases for other methods.

1.3. Waterhammer computation by the method of characteristics

The performance of modern computers made possible the numerical integration of waterhammer equations in almost all practical cases, i.e. in complex hydraulic pressure systems with a very large variety of boundary conditions. Aiming at the numerical integration of waterhammer equations, a great number of computer codes were issued, most of them based on the method of characteristics [5, 9, 23, 24]. Thus, the method of characteristics is now the most widely used to numerically simulate the waterhammer phenomenon.

Among the advantages of this method it is worth mentioning: the accuracy of the computed results, greater than with other methods (by accounting for the influence of small terms in the flow equations, the head losses can be considered either concentrated or uniformly scattered), the facility in the treatment of boundary conditions and the versatility in using complex pipe networks [28].

Using the method of characteristics, the system of flow equations

$$\frac{\partial H}{\partial x} + \frac{1}{g}\frac{\partial V}{\partial t} + \frac{\lambda}{2gD}V|V| = 0 \tag{1.11}$$

$$\frac{\partial H}{\partial t} + \frac{c^2}{g}\frac{\partial V}{\partial x} = 0 \tag{1.12}$$

which is a system of quasi-linear partial differential equations, can be transformed into a system of four equations with total derivatives [8, 24]. In the case of the waterhammer, when $V \ll c$, the four equations can be written in the form

$$\frac{\mathrm{d}x}{\mathrm{d}t} = \pm c \tag{1.13}$$

$$\pm \frac{g}{c}\frac{\mathrm{d}H}{\mathrm{d}t} + \frac{\mathrm{d}V}{\mathrm{d}t} + \frac{\lambda}{2D}V|V| = 0 \tag{1.14}$$

The plus sign in eqn. (1.14) is for the direct characteristic line (C^+), while the minus sign concerns the other characteristic line (C^-) (Fig. 1.7).

As in the usual computation practice the celerity c is constant for a given pipe, eqn. (1.13) represents straight lines in the $(x,\ t)$ plane, and along these lines the eqns. (1.14) are consistent. When the pipeline consists of several pipe runs of various diameters, the celerity is supposed to be constant on each pipe segment of constant diameter.

Using the finite difference scheme between the points P and A and the points P and B in Fig. 1.8 for eqns. (1.13) and (1.14), one obtains the following expressions [26]:

$$V_\mathrm{P} - V_\mathrm{A} + \frac{g}{c}(H_\mathrm{P} - H_\mathrm{A}) + \frac{\lambda}{2D}V_\mathrm{A}|V_\mathrm{A}|\Delta t = 0 \tag{1.15}$$

$$x_\mathrm{P} - x_\mathrm{A} = c(t_\mathrm{P} - t_\mathrm{A}) \tag{1.16}$$

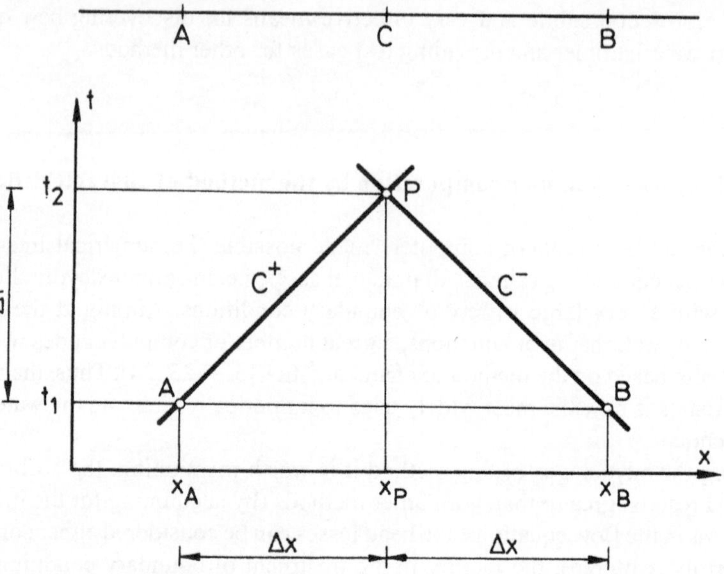

Figure 1.7. Definition of characteristic curves.

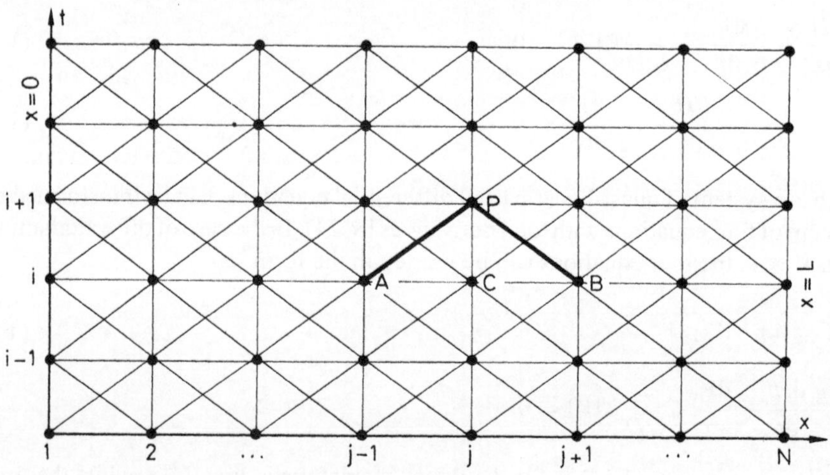

Figure 1.8. Rectangular network of characteristics.

for the *direct characteristic line* C^+, and

$$V_P - V_B - \frac{g}{c}(H_P - H_B) + \frac{\lambda}{2D}V_B|V_B|\Delta t = 0 \tag{1.17}$$

$$x_P - x_B = -c(t_P - t_B) \tag{1.18}$$

for the *indirect characteristic line* C^-. Using eqns. (1.15)–(1.18) for the entire pipe, eventually divided into segments of constant diameter, a network of characteristics is obtained, which allows solving the system of equations at each node of the network (Fig. 1.8).

Starting with a known initial condition at $t = t_o$, i.e. with known values for H and V at each point of the computation network, and defining the computation time given by eqns. (1.16) and (1.18)

$$\Delta t = \frac{\Delta x}{c} = t_P - t_A = -(t_P - t_B) \tag{1.19}$$

the characteristic lines of the initial computation network intersect at some other grid points. For example, the characteristic lines emerging from points A and B (Fig. 1.8) meet themselves at point P which belongs to the section containing the point C.

Using this scheme, to compute the velocity V_p and the head H_p at each section j, one needs to solve the system of eqns. (1.15) and (1.17), the result being [26]

$$\begin{aligned} V_{j,i+1} &= \frac{1}{2}\left[V_{j-1,i} + V_{j+1,i} + \frac{g}{c}\left(H_{j-1,i} - H_{j+1,i}\right)\right] \\ &\quad - \frac{\lambda \Delta t}{2D}\left(V_{j-1,i}\left|V_{j-1,i}\right| + V_{j+1,i}\left|V_{j+1,i}\right|\right) \end{aligned} \tag{1.20}$$

or, respectively

$$\begin{aligned} H_{j,i+1} &= \frac{1}{2}\left[H_{j-1,i} + H_{j+1,i} + \frac{c}{g}\left(V_{j-1,i} - V_{j+1,i}\right)\right] \\ &\quad - \frac{c\lambda \Delta t}{2gD}\left(V_{j-1,i}|V_{j-1,i}| - V_{j+1,i}|V_{j+1,i}|\right). \end{aligned} \tag{1.21}$$

Thus, the V and H values are computed at the next $(i+1)\Delta t$ time step as a function of the known values V and H at the previous time $i\Delta t$, starting from the initial given values at the time t_o. When performing the numerical integration the well-known [9] Courant condition

$$\Delta t = (1 - \varepsilon)\frac{\Delta x}{c} \tag{1.22}$$

with $\varepsilon = 0.01$–0.005 must be taken into account as a function of the used digit accuracy. It is worth saying that a relation like eqn. (1.22) can be sufficient for computing the largest magnitudes of waterhammer even for $\varepsilon > 0.01$, but it is not always correct when studying the hydraulic stability of operating pressure systems.

Practical computation of the waterhammer in hydraulic pressure systems revealed the following boundary conditions for example, in the layout of a pumping station (Fig. 1.9), where the following notations are used: BA water feed basin, P pump, C valve, RA air chamber, CE surge tank, BR overflow basin.

(a) *Upstream (downstream) end section with programmed valve closing* (Fig. 1.10). In this case two equations are available, one describing the valve closure law and the other, eqn. (1.15) or eqn. (1.17). The two unknowns are $V_{N,i+1}$ and $H_{N,i+1}$. From the valve closure equation the velocity $V_{N,i+1}$ is determined and from eqn. (1.15) one obtains (assuming a downstream end section):

$$H_{N,i+1} = H_{N-1,i} - \frac{c}{g}\left(V_{N,i+1} - V_{N-1,i} + \frac{\lambda}{2D}V_{N-1,i}\left|V_{N-1,i}\right|\Delta t\right) \tag{1.23}$$

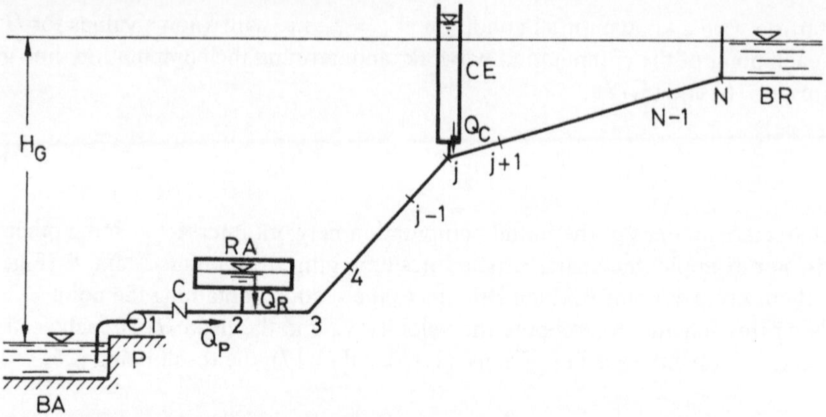

Figure 1.9. Scheme of the pumping station with a surge tank and air chamber.

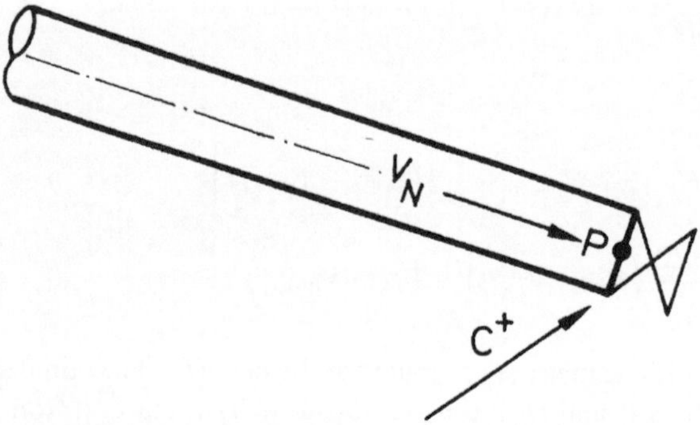

Figure 1.10. End section with valve.

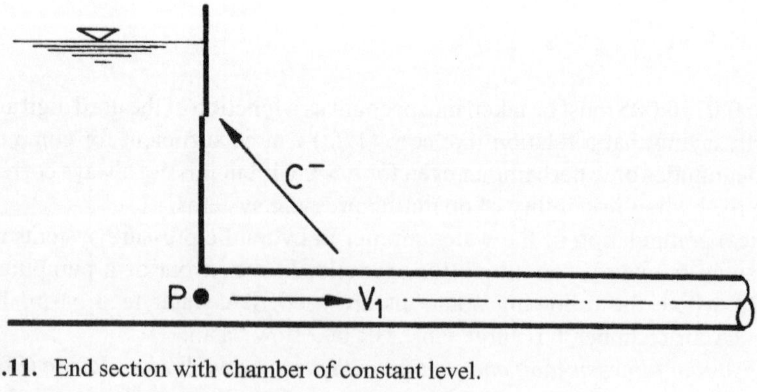

Figure 1.11. End section with chamber of constant level.

(b) *Upstream (downstream) end section with constant level reservoir* (Fig. 1.11). In this case $H_{1,i+1} = H_{am} = \text{const}$ and it represents the pressure head with respect to a reference plane. From eqn. (1.17) it follows

$$V_{1,i+1} + V_{2,i} + \frac{g}{c}\left(H_{1,i+1} - H_{2,i}\right) - \frac{\lambda}{2D} V_{2,i}\left|V_{2,i}\right| \Delta t \qquad (1.24)$$

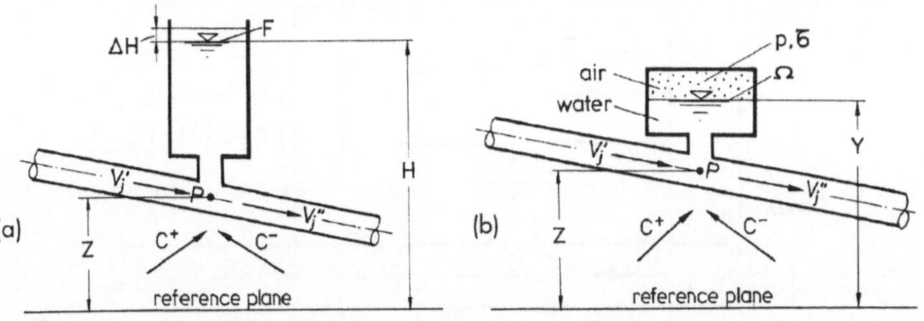

Figure 1.12. Computation section with: (a) surge tank, (b) air chamber.

(c) *Section with surge tank* (Fig. 1.12a). Such sections are intermediate and involve three unknowns, namely the upstream and downstream velocities $V'_{j,i+1}$ and $V''_{j,i+1}$ respectively and the water elevation in the tank $H_{j,i+1}$ (at this computation stage the head losses at the pipe connection are neglected). There are three equations, namely eqns. (1.15) and (1.17) and the continuity equation at the surge tank node, which in finite differences can be written as

$$\frac{\Delta H}{\Delta t} = \frac{1}{F(H)}\left(V'_{j,i+1}f_1 - V''_{j,i+1}f_2\right) \tag{1.25}$$

where $F(H)$ stands for the surge tank cross section area at elevation H and time $i+1$, and f_1, f_2 are the pipe sections upstream and downstream of the tank, respectively.

(d) *Section with air chamber* (Fig. 1.12b). In this case there are five unknowns: the velocities upstream and downstream of the computation section, $V'_{i,i+1}$ and $V''_{i,i+1}$, the piezometric head $H_{i,i+1}$, the air cushion pressure $p_{i,i+1}$ and the air cushion volume $\tau_{j,i+1}$. In addition to eqns. (1.15) and (1.17) there are the following equations specific to the air vessel

$$\begin{aligned} H_{j,i+1} &= Y_{j,i+1} + \frac{p_{j,i+1}}{\gamma} \\ p_{j,i+1}\tau^n_{j,i+1} &= p_{j,i}\tau^n_{j,i} \\ \tau_{j,i+1} &= \tau_{j,i} + \Omega(Y)[V_{\mathrm{r},i+1} + V_{\mathrm{r},i}]\frac{\Delta t}{2} \end{aligned} \tag{1.26}$$

where the following notation is involved: p and i are the pressure and the volume of the air cushion, respectively, Ω is the area of the horizontal section of the air tank at elevation Y, V_r is the rising water velocity in the tank and n is the polytropic coefficient ($n = 1$–1.4).

For the sections with surge tank or air vessel, where usually the pressure losses in the pipe connections are also taken into account, the usual relation

$$H_{i,i+1} - H_{\mathrm{d},i+1} = MQ_\mathrm{d}\,|Q_\mathrm{d}|$$

is available, where M is the head loss coefficient in the pipe connection, and H_d, Q_d are the piezometric head and the inlet/outlet rate of flow of surge tank or of the air tank respectively.

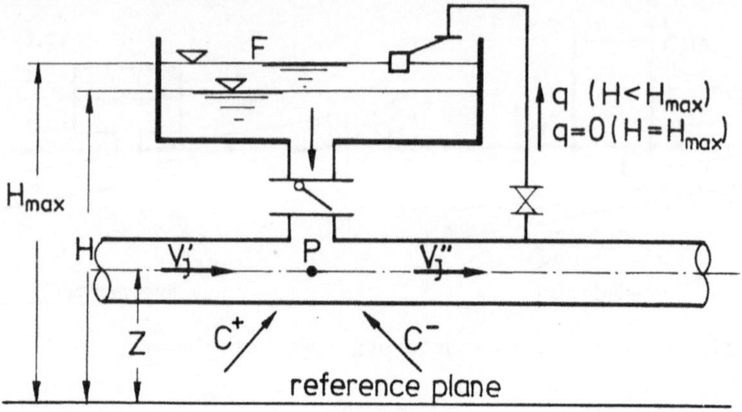

Figure 1.13. Computation section with one-way reservoir.

(e) *Section with one-way reservoir*. This kind of protection system is used on the discharge pipes of pumping stations and it is recommended especially for use at high altitudes, where abrupt changes in the longitudinal slope of the discharge pipes occur.

In addition to the surge tank, the one-way reservoir is fitted with a check valve which acts as a non-return valve, and with a connection pipe of small diameter which ensures the filling of the reservoir with water (after its dewatering) until a certain maximum level is reached (Fig. 1.13).

The one-way reservoir operates as soon as the pressure in the conduit is below a certain limit $p_P/\gamma < (H - z)/\gamma$, and during its operation it acts as a surge tank (the first phase). Immediately as the pressure in the duct rises over the imposed limit, the check valve is closed, and the operation of the one-way reservoir is stopped (the second phase). As far as the computation of a section with one-way reservoir is concerned, two algorithms are used: for the first phase the algorithm of the surge tank, and for the second phase, the algorithm of an ordinary section.

(f) *Section with over-pressure valve*. The over-pressure valve is used to limit the water pressure into the duct, when fitted. The greater the effect of pressure limitation, the higher the flow for the same value of the over-pressure.

The valve is characterized by a discharge-pressure law of the general form $H = \Phi(\Delta Q)$, and the condition $H_{j,i+1} > H_{\min}$ is satisfied during operation. The efficiency of the valve greatly decreases as the pressure rises over the pressure value which fully opens it, and starts acting as an ordinary hydraulic gate in that section.

Such sections, which are intermediate sections (Fig. 1.14), involve four unknowns: the velocities $V'_{j,i+1}$ and $V''_{j,i+1}$ in the sections upstream and downstream, the piezometric level $H_{j,i+1}$ and the discharge evacuated through the valve ΔQ. Thus, there are four equations, namely (1.15), (1.17) and the following two equations due to the specific action of the valve

$$\begin{aligned} \Delta Q &= V'_{j,i+1} f_1 - V''_{j,i+1} f_2 \\ H_{j,i+1} &= \Phi\left(V'_{j,i+1} f_1 - V''_{j,i+1} f_2\right) \end{aligned} \tag{1.27}$$

(g) *Section with pipe branching* (Fig. 1.15). Using the same notation as in the above lines, the values $V_{j1,i+1}$, $V_{j2,i+1}$, $V_{j3,i+1}$ and $H_{j,i+1}$ of this section are computed from the

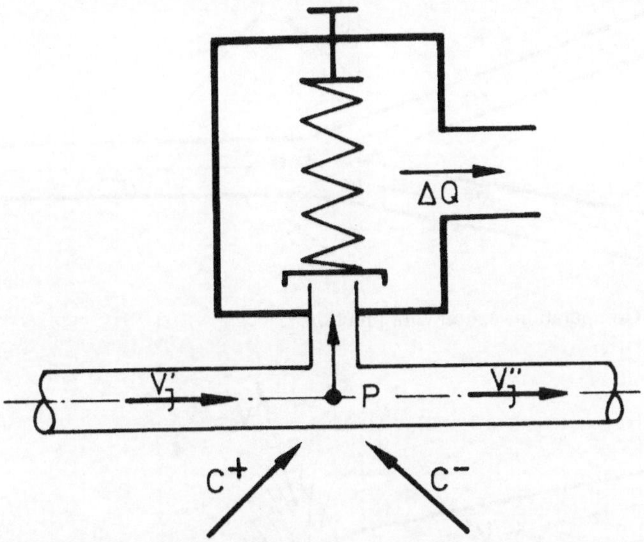

Figure 1.14. Computation section with over-pressure valve.

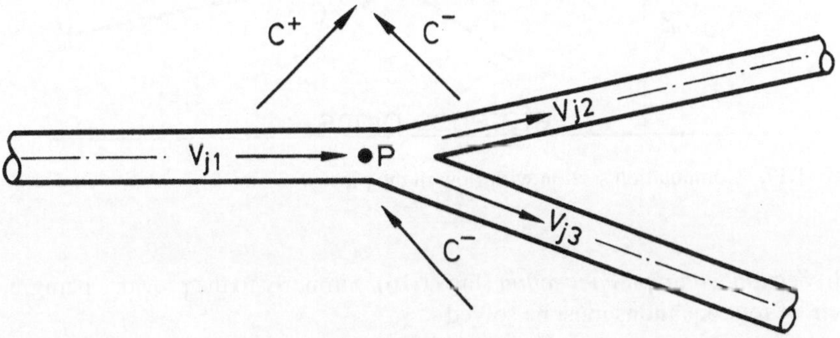

Figure 1.15. Computation section with branching.

following system of equations

$$
\begin{aligned}
&V_{j1,i+1} - V_{j1-1,i} + \frac{g}{c}\left(H_{j,i+1} - H_{j1-1,i}\right) + \frac{\lambda \Delta t}{2D} V_{j1-1,i}\left|V_{j1-1,i}\right| = 0 \\
&V_{j2,i+1} - V_{j2+1,i} - \frac{g}{c}\left(H_{j,i+1} - H_{j2+1,i}\right) + \frac{\lambda \Delta t}{2D} V_{j2+1,i}\left|V_{j2+1,i}\right| = 0 \\
&V_{j3,i+i} - V_{j3+1,i} - \frac{g}{c}\left(H_{j,i+1} - H_{j3-1,i}\right) + \frac{\lambda \Delta t}{2D} V_{j3+1,i}\left|V_{j3+1,i}\right| = 0 \\
&V_{j1,i+1} f_1 = V_{j2,i+1} f_2 + V_{j3,i+1} f_3
\end{aligned}
\tag{1.28}
$$

where f_1, f_2, f_3 are the cross-section areas of the ducts at node j.

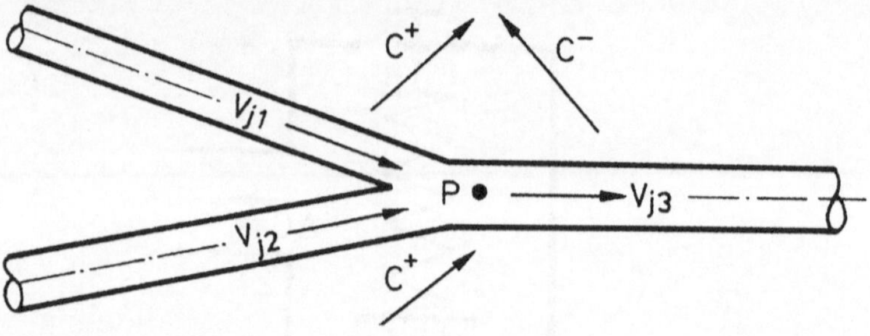

Figure 1.16. Computation section with junction.

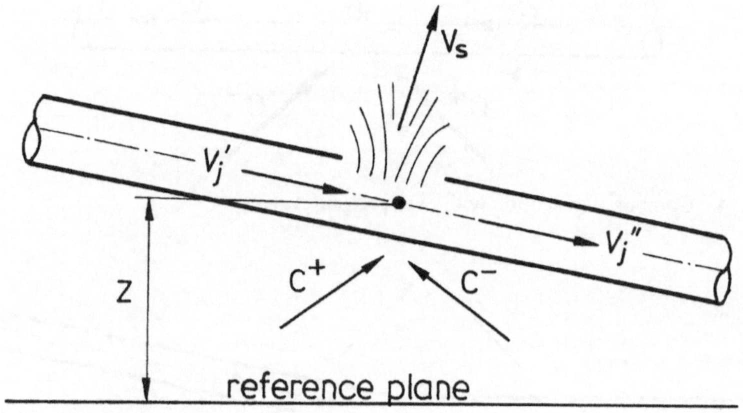

Figure 1.17. Computation section with flow in the pipe.

(h) *Section with pipe-connection* (Fig. 1.16). Similarly to the previous paragraph, a system of four equations must be solved

$$
\begin{aligned}
&V_{j1,i+1} - V_{j1-1,i} + \frac{g}{c}\left(H_{j,i+1} - H_{j1-1,i}\right) + \frac{\lambda \Delta t}{2D} V_{j1-1,i}\left|V_{j1-1,i}\right| = 0 \\
&V_{j2,i+1} - V_{j2-1,i} + \frac{g}{c}\left(H_{j,i+1} - H_{j2-1,i}\right) + \frac{\lambda \Delta t}{2D} V_{j2-1,i}\left|V_{j2-1,i}\right| = 0 \\
&V_{j3,i+1} - V_{j3+1,i} - \frac{g}{c}\left(H_{j,i+1} - H_{j3+1,i}\right) + \frac{\lambda \Delta t}{2D} V_{j3+1,i}\left|V_{j3+1,i}\right| = 0 \\
&V_{j1,i+1} f_1 + V_{j2,i+1} f_2 = V_{j3,i+1} f_3
\end{aligned}
\tag{1.29}
$$

(i) *Section in a draining duct* (Fig. 1.17). This is the case of a duct with leakages due to a broken wall or to a controlled outflow. To compute the values $V'_{j,i+1}$, $V''_{j,i+1}$, V_S and $H_{j,i+1}$, the eqns. (1.15), (1.17) and the following relations specific to this case

$$
\begin{aligned}
V_{s,i+1} &= \sqrt{2g\left(H_{j,i+1} - Z_j\right)} \\
V'_{j,i+1} f_1 &= V''_{j,i+1} f_1 + V_{s,i+1} f_s
\end{aligned}
\tag{1.30}
$$

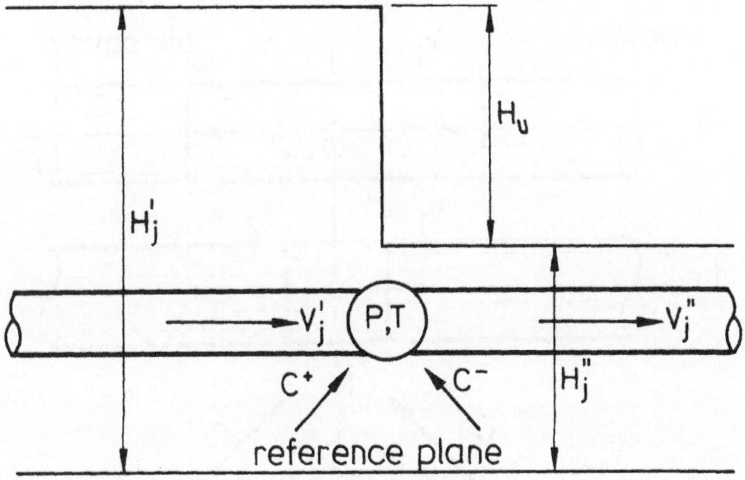

Figure 1.18. Computation section with pump of turbine.

are used, where V_s is the speed in the draining section and f_s is the area of the draining section.

(j) *Section with pump or turbine* (Fig. 1.18). To compute the values $V'_{j,i+1}$, $V''_{j,i+1}$, $H'_{j,j+1}$, $H''_{j,i+1}$ and $\Delta\bar{\omega}$, eqns. (1.15), (1.17) and the following three relations

$$
\begin{aligned}
V'_{j,i+1} f_1 &= V''_{j,i+1} f_2 \\
\Delta\omega &= \pm \frac{M}{J} \Delta t \\
H_{\text{useful},i+1} &= H'_{j,i+1} - H''_{j,i+1} = f(Q, n)
\end{aligned}
\tag{1.31}
$$

are applied, where the usual notation is used: $\Delta\bar{\omega}$ is the decrease (increase) of the rotational speed of the pump (turbine), M is the hydraulic friction (active) couple and $J = GD^2/4g$ is the inertia couple of the rotating shaft. Let us note that the characteristics of the rotating pump are used in the form $H = f(Q, n)$, by taking into account the speed n corresponding to the computation time. This equation is actually a family of characteristic curves belonging to the given pump (turbine).

The system of eqns. (1.15), (1.17) and (1.31) is usually solved by a simple iterative procedure.

(k) *Section with cavitation*. At the nodes where the pressure in the duct decreases below the vaporization pressure (H_v), cavitation phenomena take place as a result of giving off the gases dissolved in water. Whenever the stress occurring inside the liquid exceeds a certain limit, and the hole made by water vaporization and water-dissolved gas is located, for example at a higher position of the duct, the water column breaks. Inside the hole a vacuum sets in. In the first stage (Fig. 1.19a) the cavitation hole increases its volume up to a maximum value. During the second stage, its volume decreases and the water column forms again (Fig. 1.19b) and a huge over-pressure appears inside the duct (often accompanied by a big noise heard as a metallic kick).

As far as the computation of such a section with cavitation is concerned, eqns. (1.15), (1.17) and a relation of limitation of the minimum pressure in the duct to $H = H_v$ are

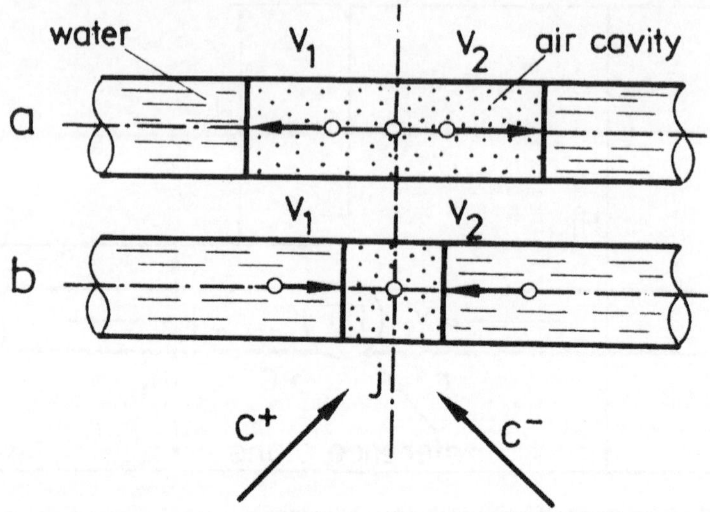

Figure 1.19. Computation section with cavitation.

considered in a first approximation. For a more accurate computation, the volume of hole (V_{cav}) is computed after water column breakdown; this volume is controlled until the water column settles again ($V_{cav} = 0$)

$$\Delta V_{cav} = \pm(V_1 f_1 + V_2 f_2)\Delta t \tag{1.32}$$

Recently, theoretical and experimental studies have attempted to take into account the complexity of waterhammer phenomena in the presence of cavitation [11, 12, 25, 26].

If the propagation time of the pressure wave between two computing sections is not the same (e.g. pipes of different lengths and diameters), calculations are performed by searching new sections through interpolating H and V with respect to position x at time $i\Delta t$, in such a manner that at time $(i+1)\Delta t$ the direct and inverse characteristics interact in the imposed computation sections. A grid of *fixed and adjustable nodes* is thus obtained (Fig. 1.20). In this case, eqns. (1.15)–(1.18) take the form

$$V_P - V_R + \frac{g}{c}(H_P - H_R) + \frac{\lambda}{2D} V_R\,|V_R|\,(t_P - t_R) = 0 \tag{1.33}$$

$$x_P - x_R = (V_R + c)(t_P - t_R) \tag{1.34}$$

for the direct characteristic C^+ and

$$V_P - V_S + \frac{g}{c}(H_P - H_S) + \frac{\lambda}{2D} V_S\,|V_S|\,(t_P - t_S) = 0 \tag{1.35}$$

$$x_P - x_S = (V_S - c)(t_P - t_S) \tag{1.36}$$

for the inverse characteristic C^-.

The values V_R, V_S, H_R and H_S are computed by linear interpolation (Fig. 1.21)

$$\begin{aligned} V_R &= V_A + k_R(V_C - V_A); \qquad V_S = V_B + k_S(V_C - V_B) \\ H_R &= H_A + k_R(H_C - H_R); \qquad H_S = H_B + k_S(H_C - H_B) \end{aligned} \tag{1.37}$$

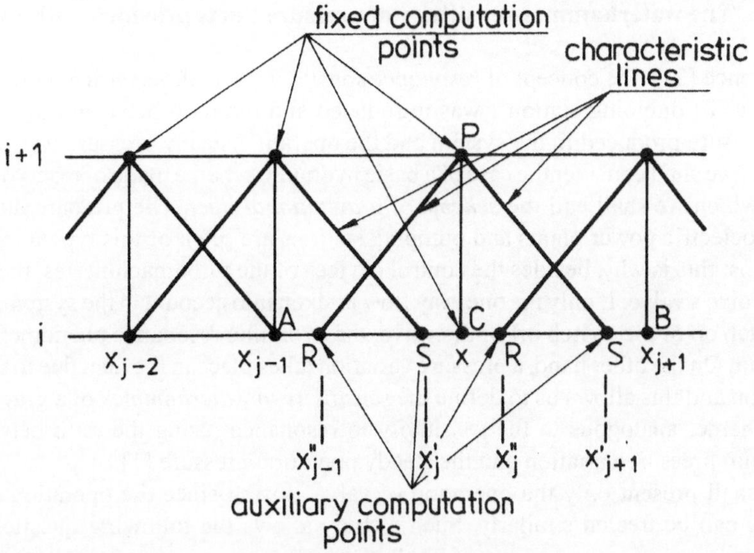

Figure 1.20. Grid with fixed nodes and auxiliary nodes.

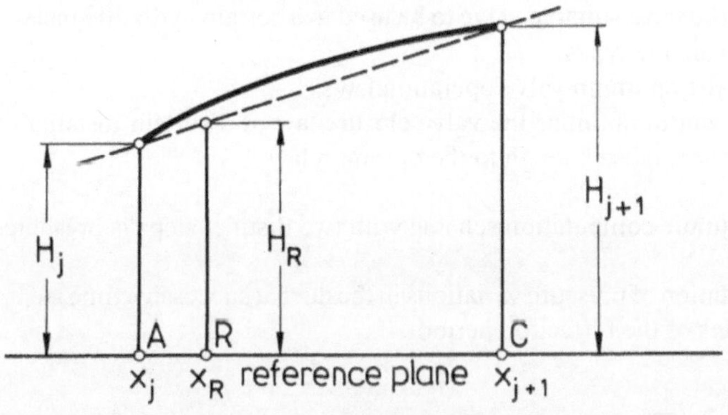

Figure 1.21. Computation scheme for the interpolation of H and V.

where k_R and k_S are the interpolating coefficients

$$k_R = \frac{x_R - x_A}{x_C - x_A}; \quad k_S = \frac{x_S - x_B}{x_C - x_B} \tag{1.38}$$

Finally, in this case, in order to compute $V_{j,i+1}, H_{j,i+1}$ in each section j at any time $(i+1)$, the system of equations

$$V_P = \frac{1}{2}\left[V_R + V_S + \frac{g}{c}(H_R - H_S)\right] - \frac{\lambda \Delta t}{2D}(V_R|V_R| + V_S|V_S|) \tag{1.39}$$

$$H_P = \frac{1}{2}\left[H_R + H_S + \frac{c}{g}(V_R - V_S)\right] - \frac{c\lambda \Delta t}{2gD}(V_R|V_R| + V_S|V_S|) \tag{1.40}$$

must be solved by taking into account eqns. (1.37) and (1.38).

1.4. The waterhammer sensitivity of pressure ducts provided with valves

In reference [21], the concept of *resonance sensitivity* of hydroelectrical power plants, as well as of pumping stations, was introduced and the ratio between the maximum over-pressure produced in the system and the operating steady pressure was defined.

Here, we shall consider the case of a basic hydraulic scheme made of reservoir–duct–valve, which we shall call the *scheme of gravitational duct*. The pressure ducts used in hydroelectric power plants and pumping stations are often of this type in operation situations; this is why, besides the control devices of the turbomachineries, these ducts also involve a valve. If only the one-way flow is taken into account in the system, namely the switch on or the switch off of the valve, the hydraulic resonance phenomenon does not occur. On the other hand, a pressure variation takes place in the duct due to the valve operation and this allows us to define the *sensitivity to waterhammer* of a gravitational duct scheme, analogous to the sensitivity to resonance, using the ratio between the maximum pressure variation and the steady operation pressure [1].

We shall present only the operation of valve closing since the operation of valve opening can be treated similarly. Such a study allows the following questions to be answered:

- what is the sensitivity to waterhammer of a certain installation;
- which is the most suitable valve to be used in a certain hydraulic installation, among several available types;
- which is the optimum valve operation law;
- is it efficient to optimise the valve closure law of a certain installation, or its own closing law is close enough to the optimum law.

First, a simple computation scheme with two distinct steps is presented:

(a) computation of pressure variations in the duct at successive time steps representing multiples of the reflection period

$$t_i = i\tau = i\frac{2L}{c}$$

(b) determination of optimum valve operation law.

Obviously, from the point of view of the *sensitivity to waterhammer*, the most disadvantageous valves are those whose operating time is smaller than the reflection time (the so-called instantaneous closing valves). For these valves, the maximum increase in pressure is given by the Jukovski's formula

$$\Delta p_{\max} = \rho c v_o; \quad \text{that is} \quad \Delta H_{\max} = \Delta H_j = \Delta p_{\max}/\gamma = c v_o/g \tag{1.41}$$

and the maximum sensitivity to waterhammer is

$$S_{\max} = S_j = \frac{\Delta H_j}{H_o} \tag{1.42}$$

where the usual symbols are used.

The computation of pressure variations

Bernoulli's equation is written in the steady state form at the sections upstream and downstream of the valve

$$H_o - h_{rco} = \frac{p_o}{\gamma} = \zeta_{v_o}\frac{v_o^2}{2g}; \quad h_{vco} = \zeta_c\frac{v_o^2}{2g} \tag{1.43}$$

where ζ_c represents the coefficient of hydraulic head losses in the duct, considered to be constant in the unsteady state, in the first approximation, and ζ_{v_o} is the coefficient of local hydraulic head losses of the valve at the initial time.

Starting with this initial flow, the valve is considered to begin closing—the closing law being approximated by a sequence of steps in time—and generates a set of primary waves, each separated by an equal time interval which is equal to the reflection time $\tau = 2L/c$. After the first step, an increase in the valve drag coefficient and in the pressure are noticed, together with a decrease in the speed. The following two relations, which form the basis of the computation model, can be written.

- Jukovski's formula, which gives the speed variation due to pressure variation

$$\Delta h_1 = c(v_o - v_1)/g \quad \text{or} \quad v_1 = v_o - g\Delta h_1/c \tag{1.44}$$

- Bernoulli's equation, written between the upstream and downstream sections of the valve

$$\begin{aligned} H_o - h_{rc1} + \Delta h_1 &= \zeta_{v1}\frac{v_1^2}{2g} \\ h_{rc1} &= \zeta\frac{v_1^2}{2g} \end{aligned} \tag{1.45}$$

From these equations, the speed v_1 is eliminated, and an algebraic equation of second degree is obtained, the solution of which is

$$\begin{aligned} \beta_1 &= \frac{\Delta h_1}{\Delta H_J} = 1 + \frac{\zeta_o S_J}{2\zeta_1} - \sqrt{\frac{\zeta_o S_J}{\zeta_1}\left(1 + \frac{1}{S_J} + \frac{\zeta_o S_J}{4\zeta_1}\right)} \\ \zeta_o &= \zeta_c + \zeta_{v_o}; \quad \zeta_1 = \zeta_c + \zeta_{v1} \end{aligned} \tag{1.46}$$

When computing the second step of pressure variation, which is produced when the first wave returns to the valve by reflection, care must be taken of the fact that, due to the reflection at the end of the duct, the sign of the pressure step has changed.

In this case, the two basic equations are:

- Jukovski's formula

$$\Delta h_2 = \frac{c}{g}(v_1 - v_2) \quad \text{or} \quad v_2 = v_o - \frac{g(\Delta h_1 + \Delta h_2)}{c} \tag{1.47}$$

- Bernoulli's equation

$$H_o - h_{rc2} - \Delta h_1 + \Delta h_2 = \zeta_{v2}\frac{v_2^2}{2g} \tag{1.48}$$

In the same manner it follows

$$\beta_2 = \frac{\Delta h_2}{\Delta H_J} = 1 + \frac{\zeta_o S_J}{2\zeta_2} - \beta_1 - \sqrt{\frac{\zeta_o S_J}{\zeta_2}\left(1 + \frac{1}{S_J} + \frac{\zeta_o S_J}{4\zeta_2} - 2\beta_1\right)} \tag{1.49}$$

Passing to the next pressure variation steps and performing the same calculations, the following general relations are obtained:

$$\beta_{2p} = \frac{\Delta h_{2p}}{\Delta H_J} = 1 + \frac{\zeta_o S_J}{2\zeta_{2p}} - \sum_{k=1}^{2p-1} \beta_k - \sqrt{\frac{\zeta_o S_J}{\zeta_{2p}}\left(1 + \frac{1}{S_J} + \frac{\zeta_o S_J}{4\zeta_{2p}} - 2\sum_{k=1}^{p} \beta_{2k-1}\right)}$$

$$\beta_{2p+1} = \frac{\Delta h_{2p+1}}{\Delta H_J} = 1 + \frac{\zeta_o S_J}{2\zeta_{2p+1}} - \sum_{k=1}^{2p} \beta_k - \sqrt{\frac{\zeta_o S_J}{\zeta_{2p+1}}\left(1 + \frac{1}{S_J} + \frac{\zeta_o S_J}{4\zeta_{2p+1}} - 2\sum_{k=1}^{p} \beta_{2k}\right)} \tag{1.50}$$

Finally, for the last (n) step of pressure variation, as $\zeta_{v_n} \to 0$, the computation formula reduces to

$$\beta_n = \frac{\Delta h_n}{\Delta H_J} = 1 - \sum_{k=1}^{n-1} \beta_k$$

After computing all the β_k coefficients, which express in dimensionless form the pressure variations connected to every primary wave, the pressure variation at each $k\tau$ moment can be determined by the relations:

$$\frac{\Delta H_K}{\Delta H_J} = (-1)^{k+1}\left[\beta_1 - \beta_2 + \beta_3 - \beta_4 + \cdots + (-1)^{k+1}\beta_k\right]; \quad k = 1, 2, \ldots, n \tag{1.51}$$

According to the definition, the sensitivity to waterhammer is given by the largest of the following ratios

$$S = \max \frac{\Delta H_K}{H_o} = \max \frac{\Delta H_K}{\Delta H_J} S_J; \quad k = 1, 2, \ldots, n \tag{1.52}$$

Some considerations concerning optimisation
In the theoretically important case of a valve closing according to a linear law for the flow, by using the formula of Michaud [16], the following expression of the sensitivity to waterhammer is obtained

$$S_M = \frac{\Delta H_{\max M}}{H_o} = \frac{2Lv_o}{gH_o T_m} = \frac{2Lv_o}{gT_m \Delta H_J} S_J = \frac{\tau}{T_m} S_J = \frac{1}{n} S_J \tag{1.53}$$

where T_m is the valve operation time. It should be noticed that the linear law for closing the valve is rather advantageous but hardly ever met in practice. The problem if the linear closing law is optimum might be raised, in the sense that the minimum sensitivity to waterhammer is desired (for the same operating time, of course).

If we consider the usual case where during the entire closing of the valve the pressure in the duct overpasses the steady-state pressure, then by grouping two by two relations (1.51) we obtain the following inequalities, expressing the optimisation condition:

$$\beta_1 \le \beta_{\min}, \quad \beta_k \le 2\beta_{\min}; \qquad k = 2, 3, \ldots, n \tag{1.54}$$

By adding all these inequalities and by considering the sum of the coefficients of all the primary waves (β_k) equal to unity, the value $\beta_{\min}$

$$\sum \beta_k = 1 \leq (2n-1)\beta_{\min}, \quad \text{or} \quad \beta_{\min} = \frac{1}{2n-1} \tag{1.55}$$

and the optimum closing law

$$\beta_1 = \beta_{\min}; \quad \beta_k = 2\beta_{\min}; \quad k = 2, 3, \ldots, n \tag{1.56}$$

are obtained.

Now, the minimum sensitivity to waterhammer can also be evaluated, and the result shows that it is quite near to half of the value obtained for the linear law

$$S_{\min} = \frac{1}{2n-1} S_{\mathrm{J}} \tag{1.57}$$

It follows that, generally, the *sensitivity to waterhammer* for a gravitational duct varies between two extreme values

$$S_{\min} = \frac{1}{2n-1} S_{\mathrm{J}} \leq S \leq S_{\mathrm{J}} = S_{\max} \tag{1.58}$$

It is worth mentioning that these considerations on the optimisation, namely the values of the β_k coefficients of the primary waves, are independent of the way these coefficients are computed.

To enable a closer look to the *sensitivity to waterhammer* the introduction of some degrees of sensitivity is proposed, defined by a corresponding interval ΔS for every degree of sensitivity

$$\Delta S = \frac{S_{\max} - S_{\min}}{10} \tag{1.59}$$

Thus, the first degree of sensitivity belongs to those gravitational ducts where $S_{\min} \leq S \leq S_{\min} + \Delta S$ and the tenth degree belongs to those having $S_{\max} - \Delta S \leq S \leq S_{\max}$. This procedure allows us to eliminate the influence of the valve operating time because no matter the operating time in some given practical cases of hydraulic schemes of the type reservoir–duct–valve, this scheme will belong to a certain class of sensitivity, which somehow proves how suitable the valve is for the given installation. From this point of view, it seems necessary to forbid the use of valves involving large degrees of sensitivity in large-diameter ducts.

Numerical example

Consider a gravitational duct with diameter $d = 2$ m, wall thickness $e = 20$ mm, fitted with expansion sleeve and having a length $L = 1{,}000$ m. The water velocity is $v_o = 2$ m/s, the piezometric head is $H = 20$ m and the roughness coefficient of the duct is $n = 0.012$.

The computed celerity is $c = 1{,}015$ m/s and the discharge modulus of the duct is $K = 96.6\,\mathrm{m^3/s}$. Successively the other parameters concerning the duct in steady-state flow are computed as $J_o = 4.22 \times 10^{-3}$; $h_{ro} = 4.23$ m ; $\zeta_c = 20.74$. For the unsteady flow, due to the valve closing, the following values are obtained: the reflection time $\tau = 1.95$ s and the maximum sensitivity to waterhammer $s_j = 103.5$. In this case, let us suppose the possibility to use a plane valve or a butterfly valve, both having the operating time $T_m = 14\tau = 27.6$ s. The minimum sensitivity to waterhammer follows to be $S_{min} = 3.83$, meaning that the sensitivity of the hydraulic scheme ranges between the two limits: $3.83 \leq S \leq 103.5$.

Considering a uniform closing of the valve in agreement with the appendix in reference [3], the closing laws of the two valves are shown in Table 1.1 containing the values for the hydraulic losses ζ_{vk} $(k = 0, 1, \ldots, 15)$ of the valve. In this table, the values τ_k, assumed to be constant, and obtained by summing the head-loss coefficients of the valve with the head-loss coefficient of the pipe, are shown as well. Also, the value of the β_k coefficients for the primary waves and the dimensionless values for the resulting pressure $\Delta H_k/\Delta H_j$ are shown.

From these computation it follows

- for the plate valve, the sensitivity to waterhammer is

$$S = 0.386 \times 103.6 = 40$$

- for the butterfly valve, the sensitivity to waterhammer is

$$S = 0.263 \times 103.6 = 27.3$$

From the point of view of the sensitivity to waterhammer, for this hydraulic scheme, the use of a butterfly valve is recommended. According to the proposed terminology, for this installation and for the valves with operating time $T_m = 14\tau$, the successive sensitivity degrees differ by a step $\Delta S = 9.975$. That is why, by using a butterfly valve the sensitivity degree is three, while by using a plane valve the sensitivity degree is four.

Table 1.1. Hydraulic calculations for plane and butterfly valves.

k	Plane valve				Butterfly valve			
	ζ_{vk}	ζ_k	$\beta_k = \frac{\Delta h_k}{\Delta H_j}$	$\frac{\Delta H_k}{\Delta H_j}$	ζ_{vk}	ζ_k	$\beta_k = \frac{\Delta h_k}{\Delta H_j}$	$\frac{\Delta H_k}{\Delta H_j}$
0	0.01	20.753	–	–	0.05	20.793	–	–
1	0.02	20.763	0.0000	0.0000	0.2	20.943	0.0001	0.001
2	0.05	20.793	0.0000	0.0000	0.5	21.243	0.0003	0.0002
3	0.1	20.843	0.0000	0.0000	1	21.743	0.0006	0.0004
4	0.2	20.943	0.0001	0.0001	1.8	22.543	0.0011	0.0007
5	0.5	21.243	0.0003	0.0002	2.8	23.543	0.0019	0.0012
6	1	21.743	0.0006	0.0004	4	24.743	0.0029	0.0017
7	2	22.743	0.0013	0.0009	7.5	28.243	0.0049	0.0032
8	5	25.743	0.0031	0.0022	10	30.743	0.0073	0.0041
9	10	30.743	0.0065	0.0043	20	40.743	0.0122	0.0081
10	15	35.743	0.0105	0.0062	32	52.743	0.0321	0.0240
11	40	60.743	0.0223	0.0161	60	80.743	0.0442	0.0202
12	100	120.743	0.0523	0.0362	125	145.743	0.0577	0.0375
13	400	420.743	0.1707	0.1345	280	300.743	0.1364	0.0989
14	1000	1020.743	0.2400	0.1055	700	720.743	0.2669	0.1680
15	∞	∞	0.4923	0.3868	∞	∞	0.4314	0.2634
16	–	–	0.0000	0.3868	–	–	0.0000	0.2634

1.5. Computation examples for hydroelectric power plants and pumping stations

The examples show how numerical methods can be used to study the evolution with respect to time of the waterhammer phenomenon produced in the galleries and pipes of hydroelectric power plants and pumping stations.

These examples have been taken from hydraulic studies conducted at the Hydraulic Engineering Research Institute of Bucharest (Institutul de Cercetari Hidrotehnice I.C.H.) in order to solve some practical problems related mainly to waterhammer and less to the solutions to minimize this phenomenon. The ways to minimize waterhammer are presented in the next chapters [18, 20].

1.5.1. The waterhammer computation for hydroelectric power plants

The hydroelectric power plant Sadu V

The water conveyance structure of this plant is made of a gallery, 5,240 m long, with a diameter of 1.8 m, a surge tank with an upper chamber and a lower chamber, a 106 m long penstock with a diameter of 1.8 m, the valve unit, the 940 m long penstock from the valve unit to the power plant, with a diameter varying from 1.5 m to 1.2 m, the power plant equipped with two Pelton turbines and one Francis turbine, all three having a power of 27 MW for a constant rated flow of 8.3 m^3/s and the second water intake from the river Sadurel with a constant flow of 1.7 m^3/s, captured by a vertical pipe of 42 m, into the main intake (Fig. 1.22).

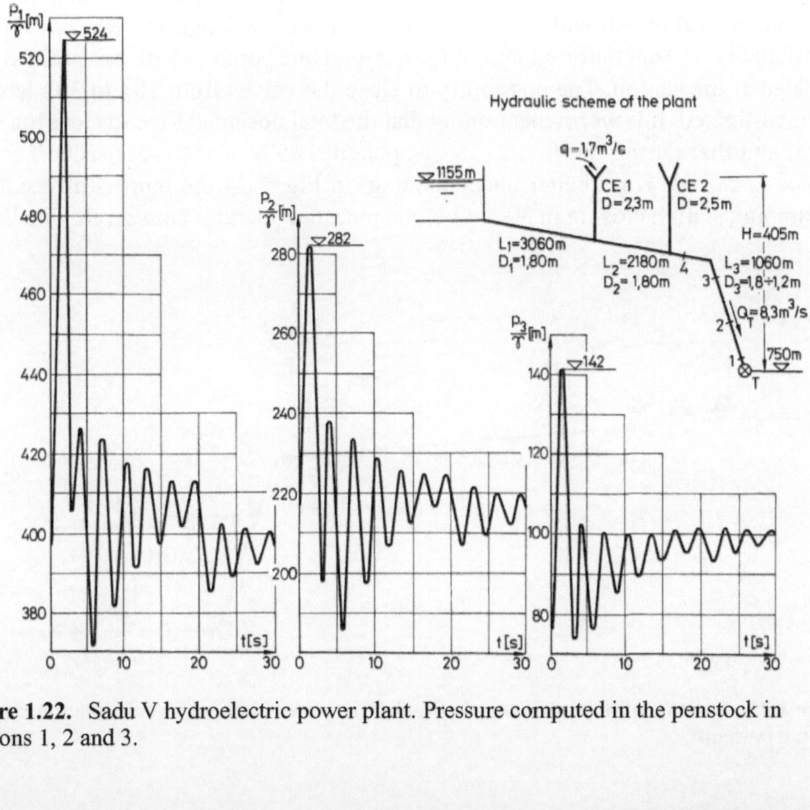

Figure 1.22. Sadu V hydroelectric power plant. Pressure computed in the penstock in Sections 1, 2 and 3.

The waterhammer in the penstock for a complete halt of the power plant has been computed and the results are graphically plotted as functions of time in three characteristic sections (Fig. 1.22). The closing times for the valves at the turbines are 3 s for the Pelton turbines and 20 s for the Francis turbine.

The hydroelectric power plant with pumped storage Sebes–Gilceag
The main characteristics of this installation are: Oasa reservoir with a maximum level of 1,259 m and a minimum level of 1,205 m; the water pressure tunnel, 8,500 m long and 3.8 m in diameter; the surge tank with chambers, the lower chamber being shaped as an inclined gallery of 120 m in length and 6 m in diameter, the surge tank well being 83 m in height and 5 m in diameter; the conical upper chamber of the surge tank 22 m high and 12 m/22 m in diameter; the penstock 2.8 m in diameter; the discharge pipe 340 m long and 1.3 m in diameter (pipeline station penstock); the tail race 481 m long; the Tau Lake, with a minimum level of 785.5 m and maximum level of 790 m. The pumping station of the power plant is equipped with two pumps of type PRO-9-245, with the nominal characteristics $Q_o = 3\ m^3/s$, $H_o = 247$ m coupled to electric motors having a power $N = 8.6$ MW and a speed 1,000 rot/min (Fig. 1.23). Along the 600 mm diameter pump discharge pipes spherical valves with programmed closing are provided. The adopted computation scheme involves 30 computation points, out of which four are along the way from the pumping station to the surge tank and the other 26 points are located from the surge tank to lake Oasa [20, 31].

As the surge tank ensures the protection against waterhammer of the water conveyance gallery (the results of the computations show a slowly-varying flow within the gallery–surge tank system) special attention was paid to the pressure variation in transient flow at the characteristic points of the penstock. An accidental stop of one or of two pumps was numerically simulated. To find the most dangerous way of operating the installation, 10 operation situations (either with one pump or with two pumps) were simulated at the station. The possibility to close the valves from 10 s to 30 s has also been investigated. It is worth mentioning that the total designed time to close the valve is 24 s, and that during the first 2–3 s of operation 85% of the cross-section area of the pipe is closed. For a better understanding, in Fig. 1.24 the numerical results for the pumping station closing in 30 s, with one pump in operation are given. Finally, the

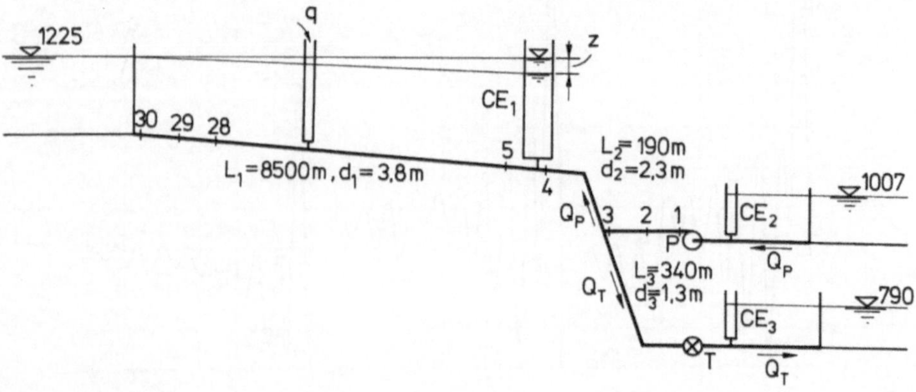

Figure 1.23. Scheme of Sebes–Gilceag hydroelectric power plant with accumulation through pumping (variant).

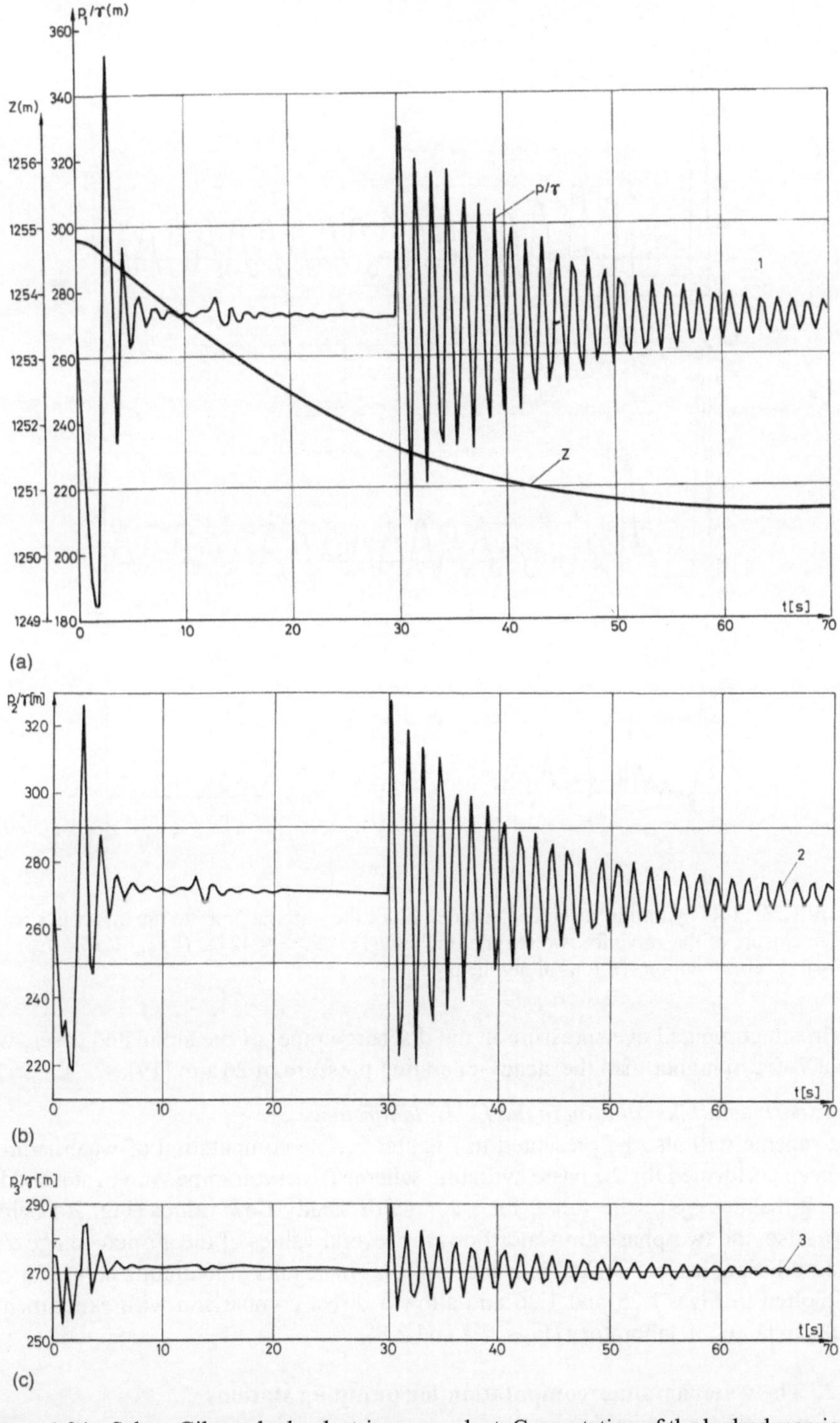

Figure 1.24. Sebes–Gilceag hydroelectric power plant. Computation of the hydrodynamical pressures in the pipe of the station (p/γ) and of the water level in the tank (Z); 1, 2, 3 computation sections.

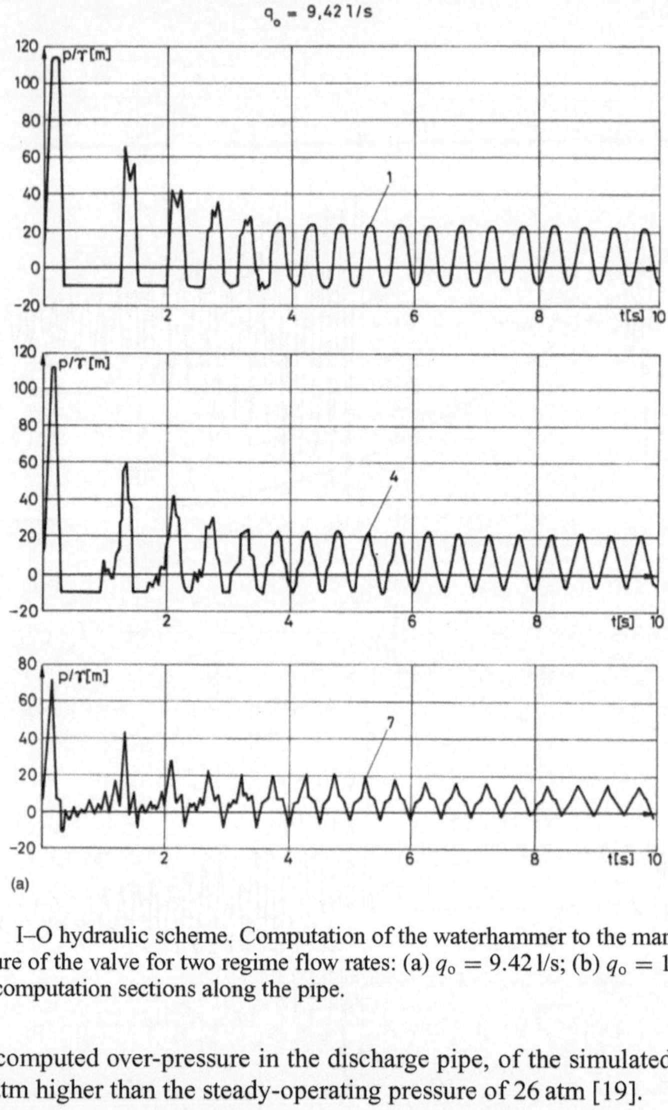

Figure 1.25. I–O hydraulic scheme. Computation of the waterhammer to the manoeuvre of sudden closure of the valve for two regime flow rates: (a) $q_o = 9.42$ l/s; (b) $q_o = 16.4$ l/s; (1), (4), (7) computation sections along the pipe.

maximum computed over-pressure in the discharge pipe, of the simulated cases, was about 7–9 atm higher than the steady-operating pressure of 26 atm [19].

The experimental installation in the I.C.H. laboratory
This scheme will also be presented in Chapter 2. The computation of waterhammer has been performed for the basic hydraulic scheme (reservoir–pipe–valve) at a sudden closing manoeuvre of the valve, for a couple of steady flow values (Fig. 2.4 below, I–O). Also, the two-phase air–water flow, for several values of the air percentage α of the air dilution, has been taken into account [6]. The results of hydraulic computations are plotted in Figs. 1.25 and 1.26 and allow a direct comparison with experimental results obtained in laboratory (Figs. 2.7 and 2.8).

1.5.2. The waterhammer computation for pumping stations

The Sinoe SRPA 3-1 pumping station
The station has a flow rate of 2.8 m^3/s for a geodetic height of 32.5 m and is provided with four pumps 18 NDS, with the following main technical characteristics:

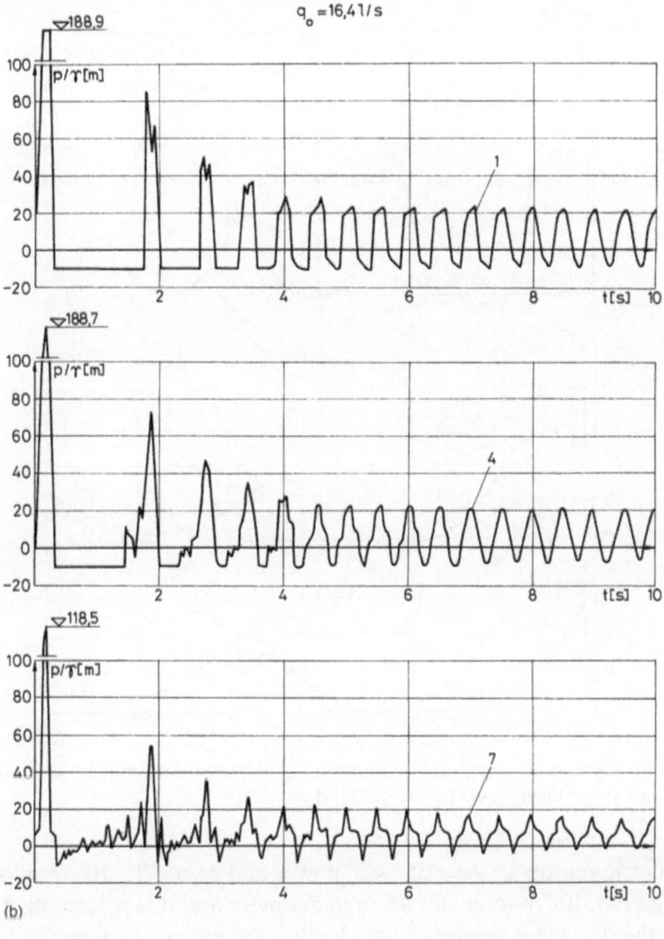

Figure 1.25. (*continued*).

$Q_o = 0.7\,m^3/s$ and $H = 36\,m$. The pumps are driven by electric motors rated to $N = 400\,kW$ power and $n = 1{,}500$ rot/min speed.

Along the discharge pipes of 0.5 m in diameter, several non-return valves are fitted. The discharge pipe of the station, made of metal, 1,000 m long and 1.25 m in diameter, is fitted with a non-return valve located near the discharge reservoir to avoid the return flow when the pumping station is shut down (Fig. 1.27). The *waterhammer* computation has been performed by assuming a cut-off in the electric power supply at maximum flow operation. The sketch presents the time variation plots of the pressures in the discharge pipe, at three characteristic sections.

Sometimes, engineering practice resorts to the presentation of the waterhammer results in a pressure installation under the form of envelope curves of the maximum and minimum pressure along the analysed pipe. For example, for this pumping station, these envelope curves are those in Fig. 1.28.

The Daeni SRP-4 pumping station

This station has a flow rate of $1.4\,m^3/s$ for a geodetic height of 73 m and is equipped with 7 pumps MV-303 × 2 with main characteristics: $Q_o = 0.2\,m^3/s$ and $H_o = 84\,m$,

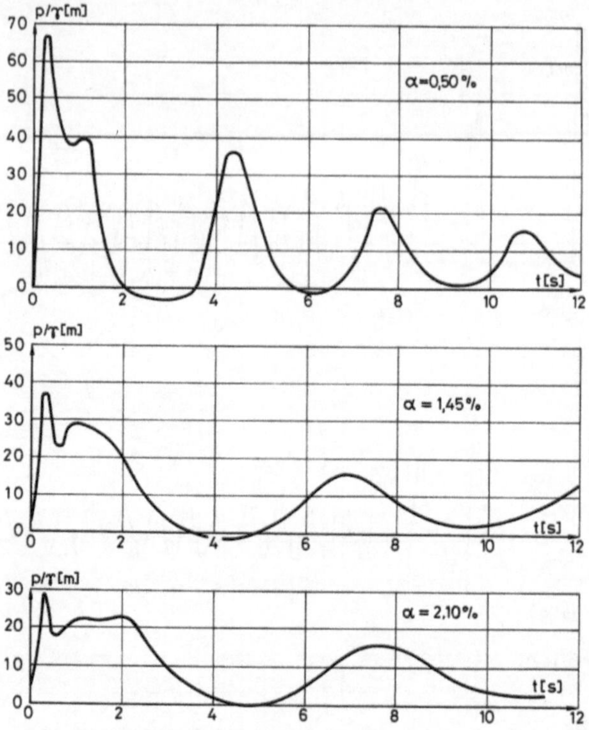

Figure 1.26. I–O hydraulic scheme. Computation of the waterhammer in Section 1, for various water–air two-phase flows ($q_o = 19.2$ l/s).

driven by electric motors of $N = 250$ kW power and of $n = 1{,}470$ rot/min speed. The discharge pipe is 1,408 m long and 0.8 m in diameter and it is fitted with a blade valve located near the discharge reservoir, which allows a non-return flow.

The scheme of the pumping station and the results of hydraulic computations performed for an accidental cut off in the electric power supply of the station, are shown in Fig. 1.29. The results exhibit both low negative pressures (cavitation) and high over-pressure ($p_{max} = 14.3$ atm), all these on the background of suddenly-varying flow (waterhammer).

The Sinoe SRPA 1-5 pumping station

The station has a flow rate of 5 m^3/s for a geodetic height of 70.5 m and is equipped with five pumps 24 NDS-g, with hydraulic parameters: $Q_o = 1$ m^3/s and $H_o = 76$ m. The pumps are driven by electric motors MAB-500 with $N = 1{,}000$ kW power, $n = 750$ rot/min speed and voltage $U = 6{,}000$ V.

Along the discharge pipes of each pump (0.6 m in diameter) non-return valves are fitted. The discharge pipe consists of two distinct pipes: the first (near the pumping station) is 1,000 m long and 1.8 m in diameter, the second is 1,000 m long, but 1.6 m in diameter. At a distance of 1,000 m from the pumping station, a 0.45 m^3/s flow is taken away from the main discharge pipe.

The scheme of the station and the results of hydraulic computations, assuming the same hypothesis as for the other pumping stations, are presented in Fig. 1.30. They

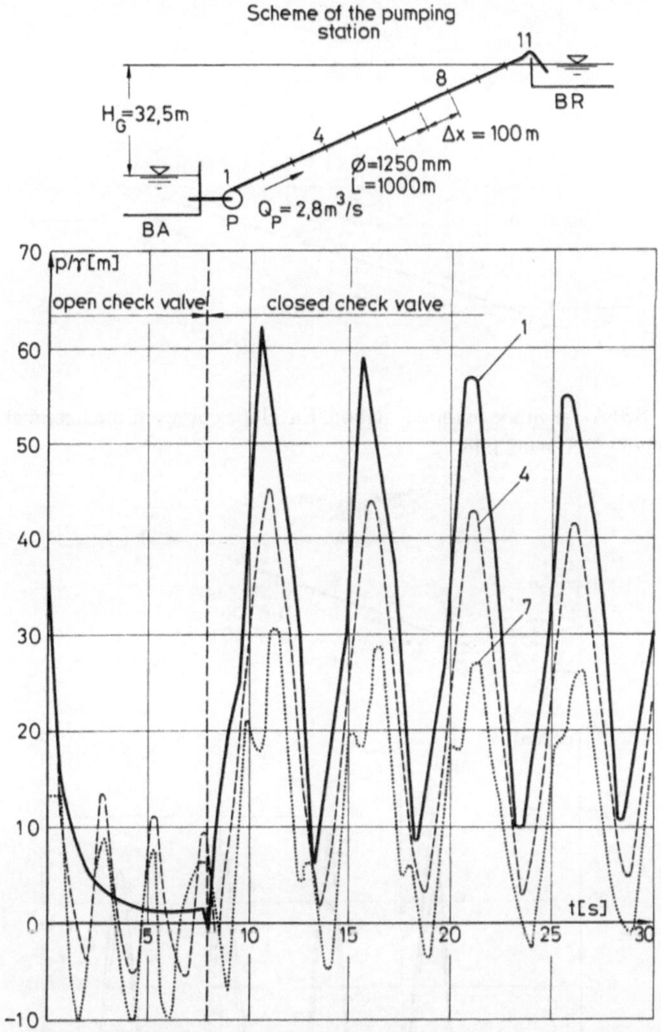

Figure 1.27. SRPA 3-1 Sinoe pumping station. Hydrodynamical pressures calculated in the pipe.

show that in the absence of devices for protection against waterhammer, the discharge pipe suffers high over-pressures ($p_{max} = 13.5$ atm) at rapidly varied flows in the pipe.

The Sinoe SPA pumping station

This is one of the largest pumping stations built in Romania. Its main characteristics are: a 46 m^3/s flow rate and a geodetic height of 50.5 m. The discharge pipe consists of two independent metallic pipes with a length of 810 m and a diameter of 2.8 m. The station is equipped with eight centrifugal pumps with vertical shaft 52-B-17, with the hydraulic characteristics: $Q_o = 6.5$ m^3/s and $H = 58.5$ m, driven by synchronous electric motors BDC 325/49-16 with $N = 5$ MW power and $n = 375$ rot/min speed. Along the discharge pipes of each pump hydraulically operated and programmed closing valves are fitted with closing times $t_s = 50$–54 s.

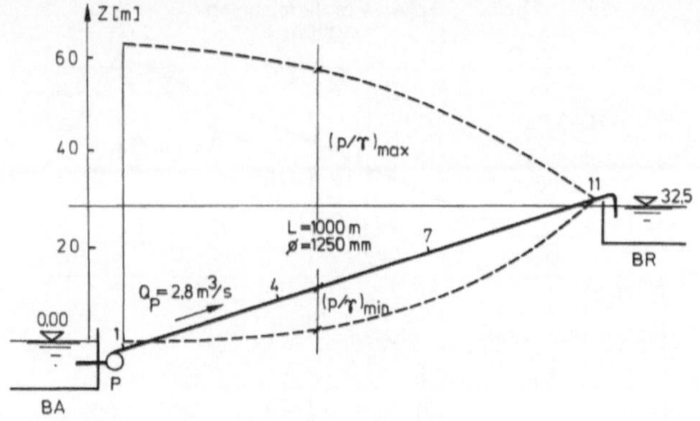

Figure 1.28. SRPA 3-1 Sinoe pumping station. Envelope curves of the maximal and minimal pressure along the repressing pipe.

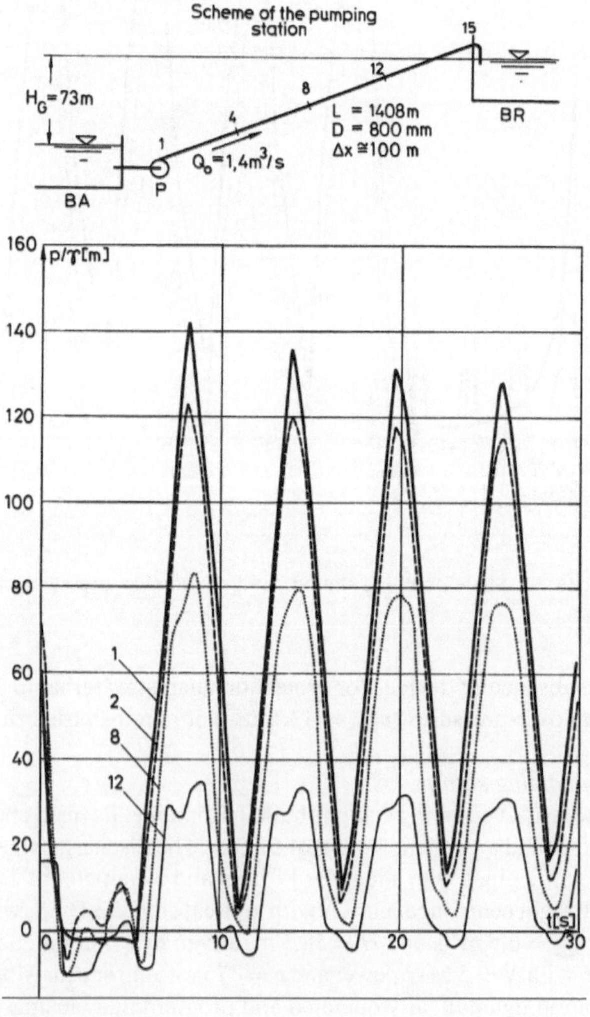

Figure 1.29. SRP-4 Daeni pumping station. Curves of pressure variation in pipe.

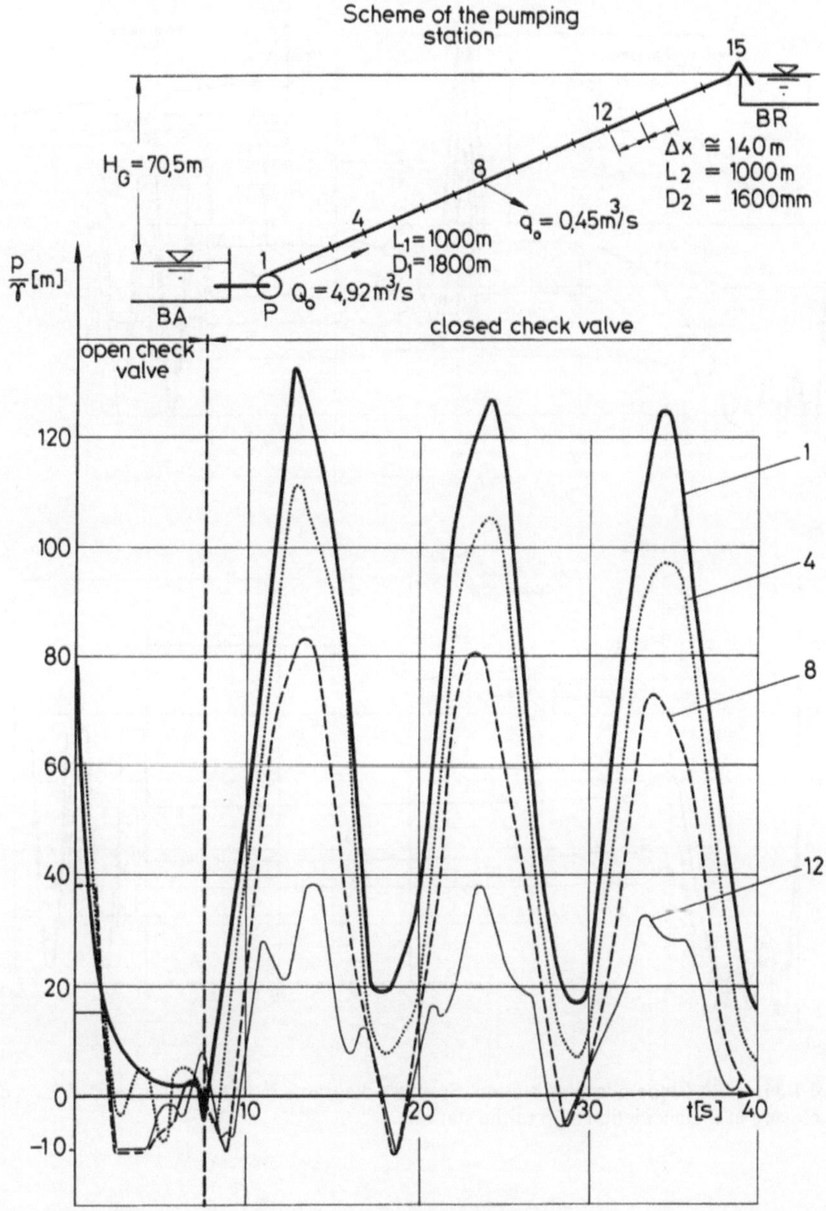

Figure 1.30. SRPA 1-5 Sinoe pumping station. Hydrodynamical pressures calculated in the pipe.

A special study on the effect of valve closing on the maximum pressure in the discharge pipes has been performed for this pumping station. As the discharge pipes are independent, the hydraulic computations have been performed for one pipeline. As it is well known, the law of valve closing greatly influences the dynamic pressure regime inside a discharge pipe without protection devices against waterhammer caused by accidental shut-down of the pumping station [2, 8]. This is shown in the plots of

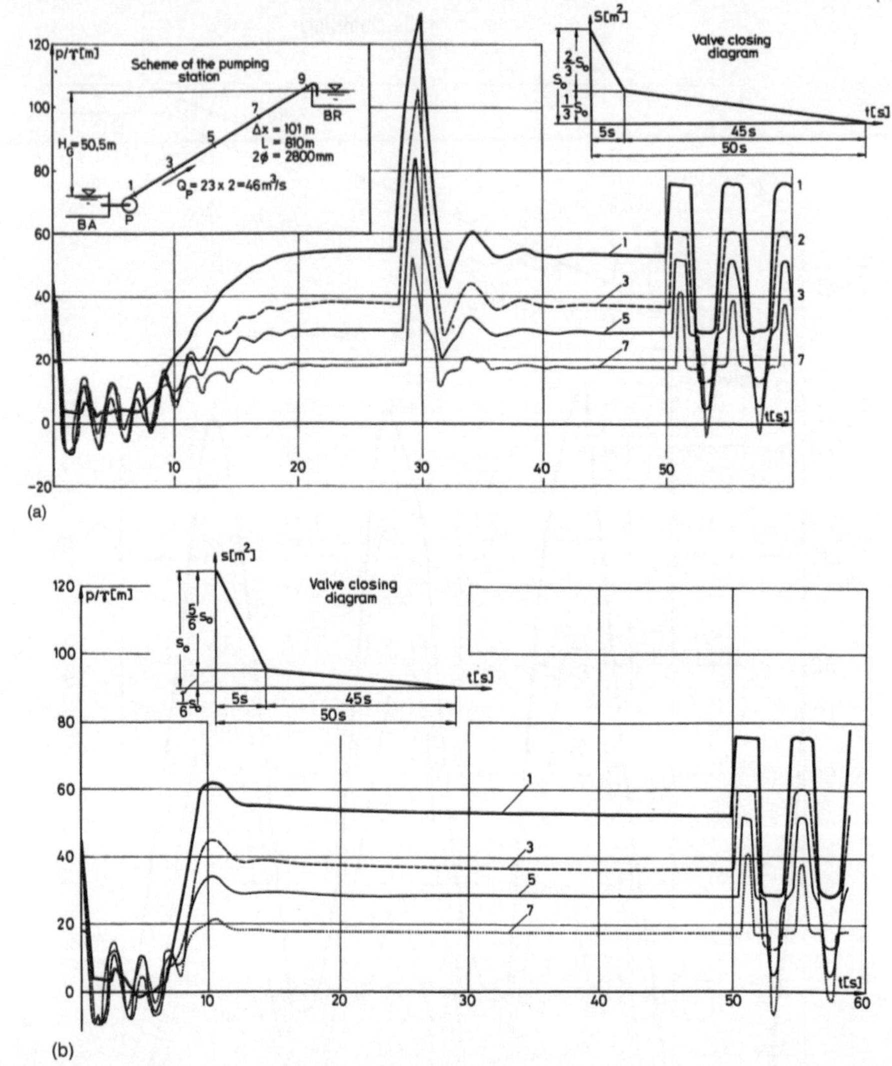

Figure 1.31. SPA Sinoe pumping station. Pressure diagrams in the pipe for various laws of valve closure at the accidental stop of the station.

the computed pressures inside the discharge pipe of the Sinoe SPA pumping station (Fig. 1.31), which prove that small changes into the closing law of the valves entail large pressure being produced in the system (S_o is the flow area of the completely opened valve). From Fig. 1.31b, it seems that the choice of an optimum law of valve closing is sufficient to ensure the protection against waterhammer of the pumping station and, as a consequence, it might be possible to eliminate the usual protection device i.e. the reservoir with air-chamber, or surge tank.

This effect of programmed valve closing, shown above, has mainly a theoretical importance, because in practice, at least in our opinion, it is not recommended to expose the pumping station to the risk of violent waterhammer and thus to a possible

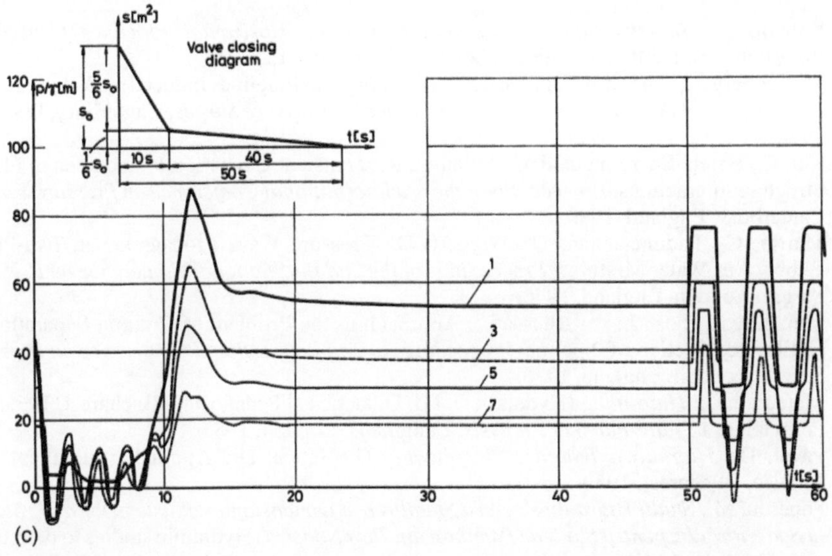

Figure 1.31. (*continued*).

accident, due to a mistake in operating the valve recording to the required law of closing. However, in some papers descriptions of pumping stations operating under optimum programmed valve closing law without the usual protection devices against waterhammer, are given. Nevertheless, in practice, the most intelligent way to avoid the risk of waterhammer is to use the optimum law of valve closing, as well as the usual protection devices against waterhammer, such as surge tanks, air chambers and one-way reservoirs.

References

1. Arsenie, D.: Sensibilitatea la Soc Hidraulic a Conductelor sub Presiune Prevazute cu Vane (The Sensitivity to Waterhammer of Pressure Pipes Provided with Valves). *Hidrotehnica*, nr. 3, 1985.
2. Chaudhry, H.: *Applied Hydraulic Transients*. Van Nostrand Reinhold Company, New York, 1987.
3. Cioc, D.: *Contributii la Calculul Miscarii Nepermanente in Conducte si la Teoria Sonicitatii cu Aplicatii la Pompajul Sonic* (Contributions to the Computation of Unsteady Flows in Ducts and to the Theory of Sonicity Applied to Sonic Pumping). Studii de Hidraulica I.C.H., vol. XVIII, Bucharest, 1968.
4. Cioc, D.: *Hidraulica* (Hydraulics). Ed. Didactica si Pedagocica, Bucharest, 1975.
5. Dumitrescu, D., Iamandi, C.: *Manualul Inginerului Hidrotehnician*, Vol. I, Sectiunea III Hidraulica (Handbook of Hydropower Engineering, Vol. I, Section III Hydraulics). Ed. Tehnica, Bucharest, 1969, pp. 211–224.
6. Florea, J., Panaitescu, V.: *Mecanica Fluidelor* (Fluid Mechanics). Ed. Didactica si Pedagocica, Bucharest, 1979.
7. Franke, P., Seyler, F.: Computation of Unsteady Pipe Flow with Respect to Viscoelastic Material Properties. *Journal of Hydraulic Research – IAHR*, 5 (1983).
8. Hancu, S., Popescu, M., Zaharescu, E.: *Hidraulica Aplicata* (Applied Hydraulics). Ed. Tehnica, Bucharest, 1985.
9. Iamandi, C., Petrescu, V.: *Fluid Mechanics*. Ed. Didactica si Pedagocica, Bucharest, 1978 (in Romanian).

10. Kalwijk, J.: *Investigation into Cavitation in Long Horizontal Pipelines Caused by Waterhammer*. Delft Hydraulics Laboratory, Publ. no. 115, 1974.
11. Kranenburg, C.: The Effect of Free Gas on Cavitation in Pipelines Induced by Waterhammer. *Proceedings of the 1st International Conference on Pressure Surges*, Canterbury, England, 1972.
12. Kot, C., Hsieh, B., Youngdahl, C., Valentin, R.: Transient Cavitation Phenomena in Fluid–Structure Interactions. *Proceedings of the 3rd International Conference on Pressure Surges*, Canterbury, England, 1980.
13. Martin, C., Padmanabhan, D., Wiggert, D.: Pressure Wave Propagation in Two-Phase Bubbly Air–Water Mixtures. *Proceedings of the 2nd International Conference on Pressure Surges*, London, England, 1976.
14. Marsden, N., Fox, J.: An Alternative Approach to the Problem of Column Separation in an Elevated Section of Pipeline. *Proceedings of 2nd International Conference on Pressure Surges*, London, England, 1976.
15. Mateescu, C.: *Hidraulica* (Hydraulics). Ed. Didactica si Pedagocica, Bucharest, 1963.
16. Parmakian, J.: *Waterhammer Analysis*. Longmans, London, 1985.
17. Pavel, D.: *Hidraulica Teoretica si Aplicata* (Theoretical and Applied Hydraulics). Ed. Tehnica, Bucharest, 1950.
18. Popescu, M.: *Studii Hidraulice pentru Stabilirea si Dimensionarea Sistemelor de Protectie la Soc Hidraulic pentru Statii de Pompare din Tara Noastra* (Hydraulic studies to determine and design protection devices against waterhammer for pumping stations in Romania). Studii interne I.C.H., Bucharest, 1970–1980.
19. Popescu, M.: *Studiul Functionarii la Regimuri Tranzitorii a Statiilor de Pompare de Mare Presiune din Cadrul Amenajarilor Hidroenergetice Complexe* (Study of the operation in hydraulic transient flows of high pressure pumping stations in the complex hydroelectric power plants). Internal Report, I.C.H., Bucharest, 1982.
20. Popescu, M.: *Cercetari Privind Functionarea Hidraulica la Regimuri Tranzitorii a Uzinelor Hydroelectrice si Statii de Pompare* (Researches Concerning the Hydraulic Operation in Transient Regimes of the Hydroelectric Power Plants and Pumping Stations). Sesiune Stiintifica Jubiliara, Volume edited by I.C.H., 1977, pp. 149–162.
21. Popescu, M., Halanay, A.: Resonance Sensitivity Hydropower and Pumping Stations. *Water Power and Dam Construction*, 9 (1984).
22. Provoost, G.: Investigation into Cavitation in a Prototype Pipeline Caused by Waterhammer. *Proceedings of 2nd International Conference on Pressure Surges*, London, England, 1976.
23. Racoveanu, N., Dodescu, G., Mincu, I.: *Metode Numerice pentru Ecuatii cu Derivate Partiale de Tip Hiperbolic* (Numerical Methods for Hyperbolic Partial Differential Equations). Ed. Tehnica, Bucharest, 1976.
24. Streeter, V., Wylie, E.: *Hydraulic Transients*. McGraw Hill, New York, 1967.
25. Tijsseling, A.S.: *Fluid–Structure Interaction in Case of Waterhammer with Cavitation*. Delft University of Technology, Raport No. 93–96, 1993.
26. Weyler, M., Streeter, V., Larsen, P.: An Investigation of the Effect of Cavitation Bubbles on the Momentum Loss in Transient Pipe Flow. *Journal of Basic Engineering*, 3 (1971).
27. Zaoui, J., Combes, G.: Analyse des Erreurs Introduites par l'Utilisation Practique de la Methode des Characteristiques dans le Coups de Belier. *La Houille Blanche* 2 (1967).
28. Sattler, H., Strohmer, F.: Computer Solutions to Waterhammer Problems Using the Method of Characteristics. *Water Power and Dam Construction*, 9 (1974).
29. Constantinescu, G.: *Contributii privind Protectia Instalatiilor sub Presiune Utilizind Efectele Aerului Liber* (Contributions to the protection of pressure installations using the effects of free air). PhD Thesis, Institutul Polithenic Timisoara, 1983.
30. Marinovici, D.: *Hydraulic Transients for Pipes*. Ed. Bren, Bucharest, 2000 (in Romanian).
31. Popa, G., Ion, M., Muj, S.: *The Stress and Unit Strain Analysis Study in the Part of Aspiration with Low Head*. Ovidius University Annals of Constantze, Year III–IV, Series Constructions, 2002, pp. 137–142.
32. Florea, M.: *Contributions to the Study of the Unsteady Motions in Hydraulic Surge Systems*. PhD Thesis, University of Constantze, 1988.

Chapter 2

Experimental Results Concerning the Waterhammer

2.1. Laboratory studies

2.1.1. Experimental installation and measuring devices

The experimental installation, of relatively large dimensions, designed and built at the Hydraulic Engineering Research Institute (Institutul de Cercetari Hidrotehnice–I.C.H.) in Bucharest for the study of the unsteady flow in hydraulic pressure systems has outstanding technical and operational characteristics allowing simulation and research in laboratory conditions of practically all kinds of unsteady flows encountered in hydraulic pressure systems [11].

From this point of view, the experimental setup appears like a remarkable installation as compared to similar laboratory installations, known or mentioned in papers (Figs. 2.1 and 2.2). The installation can operate in two basic variants:

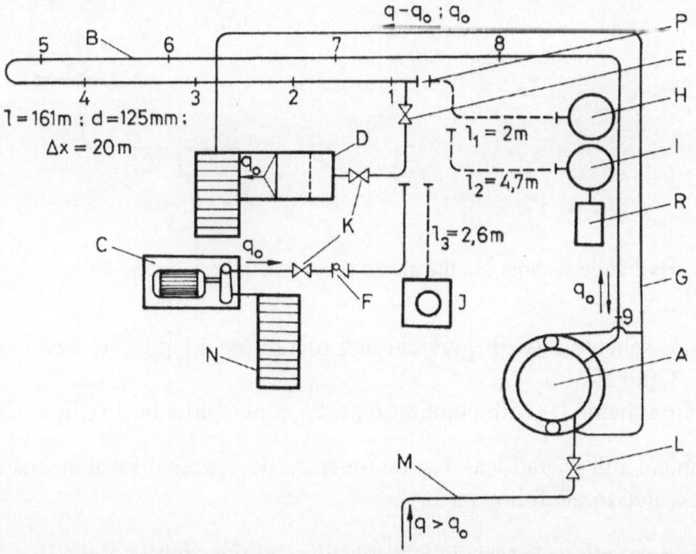

Figure 2.1. General scheme of the (ICH) laboratory experimental installation for the study of non-stationary motions inside hydraulic surge systems.

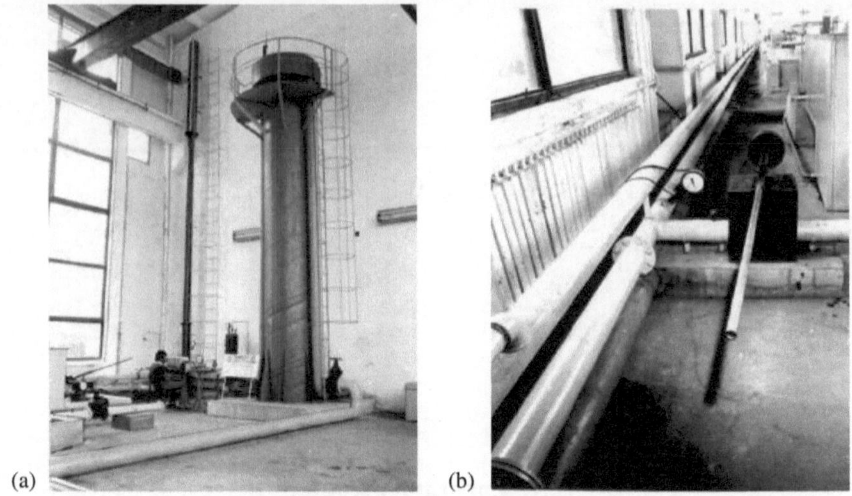

Figure 2.2. (ICH) laboratory experimental installation for the study of unsteady motions inside hydraulic surge systems—overall view: (a) supply chamber, protection system; (b) surge pipe, valve, measure chamber.

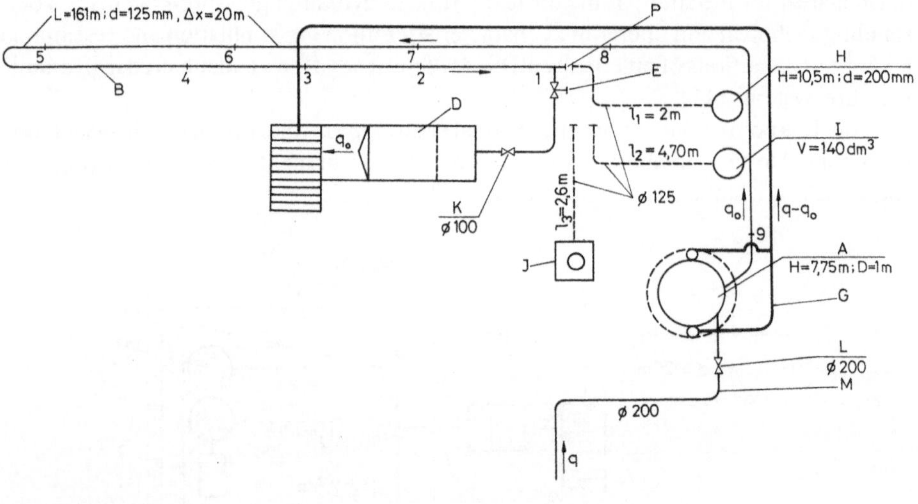

Figure 2.3. Hydraulic scheme I in the plane: supply chamber—pipe—valve.

- Hydraulic scheme I (with gravitational or turbine supply): reservoir–pipe–valve (Figs. 2.3 and 2.4).
- Hydraulic scheme II (with pumping): pump–pipe–outlet basin (Figs. 2.5 and 2.6).

Some technical and operational data on the main devices that form the installation are briefly presented in the following.

(A) *The water tank or reservoir* is a metallic cylinder with height $H = 7.75$ m and diameter $D = 1$ m. In its upper part it has an overflow with a measuring weir and two effluent ducts of 0.2 m diameter. The water supply of the tank is ensured

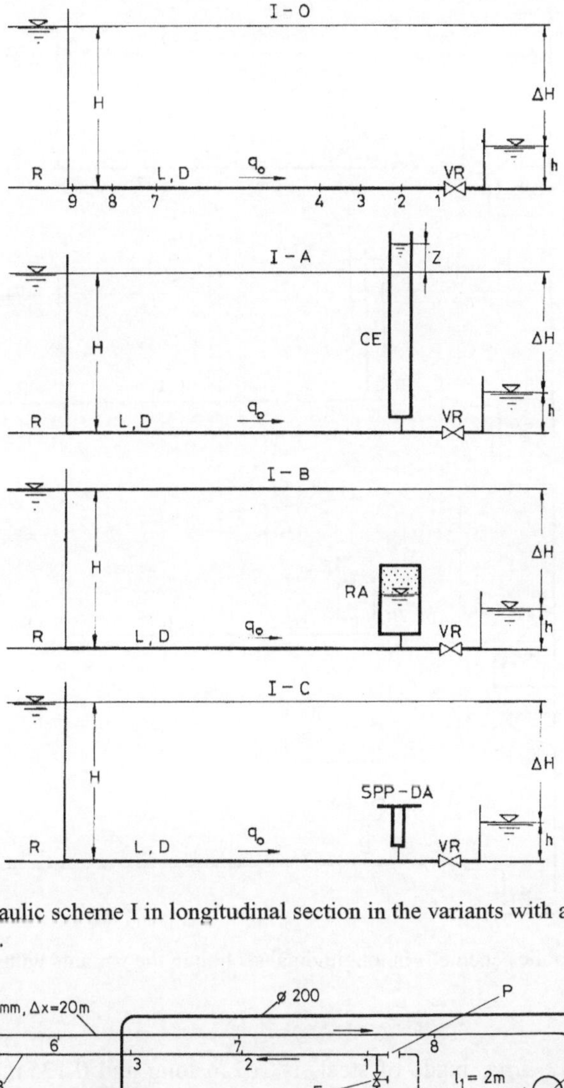

Figure 2.4. Hydraulic scheme I in longitudinal section in the variants with and without protection devices.

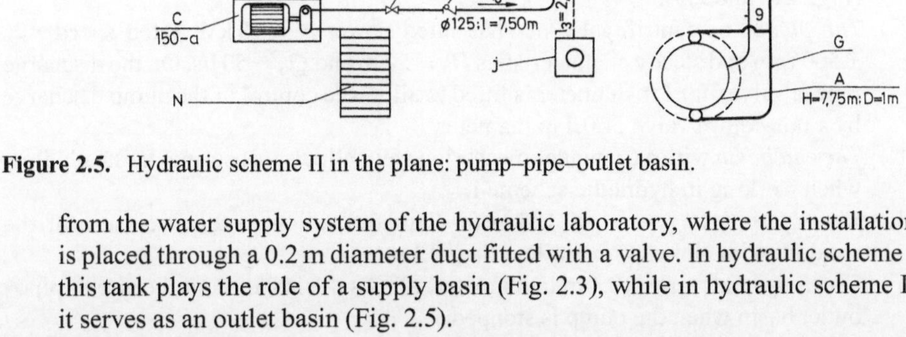

Figure 2.5. Hydraulic scheme II in the plane: pump–pipe–outlet basin.

from the water supply system of the hydraulic laboratory, where the installation is placed through a 0.2 m diameter duct fitted with a valve. In hydraulic scheme I this tank plays the role of a supply basin (Fig. 2.3), while in hydraulic scheme II it serves as an outlet basin (Fig. 2.5).

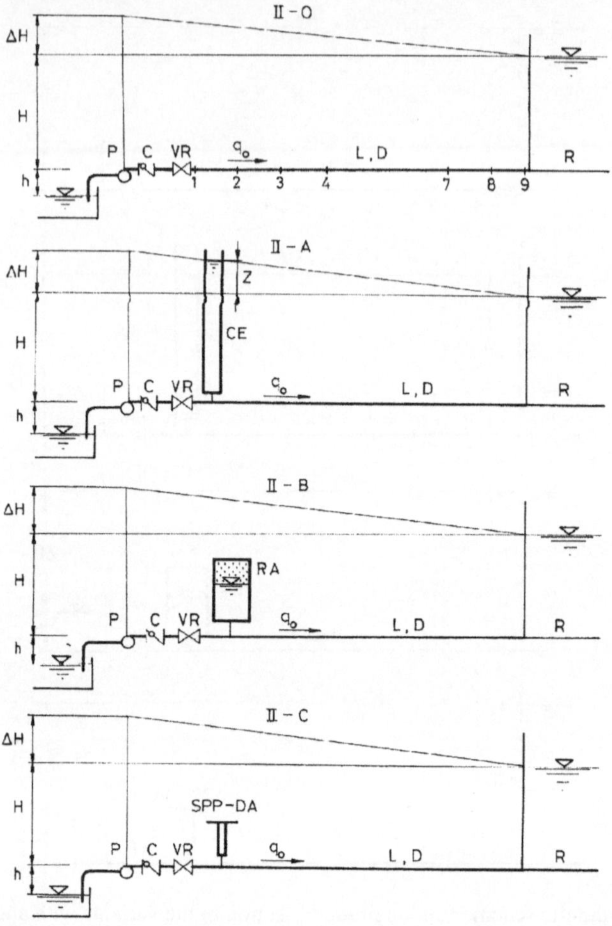

Figure 2.6. Hydraulic scheme II in longitudinal section in the variants with and without protection devices.

(B) *The pressure pipe*, made of steel, is 161 m long and 0.125 m in diameter. To measure the pressure in the pipe, eight sections equally spaced along the pipe ($\Delta x = 20$ m) were provided, where several sleeves were fitted to allow pressure transducers to be connected. The measuring points are numbered from 1 to 8 (Figs. 2.3 and 2.5).

(C) *The pump*, of centrifugal type, has rated power $N = 13$ kW, rated speed $n =$ 1,500 rpm and steady characteristics $H_o = 20$ m and $Q_o = 50$ l/s. On the discharge pipe, a valve of 0.1 m diameter is fitted to allow the control of the pump discharge by a non-return valve of 0.1 m diameter.

(D) *The end basin* with a triangular measuring weir, allows measuring the steady flow when working in hydraulic scheme I.

(E) *The rapid closing valve* (0.125 m in diameter), allows sudden closure of the conduit, with closing times t_i less than 0.1 s.

(F) *The non-return valve* (0.1 m in diameter) allows the isolation of the circuit pipe–outlet basin when the pump is stopped.

(G) *The outlet pipe* (0.2 m in diameter), ensures the discharge of the water overflow from the reservoir in both hydraulic schemes.
(H) *The surge tank*, $H = 10.5$ m in height and $D = 0.2$ m in diameter, can be connected to the installation in both hydraulic schemes.
(I) *The air chamber*, with a total volume of 140 dm^3, can ensure a controlled air cushion of volume 10–70 dm^3. It can be fitted to the installation in both schemes.
(J) *The protection device against waterhammer*. The following devices were tested:
- overpressure valve (SSP), 0.008 m diameter;
- air-device (DA), 0.15 m diameter;
- air inflow/outflow device (DAD), 0.1 m diameter.

(K) *The control valve* (0.1 m diameter), which allows control of the steady flow rate for hydraulic scheme I (Fig. 2.3) or for hydraulic scheme II (Fig. 2.5).
(L) *The control valve* (0.2 m diameter), which allows adjustment of the inflow to the constant-level reservoir for hydraulic scheme I.
(M) *The feed pipe* (0.2 m diameter) to the constant-level reservoir when hydraulic scheme I is operating (it works in an open circuit).
(N) *The suction basin of the pump*, to which the pipe coming from the overflow of the weir is connected. Thus, hydraulic scheme II operates in a closed circuit.
(P) *The section for connecting to the installation in both operational variants*, the experimental devices of protection: surge tank, air chamber, device against waterhammer.
(Q) *The air compressor*, which allows the necessary volume for the air cushion at steady flow.

The installation can operate in steady regimes with flow rates $q_o = 5$–25 l/s, corresponding to velocities $v_o = 0.4$–2 m/s in the pipe. As can be seen in Figs. 2.3 and 2.5, the systems and the devices for protection against waterhammer have been located in such a manner to be able to connect them to the experimental installation in Section 1, by using relatively short pipes.

To continuously measure and record the variable pressures in the pipe, pressure transducers, designed and manufactured by I.C.H., have been used and connected to the conventional devices of the institute. The used pressure transducers are of resistive type with *strain gauges* and belong to the class of transducers with plane elastic diaphragm. Among the technical characteristics of these transducers, it is worth mentioning the range, 1–50 atm, and the sensitivity of 1.5%.

The electronic installation offers the possibility of simultaneous measurements of pressure at five measuring points by using an oscillographic recorder type Visicorder-1,706 with six channels, which can be fitted to the channels of the tensometric bridges. The recorder is an ultraviolet light oscillograph recording on photographic paper and allowing development under natural light. The connection of the transducers to the measuring bridge is made using some special devices and double-wire tensometric cables.

The measuring error of the entire chain is about 1.5% and can operate under the control of a computer. More information on this alternative will be given in one of the following sections.

The installation was fitted with an automated electronic system of operating, measuring and data processing. In this way it can operate in the computer-controlled variant too. This variant will be discussed in a subsequent section.

2.1.2. Experimental results

Experiments with steady flow rates $q_o = 5$–25 l/s were performed, corresponding to the following head losses in the pipe: $\Delta H = 0.5$; 1; 1.5; 2–6 m. The phenomenon of waterhammer was experimentally produced and measured, using the hydraulic schemes I and II with and without protection devices against waterhammer, by keeping unchanged the feature of suddenly varying flow.

Hydraulic scheme I without protection (I–O)

The experimental installation, which represents the main hydraulic scheme I without protection, is made of the following devices: constant level supply tank, pressure pipe, rapidly closing valve, end basin with measuring weir (Figs. 2.3 and 2.4). With this scheme, a great number of experiments, for several steady flow rates with pressure recordings at many points of the pipe, have been performed.

Based on these measurements, a careful analysis of waterhammer for the case of cavitation appearing in the system has been made. The restoration of the water column into the pipe during the last stage of the cavitation phenomenon produces high noise, much like "hammer kicks in the pipe". These noises are periodic, following the periods of cavitation appearance [1, 3, 8, 18].

The results of the experiments are shown in a plot (Fig. 2.7) of the pressures measured simultaneously at points 1 and 7, for a steady flow-rate $q_o = 16.5$ l/s after sudden closing of the valve during $t_i < 0.1$ s [1]. It can be seen that at point 1 several pressure peaks appear immediately after the phenomenon of cavitation and they are superposed over the maximum amplitudes of the waterhammer.

Taking into account the practical and theoretical importance of this experimental result, which is also mentioned in some references [3], the I.C.H. intends to continue their own theoretical and practical studies on this topic. It should be mentioned that several papers [5, 6] report some results closely resembling ours for the I–O scheme.

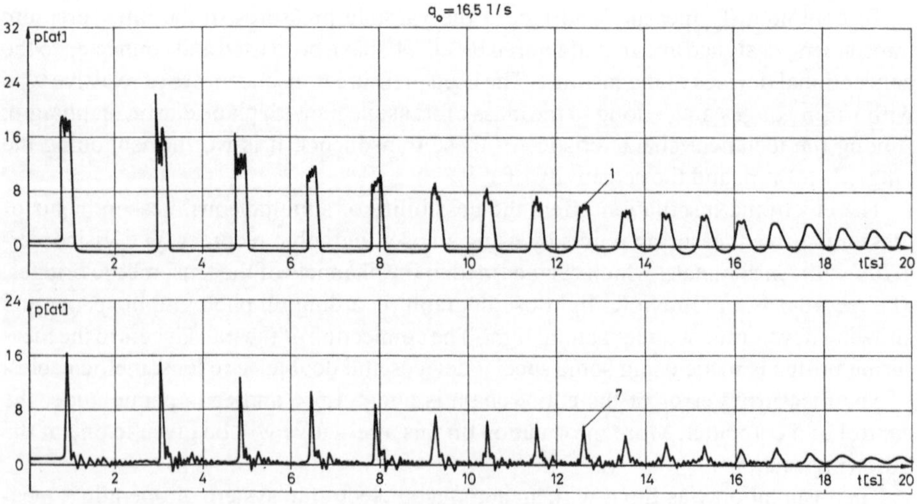

Figure 2.7. I–O: simultaneous pressure measurements at points 1 and 7 on the pipe ($q_o = 16.5$ l/s).

In reference [4], studies concerning the flow of two-phase bubbly water–air mixtures have been conducted on the same installation. The results shown in Figs. 2.8 and 2.9 show the influence induced by the air bubbles present in water on the extreme values of pressures. Thus, for an air concentration in the water of about 0.5%, the maximum pressures in the system are reduced by more than 50%; these reductions reach 70% for an air dilution of about 2%. Also, the minimum pressures in the system do not decrease below the atmospheric pressure for values of *a* higher than 1%.

For the same problem, in some other papers [7, 15, 16] quite similar results are given.

Hydraulic scheme I with protection devices against waterhammer (I–C)
Experiments have been performed using hydraulic scheme I with the following protection devices against waterhammer: overpressure valve (SSP) of 0.08 m in diameter; air chamber (DA) of 0.15 m in diameter; air inflow/outflow device (DAD) of 0.1 m in diameter.

The effect of the overpressure valve on the development of the waterhammer phenomenon within the pipe is shown by several experimental results in the plots of Figs. 2.10 and 2.11. Even if the valve characteristic (the discharge pressure) is 10.5 atm, some pressure peaks over this limit appear as a consequence of the response time of the valve in dynamic operation. Also, the elimination of cavitation phenomena in the pipe

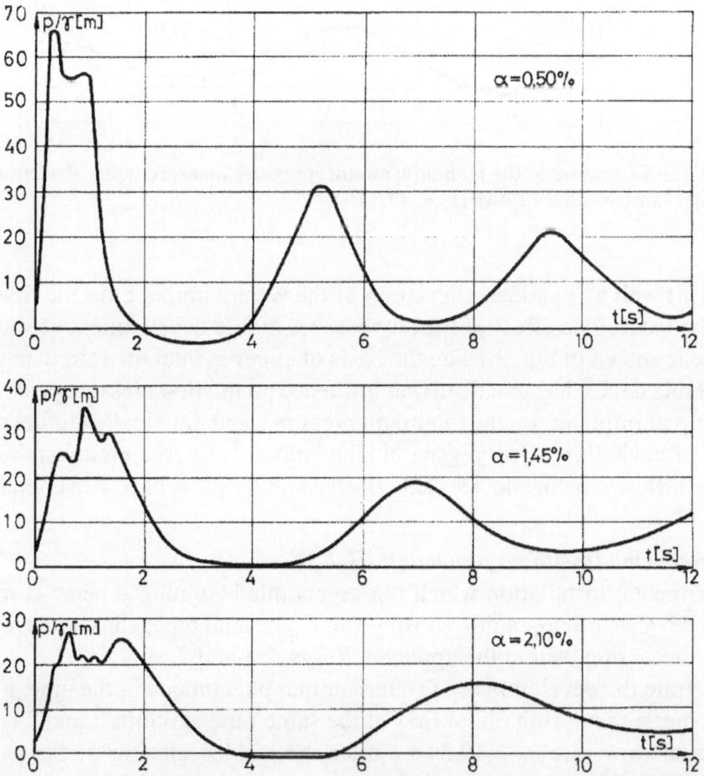

Figure 2.8. I–O: pressure measurements in Section 1 for various water–air two-phase flows.

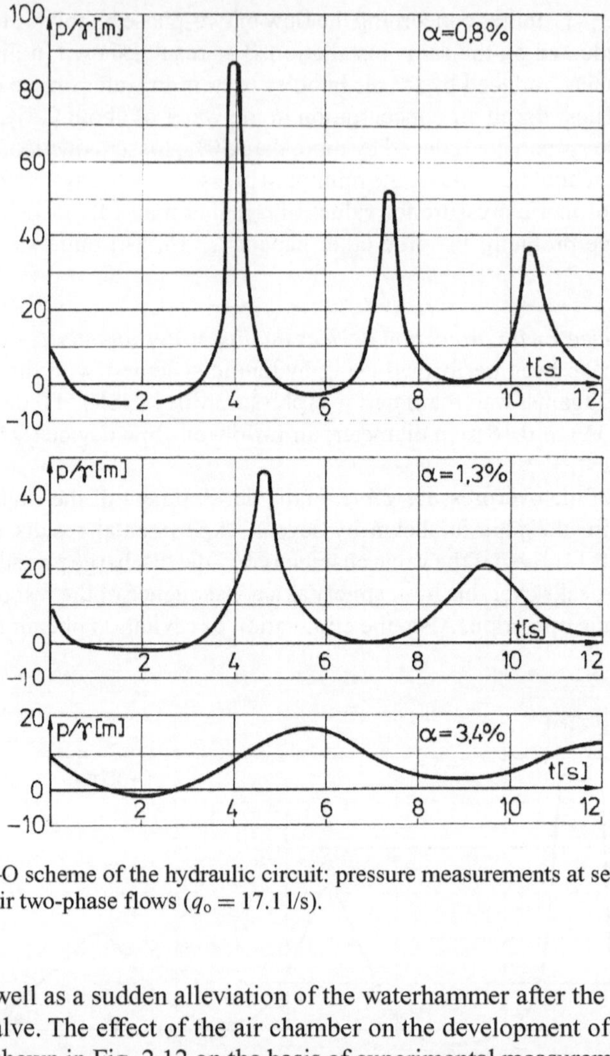

Figure 2.9. II–O scheme of the hydraulic circuit: pressure measurements at section 1 for various water–air two-phase flows ($q_o = 17.1$ l/s).

is noticed, as well as a sudden alleviation of the waterhammer after the first discharge through the valve. The effect of the air chamber on the development of waterhammer in the pipe is shown in Fig. 2.12 on the basis of experimental measurements. It is easy to see that this device has practically no influence on the first pressure peak in the pipe, but it has great influence on the minimum pressures and practically eliminates the phenomenon of cavitation. Also, a general alleviation of waterhammer can be observed. For the air inflow/outflow device (DAD) very similar plots have been obtained.

Hydraulic scheme II without protection (II–O)

The experimental installation which represents this hydraulic scheme is made of the following devices: pump, rapidly closing valve, pressure pipe, constant level discharge tank with measuring weir at the upper end (Figs. 2.5 and 2.6).

To illustrate the development of waterhammer phenomena in the pipe, a plot of the pressures measured during initial flow at the same time at points 1 and 7 is presented, for an initial flow rate $q_o = 19.2$ l/s , after the sudden closing of the valve during $t_i < 0.1$ s (Fig. 2.13). A violent waterhammer takes place in the pipe, quite similar to the one encountered in the case of hydraulic scheme I–O, which involves great peaks

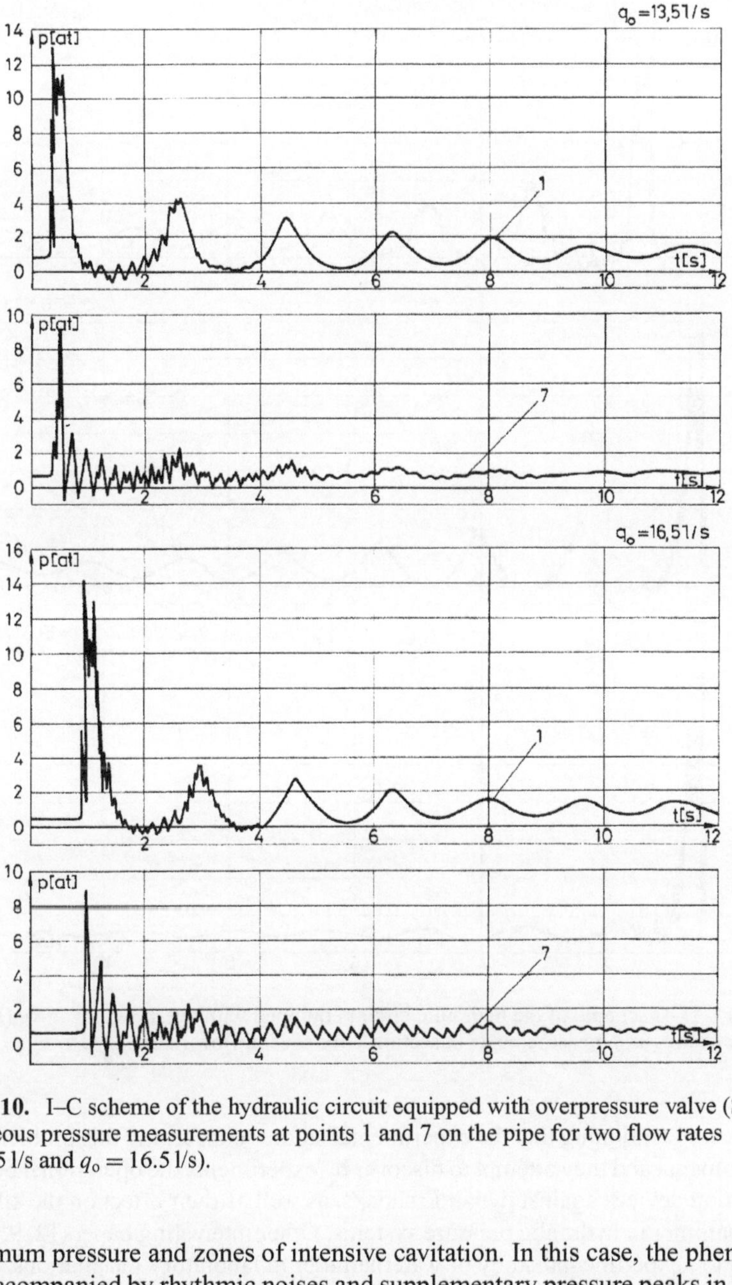

Figure 2.10. I–C scheme of the hydraulic circuit equipped with overpressure valve (SSP): simultaneous pressure measurements at points 1 and 7 on the pipe for two flow rates ($q_o = 13.5$ l/s and $q_o = 16.5$ l/s).

of maximum pressure and zones of intensive cavitation. In this case, the phenomenon is also accompanied by rhythmic noises and supplementary pressure peaks in the pipe.

Hydraulic scheme II with protection devices against waterhammer (II–C)

Experiments involving the above-mentioned three kinds of protection devices against waterhammer have been performed. Generally, the effect of these devices on the alleviation of the phenomenon of waterhammer is quite similar to that of case I–C, and several plots with experimental results are given in Figs. 2.14 and 2.15, for the overpressure valve (SSP) and for the air chamber (DA), respectively.

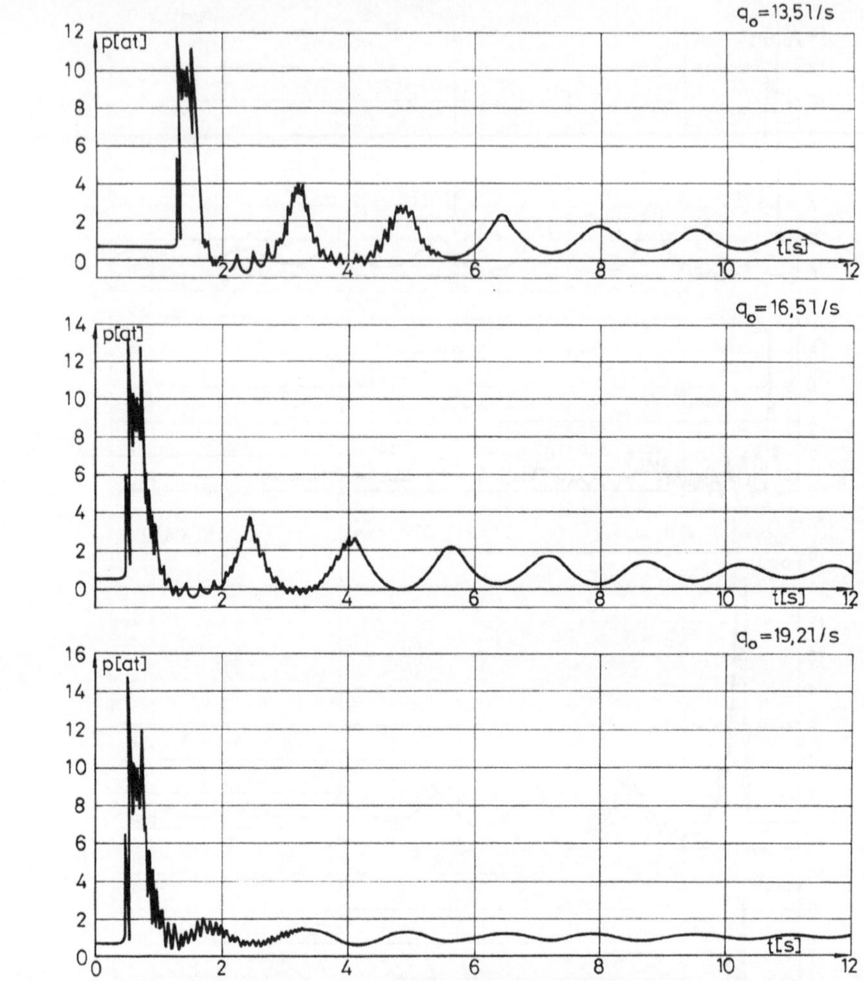

Figure 2.11. I–C scheme of the hydraulic circuit equipped with overpressure valve (SSP): simultaneous pressure measurements at section 1 for several flow rates.

It is worth mentioning that the experimental studies concerning this problem are the first in Romania and they attempt to discover by experiments the operational behaviour of protection devices against dynamic shocks, as well as their effect on the alleviation of waterhammer in hydraulic pressure systems. Other interesting papers [2, 9, 10] also deal with the experimental study of waterhammer in laboratory installations.

2.1.3. Alternative experimental installation operated by computer

A brief description of the electronic and electric equipment fitted with the installation is given below together with some explanation necessary to understand the operation of this equipment. The software (programs) for the process computer and for the computer which ensures the automatic operation of the laboratory installation is also presented (Fig. 2.16) [17].

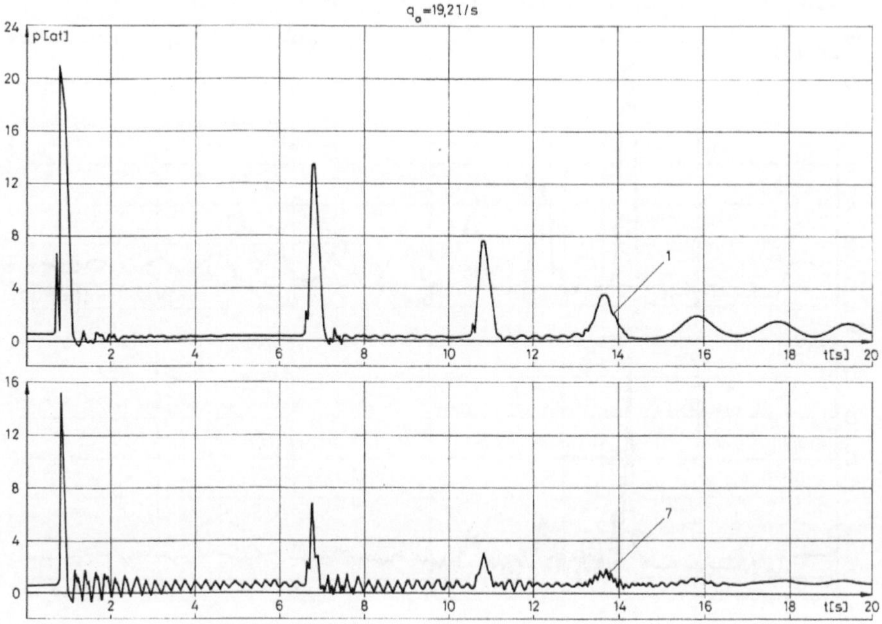

Figure 2.12. I–C scheme of the hydraulic circuit equipped with air device (DA): pressure measurements at two points on the pipe ($q_o = 19.2$ l/s).

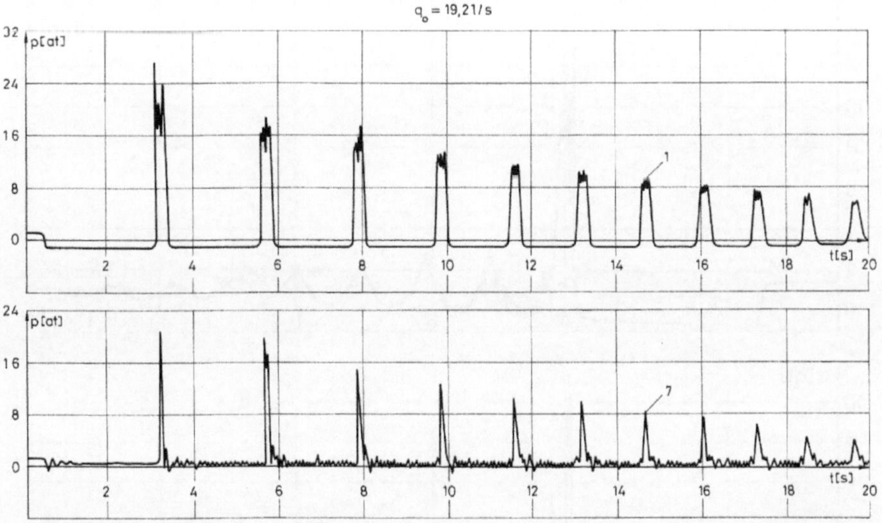

Figure 2.13. II–O scheme of the hydraulic circuit: simultaneous pressure measurements at point 1 and 7 on the pipe ($q_o = 19.2$ l/s).

Electric and electronic equipment

It should be mentioned from the beginning that the given hydraulic flow necessary for the experiment is obtained by acting upon the electrically operated valves of the hydraulic circuit of the model. These commands are given by the process computer

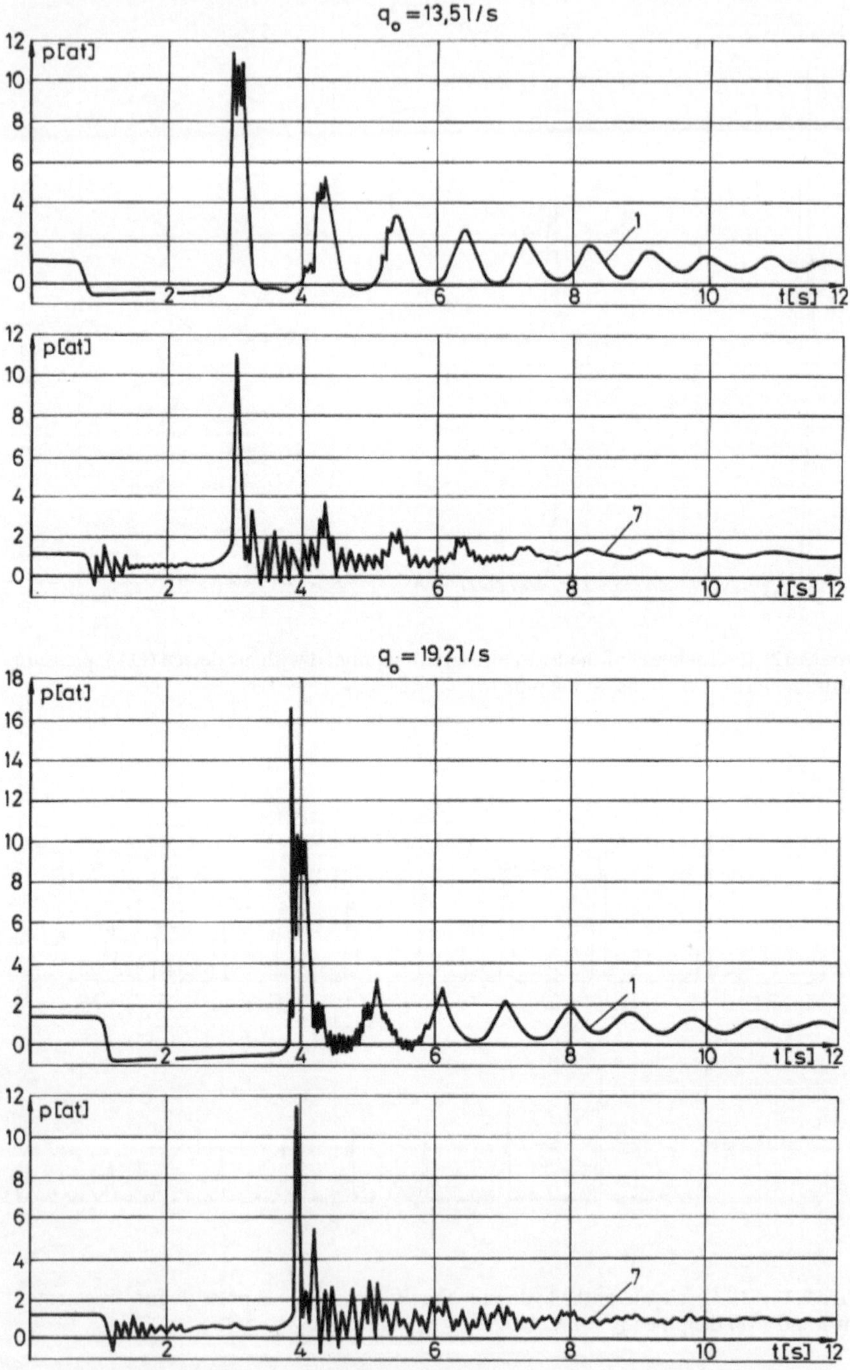

Figure 2.14. II–C scheme of the hydraulic circuit equipped with overpressure valve (SSP): simultaneous pressure measurements at points 1 and 7 on the pipe for two flow rates ($q_o = 13.5$ l/s and $q_o = 19.2$ l/s).

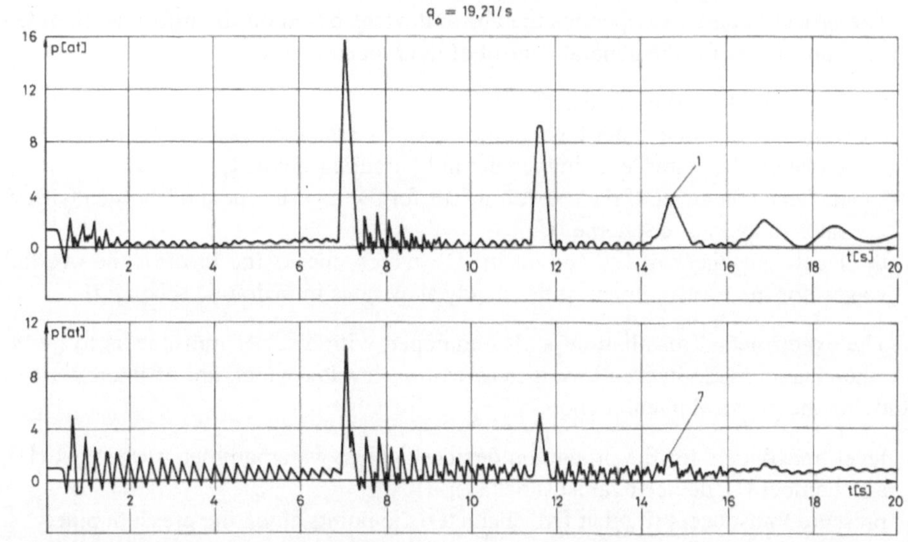

Figure 2.15. II–C scheme of the hydraulic circuit equipped with air devices (DA): pressure measurements at two points on the pipe ($q_o = 19.2$ l/s).

(a) 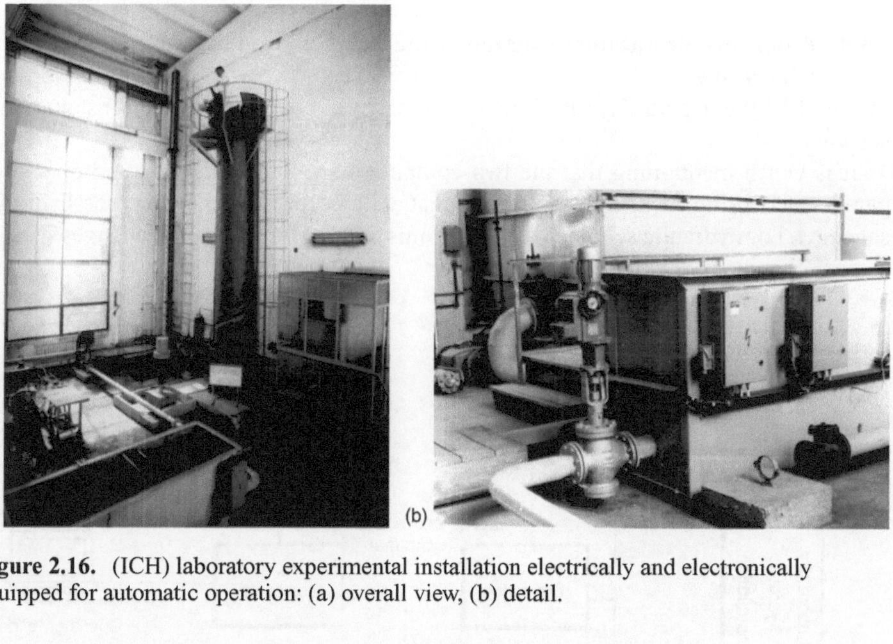 (b)

Figure 2.16. (ICH) laboratory experimental installation electrically and electronically equipped for automatic operation: (a) overall view, (b) detail.

through some control relays and contacts placed in the relay block. The actuators, electrically operated, are the following (Fig. 2.1):

- the general feed valve (L);
- the steady flow control valve in hydraulic scheme I (K);
- the steady flow control valve in hydraulic scheme II (K);
- the rapidly closing valve (E).

The process computer operates the above devices based on the information given by the transducers for the general control of the process, namely:

- the transducers for the valve position;
- the transducer for the water level in the tank (A), which plays a double role:
  - — it checks the water level in the tank in hydraulic scheme I;
  - — it checks the level of the overfall waves for the measurement of the steady flow rates in hydraulic scheme II;
- the level transducer of the end basin (D), which checks the level of the overfall waves, for the measurement of the steady flow rates in hydraulic scheme II.

The experimental installation is also equipped with a set of transducers to measure the characteristic hydraulic values (pressure, levels), which are of interest when studying the physical phenomenon:

- level transducers for the protection devices against waterhammer: surge tank (H), air chamber (I), device against shock (J);
- pressure transducers fitted at five characteristic points along the pressure pipe.

The measured values are temporarily stored in the memory of the process computer and then sent to the computer to be stored on diskettes and/or graphically processed and presented.

2.1.4. Programs for automatic action of the experimental installation in the laboratory

In Fig. 2.17 the general layout of the operation of the computer-driven installation is shown.

It is worth mentioning that the two computers are interconnected and operate in parallel according to special programs that will be presented below for hydraulic scheme I. For hydraulic scheme II the programs are similar with some changes specific to this scheme.

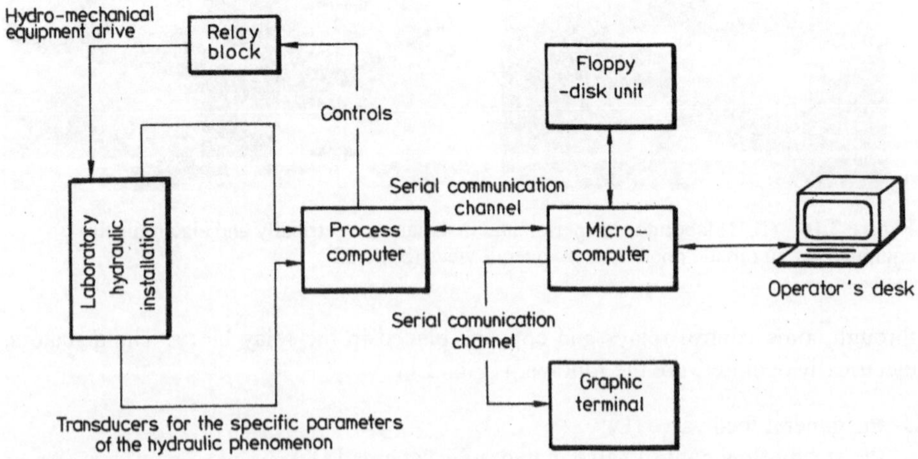

Figure 2.17. General operation scheme of the laboratory experimental installation controlled by computer.

The computer program

The main command and operation steps of this program are the following:

a) The choice of the experimental variant:
   - experiments only (in which case item j does not apply),
   - only data-processing (only item j is applied),
   - experiment and data-processing (all the following items apply).
b) Test of the communication means.
c) Loading (from diskette) of the control program for the hydraulic model.
d) Loading (from diskette) of the data characteristic for the electric and electronic control system configuration.
e) Loading from the keyboard, by the operator, of the characteristic values that define the steady flow rates from which successive experiments will be carried out.
f) Actualisation of the control program.
g) Loading of the control program and running this program on the process computer.
h) Observation of the experiment on the hydraulic model.
i) Recording of the experimental data and their storage onto disk.
j) Data processing by graphical plots of the obtained values.

The process computer program

The main command and execution steps in this program are the following:

a) Operation of the valve for the general supply of the hydraulic circuit.
b) Setting the steady flow rate to value imposed above (e).
c) Operation of the rapidly closing valve that closes the hydraulic circuit in order to produce the transient flows to be studied.
d) Measurement of the hydraulic characteristics (pressure, flow rates, levels) which are interesting for the case study.
e) Temporary storage of experimental results.
f) Transmission of the experimental results to the microcomputer.
g) Reading and execution of the commands given by the operator from the keyboard of the computer (for example, data concerning the state of the hydraulic system or the appearance of some errors in the process).
h) Repetition of the experiment with the next set of data loaded by the keyboard-operator at (e) above, until the completion of the desired program of experiments.

Finally, it should be mentioned that after two years of tests and hydraulic experiments, the laboratory installation in this operating variant—controlled by computer—proved to be an efficient tool in performing more elaborate experiments that contributed to a deeper understanding of some physical aspects of the phenomena generated by unsteady flows in hydraulic pressure systems.

2.2. Studies conducted at some pumping stations in operation

As compared to laboratory measurements, the measurements for the experimental study of waterhammer performed at pumping stations in operation are of special importance.

Indeed, the influence of certain factors involved in waterhammer (as for example, the valve shock, the inertia of the moving parts of the pump, and so on) can be better taken into account only in this kind of measurements. However, there are only a few papers [13], reporting on experiments performed on operating installations. In the following section we shall present some of the results obtained in a set of measurements performed at several pumping stations in operation in Romania, over a period of more than ten years [12]. In order to perform these measurements at the pumping stations, the conventional electronic devices already presented in the last sections were used.

2.2.1. The SRP-2 Balcescu pumping station

This station supplies a maximum flow of 5.6 m^3/s for a geodetic height of 28 m, and is fitted with four SIRET-900 electro-pumps, with the following characteristics: $Q_o = 1.4$ m^3/s and $H_o = 32$ m. The discharge of each pump consists of a metallic duct $L = 375$ m and $D = 1.6$ m (Fig. 2.18a). Along the discharge pipe of each pump ($d = 0.8$ m) some non-return and butterfly valves are fitted.

The experiment was performed under the following technical conditions:

- the station was shut down when working with one, two, or three pumps;
- the valves on the discharge pipes of the pumps did not entirely close at reverse flow, when the station was stopped.

The pressure was measured continuously at three points (1, 4, 7) along the discharge pipe when one, two or three pumps were stopped (Figs. 2.18b, c, d). The experiments

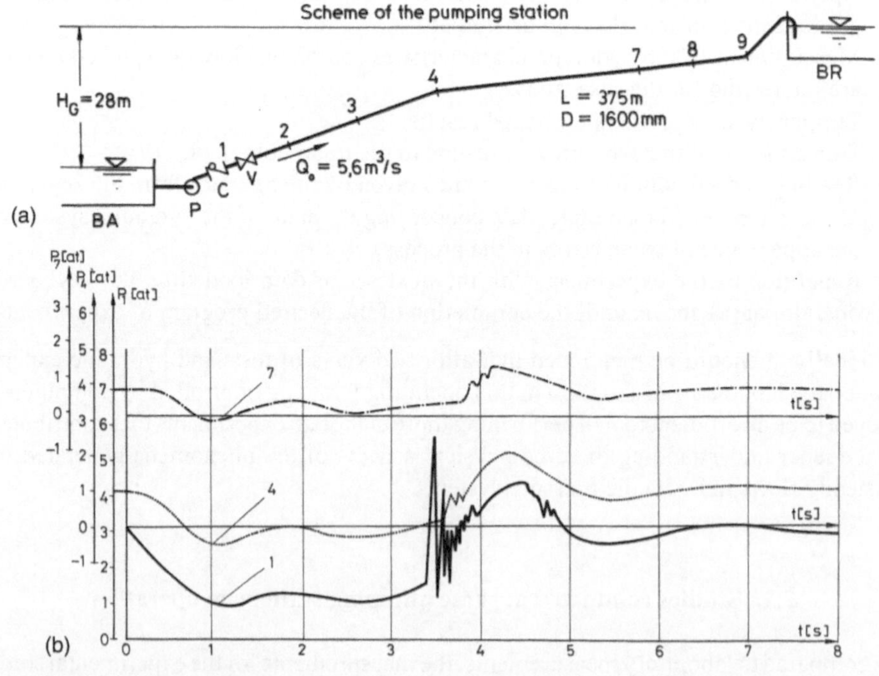

Figure 2.18. SRP-2 Balcescu pumping station: (a) station scheme, (b), (c), (d) pressure measurements at the total stop at the station with one, two and three operating pumps, respectively.

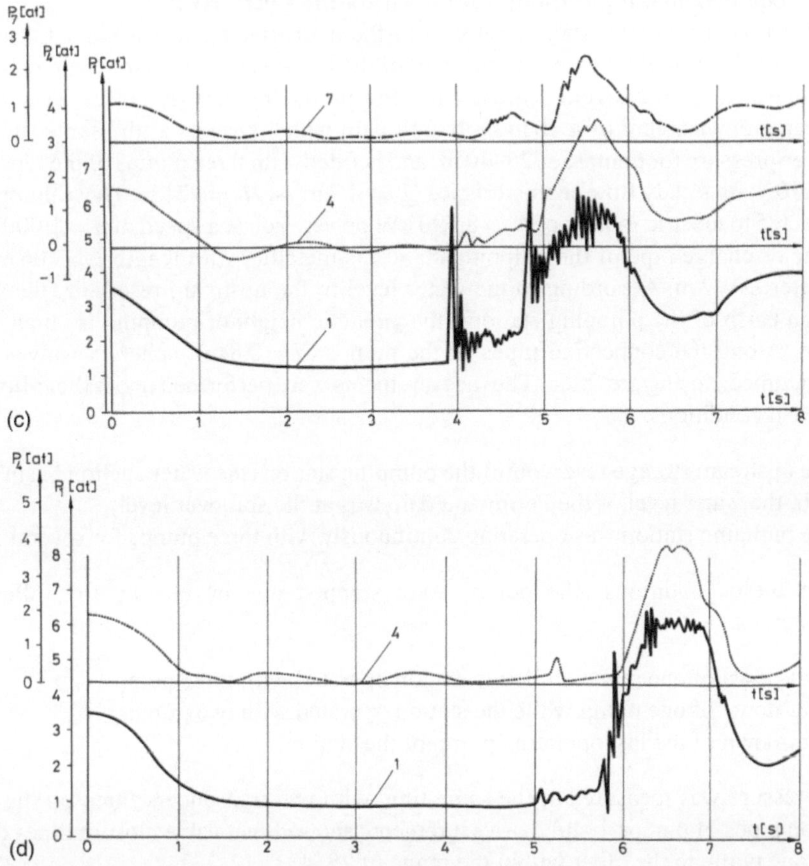

Figure 2.18. *(continued.)*

were performed while the pumps operated in different configurations (for example: pumps 2 + 3 or pumps 3 + 4, and so on) and were repeated several times.

The results reveal the appearance of rapidly varying pressures with maximum values around 6–7 atm and some negative pressures with large values in certain zones of the pipe, which can be explained by the absence of protection devices against waterhammer. The quick damping in time of the pressure oscillations is explained by the valve's leakage along the discharge pipe of the pumps, with a beneficial effect on dangerous pressure regimes.

It is certain that when the valves are closed entirely, the accidental closure of the pumps involves much higher pressure values. Even the faulty operation of some valves or their accidental crash during operation testifies to the high values of pressure due to violent waterhammer phenomena, found by us when performing later computations.

As a result of this hydraulic study two solutions were proposed for the protection of this station against waterhammer:

a) to equip the station with programmed closing values ($t_i = 5$ s) instead of the existing ones;
b) to protect the station by providing it with an air chamber of 30 m^3.

2.2.2. The Petrimanu pumping station within the Lotru HPP

The Petrimanu pumping station belongs to the hydroelectric power plant Lotru and plays the role of pumping part of the water of the Latorita river into the main reservoir of Vidraru, to be used from a greater head in producing electric power. The station is located downstream of a 40 m high arch dam which ensures at the same time the counter-pressure for pumps, of 25–40 m, and is fitted with three pumps of the type Pro-105-170, with steady flow characteristics $Q_o = 4.7\ m^3/s$, $H_o = 171$ m. These pumps are connected to electric motors of $N = 8{,}790$ kW power each at a speed of $n = 1{,}000$ rpm.

The discharge pipe of the pumping station is metallic, with length $L = 406$ m and diameter $D = 2$ m. According to the water level in the upstream reservoir (the water suction basin of the pumping station), the geodetic height of pumping is about 143–160 m. Along the connecting pipes of the pumps ($d = 0.8$ m), spherical valves with programmed closing are fitted. The measurements were performed under the following technical conditions:

- the upstream storage reservoir of the pumping station (the water suction basin) was full, the water level of the Petrimanu dam was at the spillover level;
- the pumping station was operating continuously with three pumps for several days.

Under these conditions, the pumps were stopped one by one by the following manoeuvres:

- shut-down of one pump while the station operated with three pumps;
- shut-down of one pump while the station operated with two pumps;
- shut-down of the last operating pump of the station.

The pressure was measured at the same time with two transducers fitted on the connection pipes of pumps 1 and 2. Measurements showed that valve closure times differ from one pump to the other within the range of 28–43 s [12, 13].

Some of the measurements are plotted in Fig. 2.19 and exhibit a good hydraulic operation of the station to unsteady flows, the solution of providing the station with programmed closing valves proving efficient, according to our experiments, but one should note that the pumps operated with a 40 m counter-pressure (the downstream reservoir was full, up to an elevation of 1,130 m). It is worth mentioning that starting the pumps involved insignificant pressure variations in the discharge pipe.

The same basic data of the analysed hydraulic system (the water level in the lake during suction, the effective time for closing the valves etc.) obtained by field measurements, were used to simulate by computations the manoeuvres for which measurements were performed at the station. By comparing the results of experiments and computations, one can notice that the numerical computation method satisfactorily simulated the phenomenon, and covered both the maximum and minimum pressures in the system. Within this hydraulic study, many manoeuvring alternatives for an accidental shutdown of the station which might occur during operation, were numerically simulated.

Thus, computations were performed for a total halt of the station, while operating with one, two, or three pumps, in different operation situations (empty or full water section reservoir different valve closure times etc.). The analysis of computation results shows that, during operation, minimum pressures of 9–12 atm and maximum pressures of 21–23 atm can possibly occur. As far as the maximum possible overpressures due to waterhammer are concerned, the computations showed that they range between

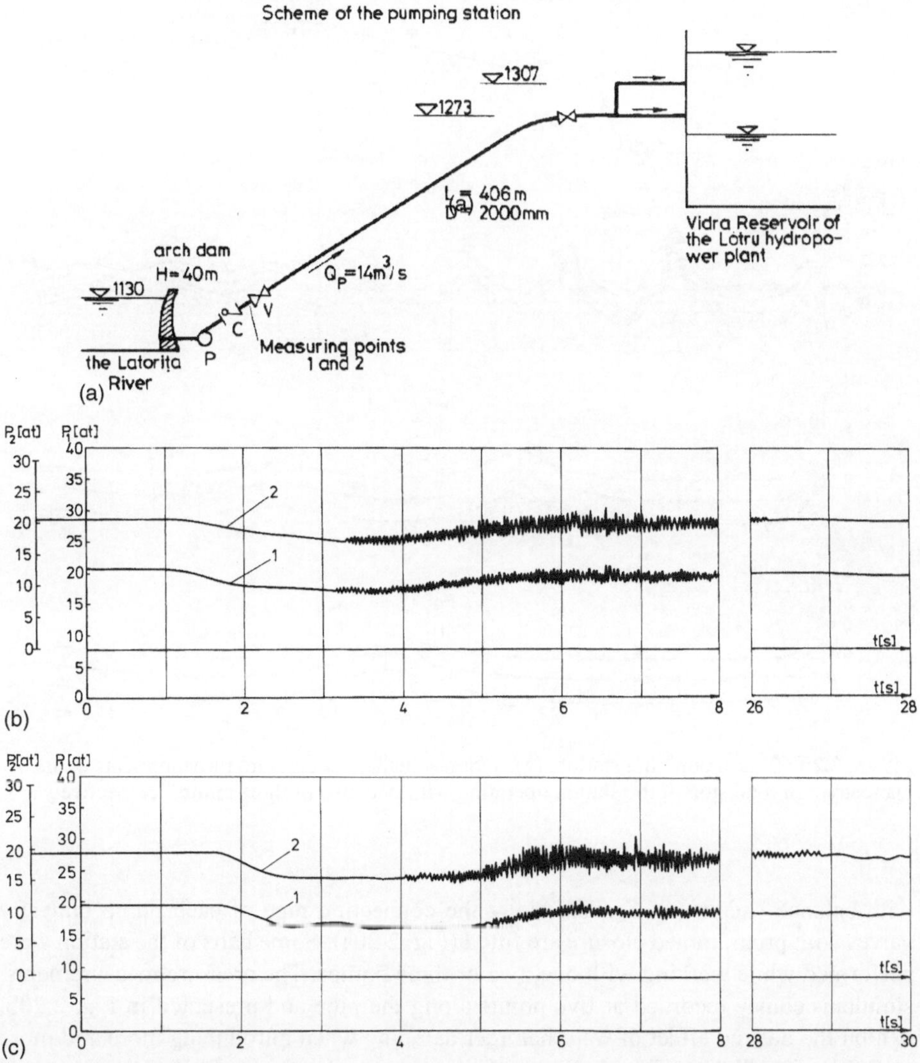

Figure 2.19. Petrimanu pumping station from Lotru hydroelectric power plant: (a) station scheme; pressure measurements at: (b) the partial stop of the station operating with two pumps (pump no.1 is stopped); (c) total stop of the station operating with one pump (pump no.2 is stopped).

3–4 atm. This indicates a good behaviour of the installation when operating in hydraulic unsteady flows.

2.2.3. The SP Tataru pumping station

The pumping station has a steady flow rate of 25 m³/s for a geodetic height of 26 m and is provided with four electro-pumps, type RV-120-170. The pumps are actuated by synchronous electric motors of $N = 3{,}150$ kW each for a speed of $n = 375$ rpm.

The discharge pipe of the station is metallic, with a length $L = 1{,}445$ m and a diameter $D = 2.8$ m. At three of its sections, the pipe is fitted with air inflow/outflow devices

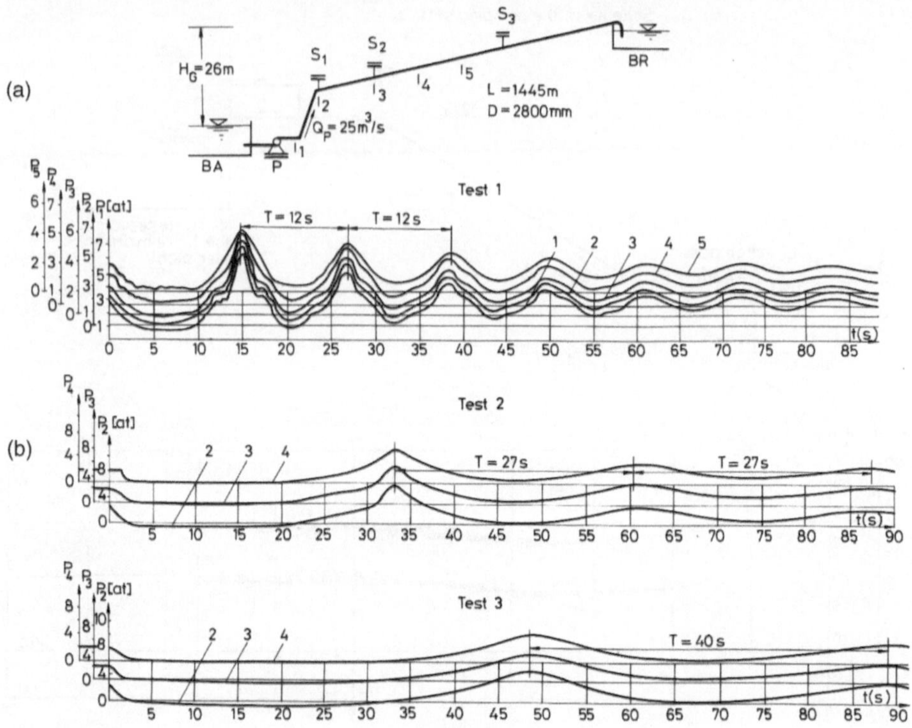

Figure 2.20. Tataru pumping station: (a) station scheme; (b) pressure measurements at the manoeuvre of total stop of the station operating with one, two or three pumps, respectively.

with balance weight (S_1, S_2, S_3). Along the connecting pipe of each pump butterfly valves with programmed closure are fitted (Fig. 2.20a). Some halts of the station were performed while working with one, two or three pumps. The pressure measurements, simultan- eously recorded at five points along the pipe and presented in Fig. 2.20b, exhibit the desired effect of waterhammer damping when introducing more and more air into the installation. The appearance of a large negative pressure during the first stage of the phenomenon, which results from the experiments, is explained by the delay in the time response of the air inflow/outflow device for negative pressures. However, the effect of air introduced into the discharge pipe of a pumping installation is not always beneficial. The outflow of the air from the pipe can produce great accelerations of the fluid particles on the separation surface of the two phases and give violent pressure variations in the entire hydraulic system [12].

The *waterhammer* into a pressure pipe, when an air cushion exists, is very complex and no satisfactory computation methods to simulate it have been developed as yet. The performed pressure measurements suggested the idea of a computation model for the waterhammer, conceived with fictitious air chambers into the system, sized in such a manner as to reproduce experiments 1 and 2. This computation model allowed finding with acceptable accuracy the measurements of experiment 3, as well as studying some new operating conditions of the hydraulic installation corresponding to several analysed versions of protection.

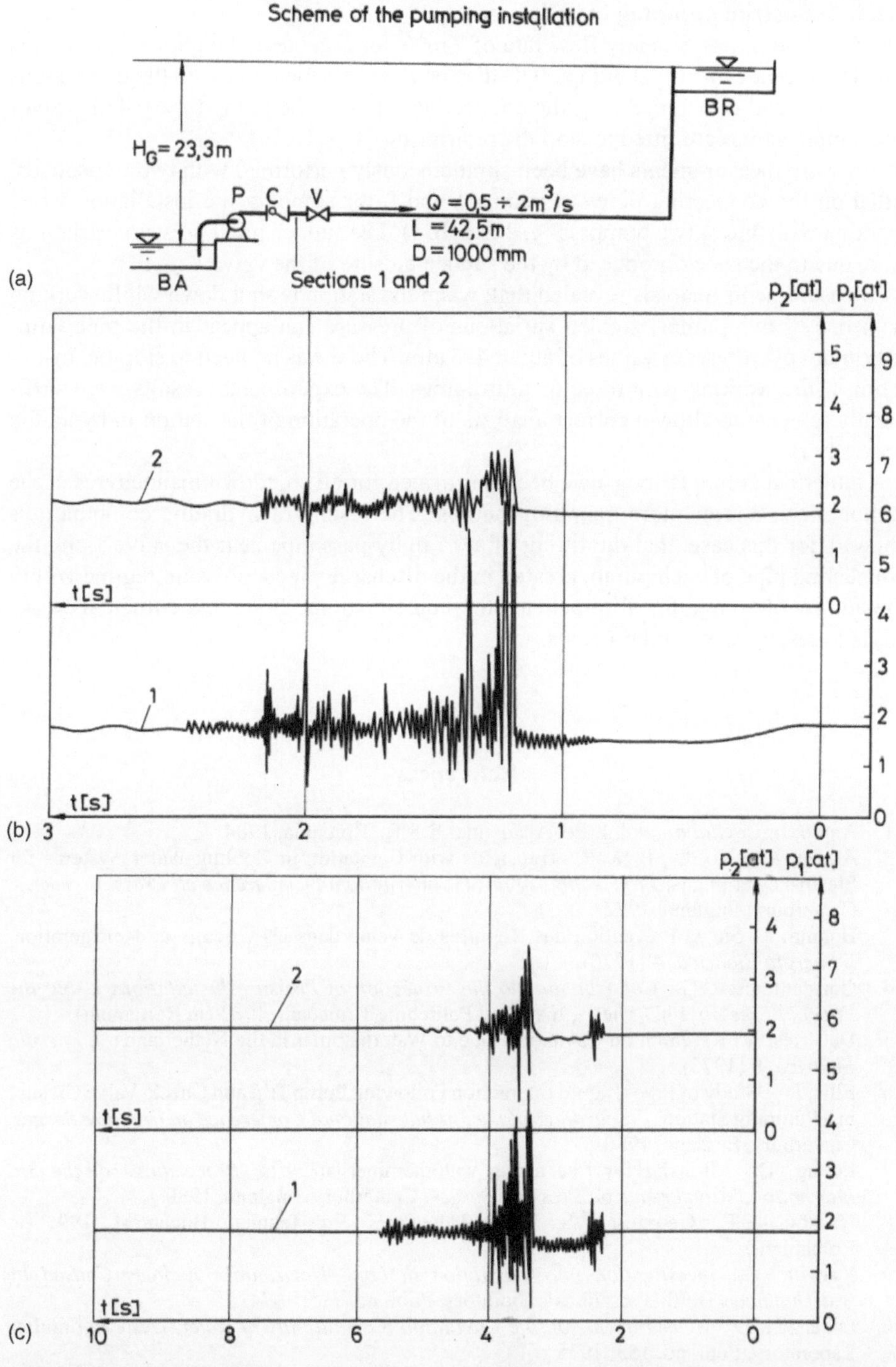

Figure 2.21. Industrial pumping installation: (a) installation scheme; pressure measurements near check valves; (b) partial stop of the station operating with two pumps; (c) total stop of the station operating with one pump.

2.2.4. Industrial pumping installation

The installation has a steady flow rate of $2\,m^3/s$ for a geodetic height of 23 m and is fitted with four centrifugal pumps. The discharge pipe is metallic, with the dimensions $L = 42.5$ m and $D = 1$ m. Along the connecting pipes of the pump ($d = 0.4$ m) valves and simple valves are fitted to stop the return flow (Fig. 2.21a).

Pressure measurements have been simultaneously performed with two transducers fitted on the connecting pipes to pumps 1 and 2 for stopping the installation while working with one or two pumps (Fig. 2.21a, b, c). The sudden great pressure variations were due to the shock produced by the sudden closure of the valve.

Measurements analysis revealed that, when the station is shut down while working with one or two pumps, sudden variations of pressure can appear in the pipe with maximum overpressure values of about 4–5 atm. There was no need to stop the installation while working with three or four pumps. The experimental results were sufficiently relevant to allow a correct analysis of the operation of the station in hydraulic unsteady flow.

Numerical computations have been performed for all shutdown manoeuvres at the station, with one or more operating pumps. The results of hydraulic computations showed, for this case, that the fitting of a 0.1 m by-pass pipe near the valve along the connecting pipe of each pump, created in the discharge pipe a pressure regime within the admissible range for all practical stopping situations. Some experimental works [6, 14] present very similar results.

References

1. Anton, I.: *Cavitation*, vol. I. Ed. Academiei R.S.R., Romania, 1984.
2. Atzeni, A., Cao, C., Piga, E.: Transients with Cavitation in Cooling-Water Systems for Steam-Power plants. *Proceedings of the 1st International Conference on Pressure Surges*, Canterbury, England, 1972.
3. Bonnin, J.: Sur la Prevention des Ruptures de Veine dans les Circuits de Refrigeration. *L'Energia Elettrica*, 4 (1975).
4. Constantinescu, G.: *Contributions to the Protection of Pressure Installations Using the Effect of Free Air.* PhD Thesis, Institutul Politehnic Timisoara, 1983 (in Romanian).
5. De Vries, A.: Research on Cavitation Due to Waterhammer in the Netherlands. *L'Energia Elettrica*, 9 (1975).
6. Ellis, J.: A Study of Pipe–Liquid Interaction Following Pump Trip and Check-Valve Closure in a Pumping Station. *Proceedings of the 3rd International Conference on Pressure Surges*, Canterbury, England, 1980.
7. Ewing, D.: Allowing for free air in waterhammer analysis, *Proceedings of the 3rd International Conference on Pressure Surges*, Canterbury, England, 1980.
8. Isbasoiu, C.E., Georgescu, S.: *Fluid Mechanics*. Ed. Tehnica, Bucharest, 1995 (in Romanian).
9. Kalwijk, C.: *Investigation into Cavitation in Long Horizontal Pipelines Caused by Waterhammer*. Delft Hydraulics Laboratory, Publ. no. 115, 1974.
10. Oldenziel, D.: *Measurements on the Cavitation Susceptibility of Water.* Delft Hydraulics Laboratory, Publ. no. 153, 1975.
11. Popescu, M.: Experimental Results in Studying the Unsteady Flow in Hydraulic Pressure Systems. *Studii si Cercetari de Mecanica Aplicata*, Ed. Academiei, 41 (1982) 453–458, 595–609 (in Romanian).
12. Popescu, M.: *Pressure Measurements at Pumping Stations in Romania Operating in Hydraulic Transient Flow.* Rapoarte interne ICH, 1974–1982.

13. Prevost, R., Tison, G.: *Coups de Belier dans les Conduites de Refoulement.* Revue C-Tijdschrift, 2 (1966).
14. Provoost, G.: The Dynamic Behaviour of Non-Return Valve. *Proceedings of the 3rd International Conference on Pressure Surges*, Canterbury, England, 1980.
15. Tullis, J., Streeter, V.: Waterhammer Analysis with Air Release. *Proceedings of the 2nd International Conference on Pressure Surges*, London, England, 1976.
16. Wylie, E.: Free Air in Liquid Transient Flow. *Proceedings of the 3rd International Conference on Pressure Surges*, Canterbury, England, 1980.
17. Popescu, M., Vlase, P.: L'etude des mouvements non-permanents des systemes hydrauliques sous pression sur une installation de laboratoire controlée par microcalculateur. *XXII Convegno di Idraulica e Construzioni Idrauliche*, Cosenza, Italy, Vol. 4, 1990, pp. 39–49.
18. Cazacu, M., Gheorghe, F.: *Fluid Suction into the Conduit of Flow-Meter Calibration Stands*. Ovidius University Annals of Constantze, Year III–IV, Series Constructions, 2002, pp. 243–248.

Chapter 3

The Protection of Pumping Stations Against Waterhammer

3.1. The classification of protection devices against waterhammer at pumping stations

When pressure oscillations due to waterhammer in hydraulic systems exceed in amplitude the extreme dangerous values (maximum pressure higher than the allowed maximum pressure or minimum pressure close to the cavitation pressure), the hydraulic systems are provided with protection devices. The protection devices alleviate or eliminate the waterhammer phenomenon in the entire hydraulic system, or in zones where the system security is threatened.

According to their nature, the protection devices can alter the nature of the unsteady flow in the hydraulic pressure system by transforming the *rapidly varying* movement into a *slowly varying* one. The protection devices of pumping stations against waterhammer can be classified in two main groups [3, 4, 6]:

I. Protection devices which alleviate or limit the amplitude of pressure oscillations in ducts while keeping the rapidly varying character of flow (waterhammer).
II. Protection devices which transform the rapidly varying movement (waterhammer) into a slowly varying one (mass oscillation) in the entire hydraulic system or, at least, at certain points of the system.

The following devices belong to group I: valves with programmed closure, the overpressure valves, the valve by-pass ducts, the inertia flywheel, etc. The following devices belong to group II: the surge tanks, the air chambers and the one-way reservoir.

Practically, these devices are used alone or in combinations. Air chambers and surge tanks have a large applicability.

The air chambers can be used in practically all pressure and flow conditions in which the pumping stations are operating; but, for economic reasons, surge tanks are generally used for geodetic heights over 20–25 m [7].

3.2. The model of compressible fluid and the model of incompressible fluid

In order to calculate the *waterhammer* in a hydraulic pressure system the compressible fluid model is used, while in the computations of *mass oscillations* either the model of compressible fluid or the model of incompressible fluid can be used.

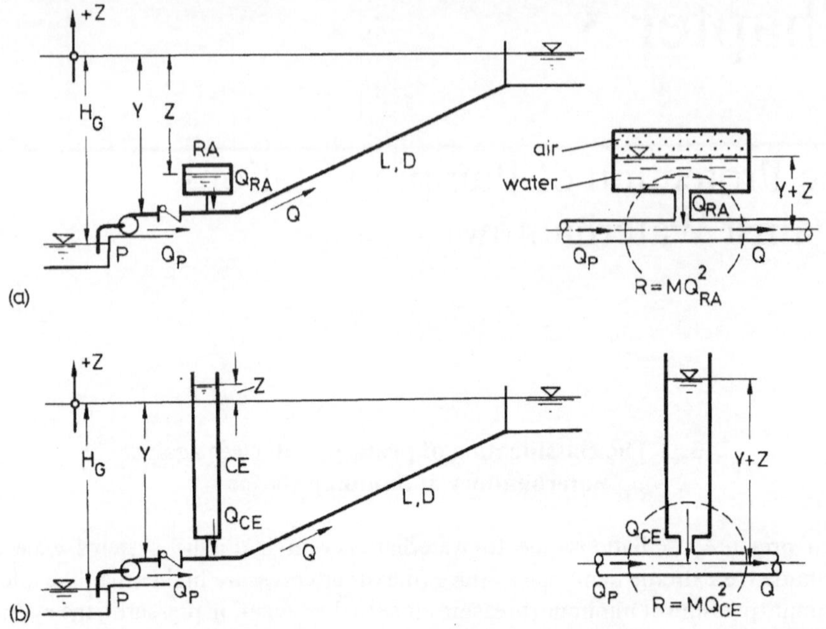

Figure 3.1. Scheme of a pumping station equipped with: (a) air chamber; (b) surge tank.

To get a clear view of the model, we shall consider a pumping station provided with an air chamber or a surge tank (Fig. 3.1) [3]. The *rapidly varying* water flow in the system (the waterhammer) is described by the equations

$$\begin{aligned} &\frac{\partial H}{\partial x} + \frac{1}{gf}\frac{\partial Q}{\partial t} + kQ\,|Q| = 0 \\ &\frac{\partial H}{\partial t} + \frac{c^2}{gf}\frac{\partial Q}{\partial x} = 0 \end{aligned} \tag{3.1}$$

where H is the piezometric height measured with respect to an horizontal reference surface, Q the flow-rate, f the cross-sectional area of the pipe, t time, g the gravitational acceleration, c the celerity, k the coefficient of longitudinal head losses in the pipe.

The *slowly varying* water flow in the system (the mass oscillation) is described by the following equations:

- For the case of a surge tank (Fig. 3.1b)

$$\begin{aligned} &\frac{L}{gf}\frac{\mathrm{d}Q}{\mathrm{d}t} - Z + P'Q\,|Q| + M'Q_c\,|Q_c| = 0 \\ &Q_P = F\frac{\mathrm{d}Z}{\mathrm{d}t} + Q \\ &Q_P = Q_P(t) \end{aligned} \tag{3.2}$$

where Z is the water level in the surge tank with respect to the water level in the outflow basin, Q the flow rate in the pipe, Q_c the inflow/outflow rate in the surge tank, Q_P the pumping flow rate, L the length of the pipe, f the cross-section area of the pipe, F the horizontal area of the surge tank at the water level, M' the local coefficient of head losses in the diaphragm of the surge tank, t time, g the gravitational acceleration.

- In the case of the station fitted with an air chamber (Fig. 3.1a)

$$\begin{aligned} &\frac{L}{gf}\frac{\mathrm{d}Q}{\mathrm{d}t} - (H+Z) + P'Q\,|Q| + M'Q_R\,|Q_R| = 0 \\ &Q_R = F\frac{\mathrm{d}Z}{\mathrm{d}t} + Q \\ &(H+h_a)V = (H_o + h_a)V_o = K \\ &Q_P = Q_P(t) \end{aligned} \tag{3.3}$$

where, moreover, the following notation is involved: Z is the level of the water in the air chamber with respect to the level in the outflow basin, (positive when bigger), Q_R the inflow/outflow rate of the air chamber, H the pressure in the air cushion, V the volume of the air cushion, h_a the atmospheric pressure in m.q.w., C the elastic constant of the air cushion (the subscript 0 stands for steady flow).

The water pressure in the discharging pipe in a section of the surge tank or the air chamber, respectively, is given by the relations

$$\begin{aligned} &\frac{p_c}{\gamma} = (Y+Z) - M'Q_c\,|Q_c| \\ &\frac{p_r}{\gamma} = (H+Y+Z) - M'Q_R\,|Q_R| \end{aligned} \tag{3.4}$$

As can be seen by comparing eqns. (3.2) and (3.3), the flow equations for the surge tank can be obtained from the equations for the air chamber by making $H = 0$.

A question arises quite normally: which are the practical applicability limits of the equations of mass oscillation (the incompressible fluid model) when the air volume in the air chamber is larger or smaller, or when the local hydraulic drag at the inflow of the surge tank or air chamber is larger ? As no strict limitation can be made between the rapidly varying and slowly varying flow of fluids in the hydraulic pressure system, there is no limitation in applying the model of compressible fluid or the model of incompressible fluid. Reference [3] shows that a limitation in the applicability of the two models can be established by accepting an admissible limiting value for the relative error (ε) which appears when computing the extreme pressures in the system by using the model of incompressible fluid instead of the model of compressible fluid.

In the long run, a *criterion for the equivalence* of the two models has been obtained. It is expressed by the following simple relation

$$T_L \geq \frac{\pi}{2\sqrt{3\varepsilon}}T_R \tag{3.5}$$

where $T_L = 2\pi\sqrt{LF/gf}$ is the period of mass oscillation, and $T_R = 4(L/c)$ is the period of waterhammer in the system.

This criterion of equivalence, which is obtained by linearising the flow equations (3.1), (3.2), (3.3) is supposed to have the same validity for the case of full non-linear flow equations, as there are no explicit drag coefficients involved.

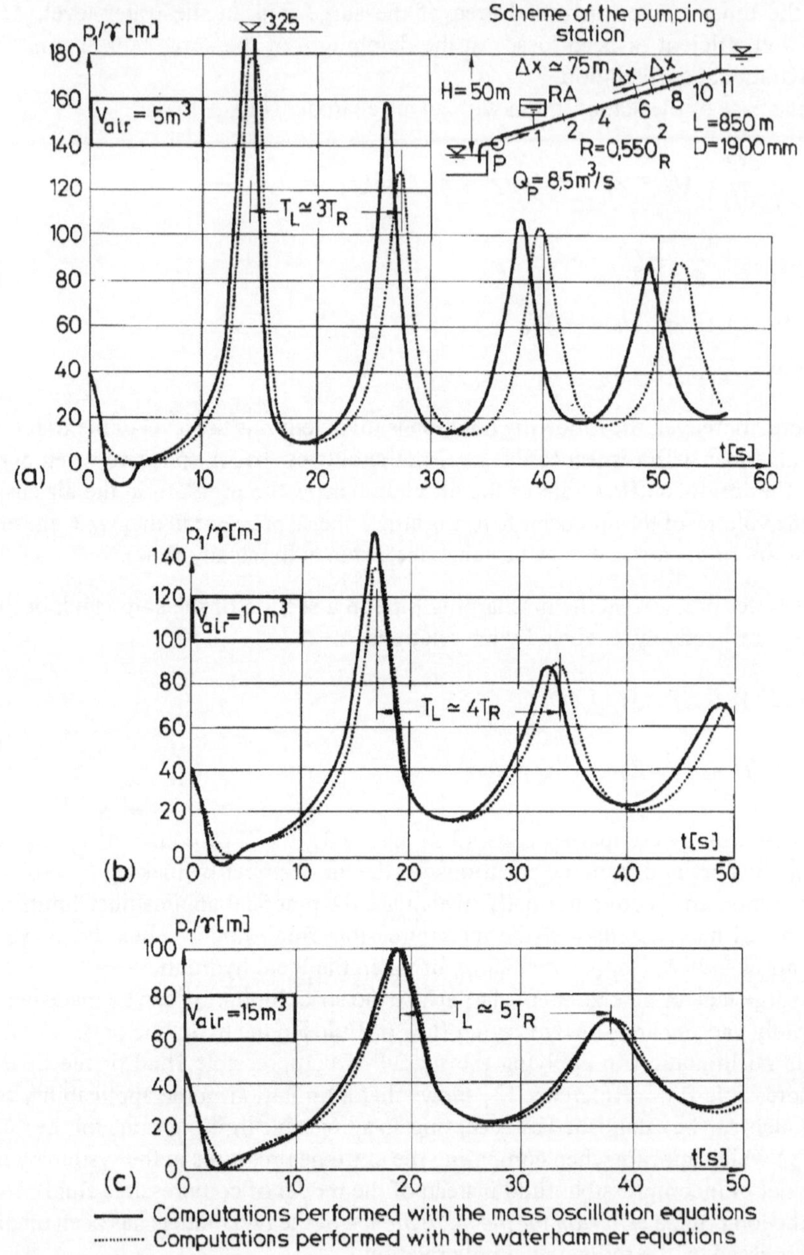

Figure 3.2. Numerical simulation of some variable motions in a pumping installation with an air chamber.

The period T in the criterion of equivalence can be the period of the forced pressure oscillations or the period of the free oscillations of the system, as only the damping factor separates the two kinds of oscillations. However, the damping factor is practically the same in both models. A confirmation of this remark can easily be obtained by numerical computations.

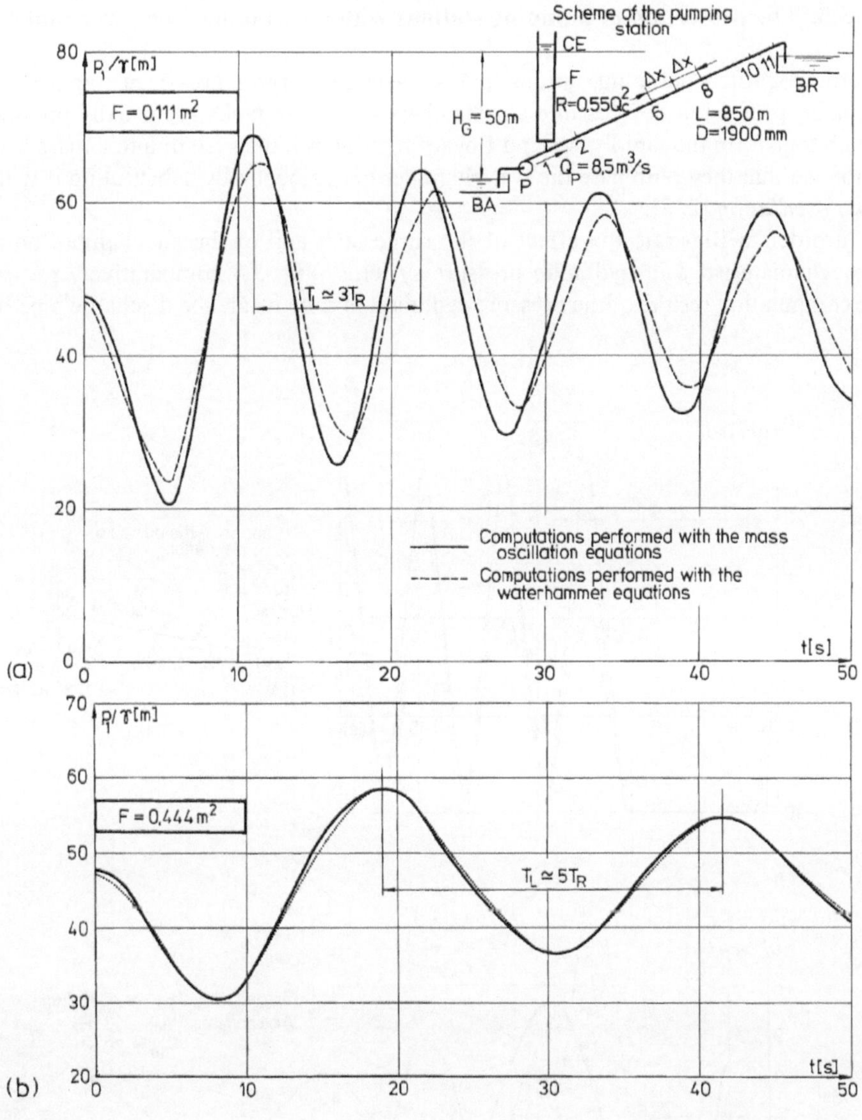

Figure 3.3. Numerical simulation of some variable motions in a pumping installation with a surge tank.

The computations have been performed for a pumping station provided with an air chamber or with a surge tank. The main characteristics of this pumping station are $H_G = 50$ m, $Q_o = 8.5$ m^3/s; the discharge pipe is $L = 850$ m long and $D = 1.9$ m in diameter. Several values of the volume V_o of the air cushion and of the area F of the horizontal section of the surge tank have been taken into account. The results of the numerical computations, plotted in Figs. 3.2 and 3.3, satisfy the criterion (3.5) of equivalence in the form $T_L \geq (5\text{–}6)T_R$, where an error $\varepsilon = 0.02$ was admitted. Knowing the applicability limits of the model of incompressible fluid, the problems of the unsteady flow in hydraulic pressure systems can be treated either by using the mass oscillation equations or the waterhammer equations.

3.3. The protection of pumping stations with air chambers or surge tanks

In order to protect pumping stations against the large overpressures occurring in waterhammer, protection devices like air chambers and surge tanks are usually provided, which transform the rapidly varying flow of the water in the system into a mass oscillation, so that they eliminate the waterhammer phenomenon by substituting it with a *mass oscillation* [2, 8].

In order to illustrate the effect of the surge tank and of the air chamber on the flow characteristics in hydraulic pressure systems, Fig. 3.4 comparatively presents the computation results of the pressure evolution in time inside the discharge pipe of a

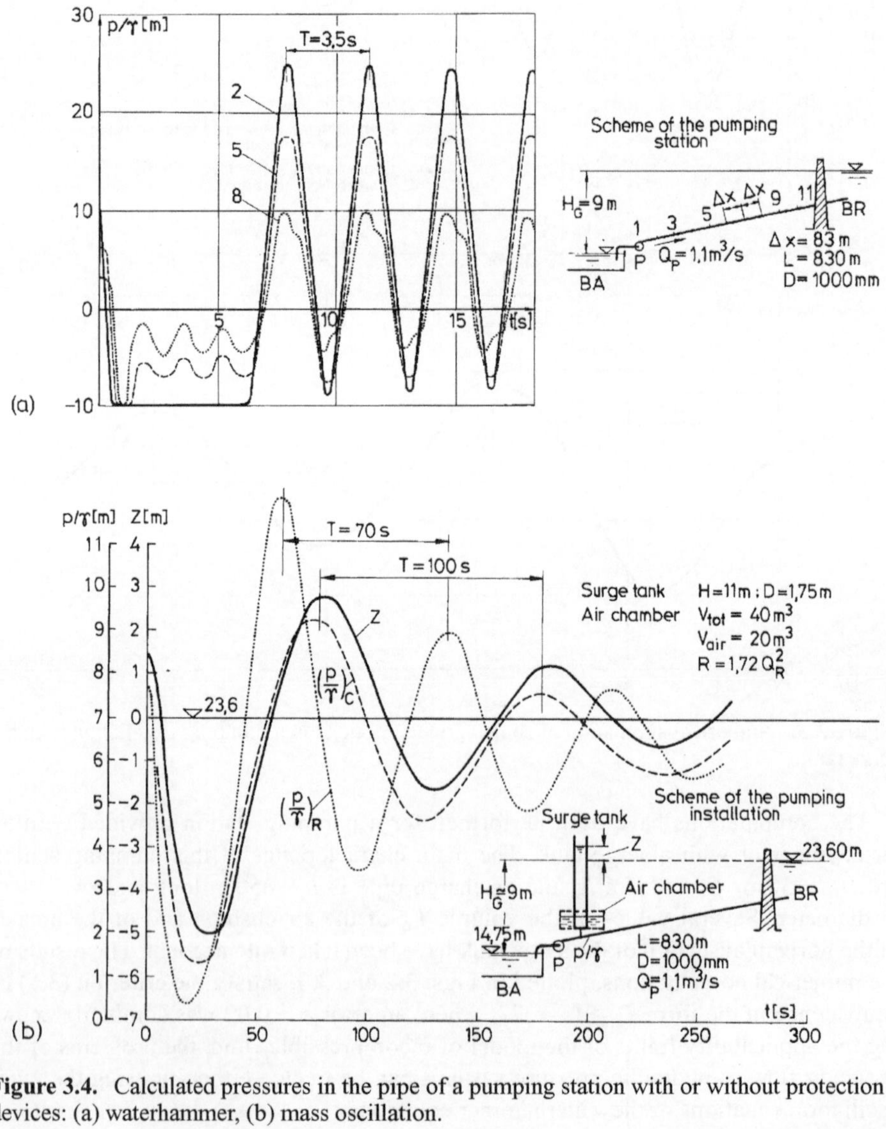

Figure 3.4. Calculated pressures in the pipe of a pumping station with or without protection devices: (a) waterhammer, (b) mass oscillation.

pumping station supposing an electric power supply cut off, in the following cases:

- pumping station with no protection devices against the waterhammer;
- pumping station equipped with surge tank;
- pumping station equipped with air chamber.

The characteristics of the pumping station are: geodetic height $H_G = 9$ m, flow rate $Q_o = 1.1$ m^3/s, length and diameter of the discharge pipe $L = 830$ m and $D = 1$ m, celerity $c = 1{,}000$ m/s respectively.

Without any protection devices, the water flow in the system is a *rapidly varying* one, the period of oscillation is $T = 3.5$ s, and the extreme pressure values are $p_{max} = 2.6$ atm and $p_{min} = -1$ atm (cavitation).

Fitted with an air chamber or surge tank, the unsteady flow in the system is *slowly varying*, and the pressure oscillations have small amplitudes and frequencies:

- for the surge tank: $p_{max} = 0.95$ atm, $p_{min} > 0$, $T = 100$ s;
- for the air chamber: $p_{max} = 1.2$ atm, $p_{min} > 0$, $T = 70$ s.

It is obvious that the surge tank as well as the air chamber alter the quality of the unsteady water flow in the system and that their presence provides protection of the discharge pipe against waterhammer. A theoretical study concerning the protection of pumping stations, fitted with one or two air chambers, against waterhammer is presented in reference [20].

In connection with the efficiency of the air chamber in attenuating the maximum pressures inside the discharge pipe, special attention should be paid to meet the very severe conditions imposed by the observance of the computed parameters in operation. For example, sometimes when the volume of the air cushion is smaller than necessary, the air chamber can generate larger maximum pressures inside the system in comparison with the pressure values appearing in the absence of any protection devices (Fig. 3.2). This is the explanation for some accidents occurring at pumping stations fitted with air chambers.

Among the Romanian achievements in the field of protection devices provided with air chambers or surge tanks, we can mention the following: *complex air chambers* and *single surge tanks* for many discharge pipes.

The complex air chamber was obtained by a combination of a classic air chamber with air valves [1]. The installation is made of a closed reservoir of small dimensions (Fig. 3.5) which contains the following supplementary devices: a number of air valves V, in the upper part, which are continued by tubes T connected to the bottom of the reservoir. A number of pipes A_{13}, for air and water, connect the reservoir and the protected pipe (their entrances to the reservoir are placed at different levels to ensure a slow air flow). The pipes S_B at the bottom of the reservoir are used mainly for the water flow. The air supply to the reservoir is made via ducts *l*, which take the naturally flowing air from the upper part of the discharge pipe or through injection inside this pipe at the pumping station from air compressors. For the description of the installation, the authors' notation [1] was preserved.

Whenever an energy breakdown occurs and the pumps stop automatically, the valves of the pumping station are closed and the protection system operates in several stages. In the first stage, immediately following the energy breakdown, the installation operates as a simple air chamber: due to the pressure of the air cushion, water is introduced into

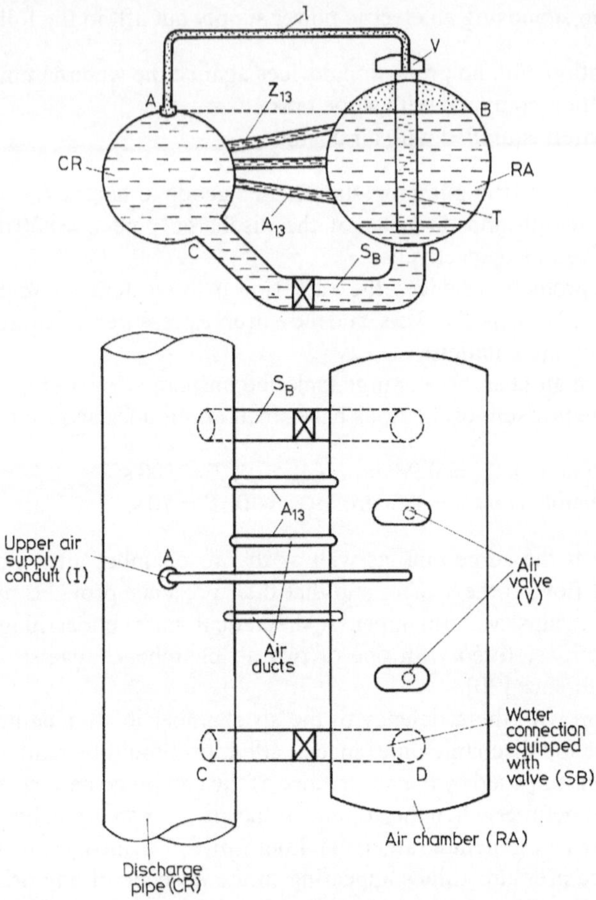

Figure 3.5. Scheme of the protection system of complex air chamber type.

the discharge pipe immediately as the pressure in the connection section decreases, unlike a simple valve which operates with a certain delay.

The second stage begins when the level inside the tank is below the inlets of ducts A_{13} (Z_{13}), when air is allowed through these entrances instead of water. Thus, a controlled decrease in head losses between the tank and the pipe is produced, resulting in a slow decrease in the connection losses until a very small value is reached.

The third stage starts when the air pressure in the tank decreases a little below the atmospheric pressure; the air valves open and atmospheric air enters the tank. At this stage, as in the second one, both water and air flow to the discharge pipe. Due to the intense airflow, pressure losses in the connection pipe are small and the pressure inside the discharge pipe can be satisfactorily controlled. At this stage, the installation operates similarly to a surge tank.

The fourth stage begins when no more water is available and only air flows inside the pipes. At this stage, the installation operates like an air valve.

The fifth stage starts when the flow in the discharge pipe and in the connection section is reversed. The inlets of tubes A_{13} and of ducts S_B to the pipe being placed at the bottom of the discharge pipe, there is almost only water flowing through the tank with a large head loss (the total section of the connection being small). This helps to control the maximum pressure in the tank and in the pipe. The fifth stage has two phases: during the first phase the complex air chamber operates as a water chamber—the air is eliminated through the tubes T, the valves V being open (the usual case of ball valves); the second phase starts when the water level exceeds the bottom inlets of the tubes T, the water flows in, closes the valves, and the air is not allowed to get out of the tank, being compressed. The installation continues to operate like a simple air chamber with large pressure losses. If one or more air valves are closed with a certain delay, no serious damage occurs, as the air cushion of the tank has a damping effect on shocks.

The bottom connections of tubes T are placed low enough, so that, in the fifth stage, an air volume is trapped into the chamber whose volume at atmospheric pressure is almost equal to the volume of the tank.

In the fifth stage, when a reverse flow takes place, the air, which in the later stages has flown into the discharge pipe, forms an elastic cushion, contributing together with the air chamber to the control of the maximum pressures.

In order to establish the dimensions of the complex air chamber, the following characteristics must be determined:

a) the total volume of the chamber;
b) the volume of the air cushion in steady flow;
c) the connection losses and, consequently, the number and dimensions of ducts A_{13} and S_B (Fig. 3.5) as well as their position on the vertical line;
d) the maximum air flow rate entering the pipe (to determine the number and dimensions of the air valves);
e) the dimensions of pipes *l* for the air supply.

The air supply pipes *l* are chosen within the diameter range 0.025–0.05 m, to allow an easy air flow so as to preserve the air cushion in the tank, but not to allow a too intense air circulation in a rapidly unsteady flow that could unfavourably alter the head losses in connection pipe.

The suitable dimensioning of the complex air chamber, as well as the hydraulic computations concerning the operation of a pumping station provided with such a kind of protection device require a special computation technique [1]. Among the pumping stations where this method was applied, the SP Celei pumped storage station should be mentioned.

The single surge tank for many discharge pipes (Fig. 3.6)
In the case of pumping stations with several independent discharge pipes, the protection solution with many surge tanks coupled into a battery—one surge tank for each discharge pipe—is well known. This solution is generally expensive. The single surge tank for pumping stations with several independent discharge pipes consists in designing a single tank connected to the discharge pipes of the station directly through pipes provided with valves (Fig. 3.6b), or through a circular duct (Fig. 3.6c), to simplify the constructive solution and reduce the investments [11].

As can be seen from the scheme, the valves along the connecting pipes of the surge tank—by their controlled opening and closing—allow the operation of the surge tank simultaneously with all the discharge pipes or with only part of them. This protection solution is applied to several pumping stations such as: SRP-3 *Hirsova*, SPA-*Salcia*, SRP-2 *Rasova*, SP-1 *Babadag* and so on.

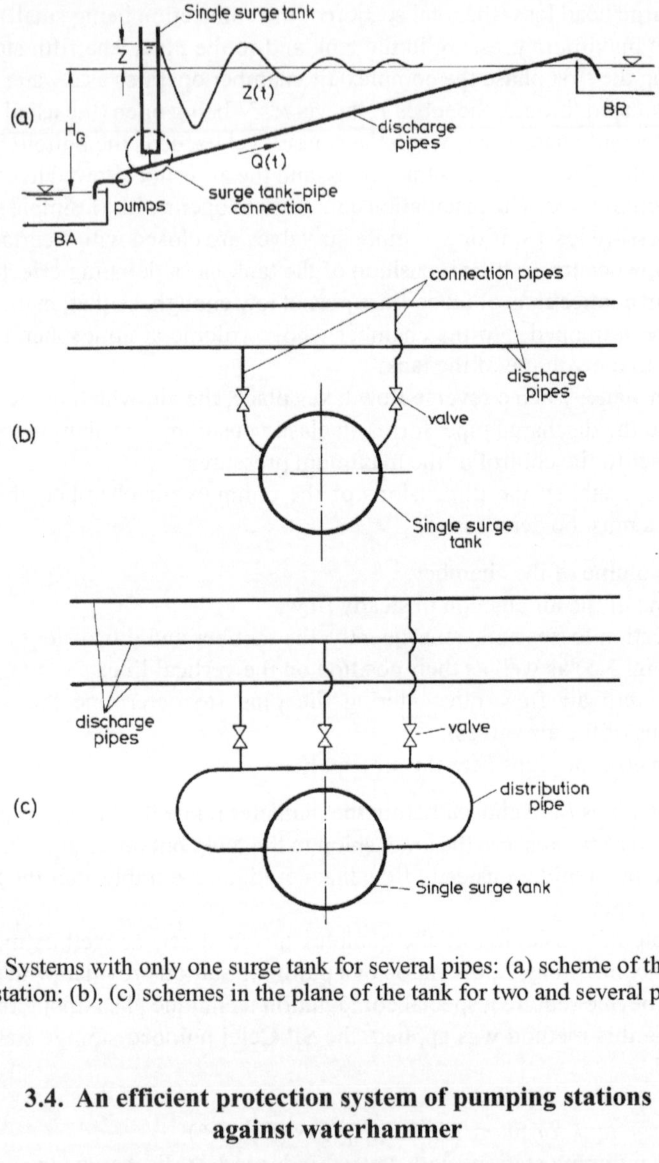

Figure 3.6. Systems with only one surge tank for several pipes: (a) scheme of the longitudinal view of the station; (b), (c) schemes in the plane of the tank for two and several pipes, respectively.

3.4. An efficient protection system of pumping stations against waterhammer

3.4.1. The effect of non-symmetrical hydraulic resistance at the entry into the air chamber or surge tank on the extreme values of pressure in the system

In order to adjust the maximum and minimum pressures in hydraulic pressure systems protected by air chambers or surge tanks, the local hydraulic resistance at the entrance to these systems is used. A local hydraulic loss can be *symmetrical* (the head loss coefficient is the same in both flow directions) or non-symmetrical (the head loss coefficient is larger in one flow direction than in the other). A special interest is paid to *non-symmetrical* resistances, as they allow different adjustment of maximum and minimum pressures in the system, thus obtaining optimum dimensioning of the hydraulic system as a whole. The non-symmetrical hydraulic losses are especially recommended for pumping stations, as the sudden loads, yielding large pressure variations, appear non-symmetrically. The halt of a pumping station can suddenly occur by accident or as a result of an electric power breakdown, while its start usually takes place slowly, under control.

The effect of non-symmetrical hydraulic resistance on the extreme pressures inside a pumping installation protected by an air chamber is shown by the numerical simulation of some unsteady hydraulic flows produced by the stop of the station when fully operating [3]. Computations were performed for the S.P. *Tiganasi* pumping station, which has the following characteristics: geodetic height $H_G = 51.5$ m, flow rate $Q_o = 8.5\,m^3/s$, length of the discharge pipe $L = 850$ m, diameter of the discharge pipe $D = 1.9$ m; several non-return valves were provided along the discharge pipe of the pumps to prevent the reverse flow. The case when the station is protected by an air chamber placed near the pumps, with symmetrical hydraulic losses ($V = 60\,m^3$, $V_{air} = 25\,m^3$; $R_1 = R_2 = 0.55\,Q_B^2$) was first considered. A maximum overpressure $\Delta p/\gamma = 33$ m followed (Fig. 3.7a).

In the case of non-symmetrical hydraulic losses with the values $R_2/R_1 = 1.78$ and $R_2/R_1 = 4$, illustratively plotted for four ducts connected to the air chamber (some of them provided with non-return valves), the maximum overpressures are reduced by 30 up to 70% (Fig. 3.7b and c) as compared to the reference case ($R_1 = R_2$) (Fig. 3.7a).

In practice, the generation of non-symmetrical hydraulic resistance with big ratios (for example, $R_2/R_1 = 4$) by means of pipes provided with valves, has the disadvantage that the moving parts can generate strong mechanical shocks in the system. For this reason, a wider applicability, especially to large pumping stations, has the non-symmetrical hydraulic resistance produced by means of fixed devices (Fig. 3.8) [19, 23].

3.4.2. Special device for non-symmetrical hydraulic resistance developed at the Hydraulic Engineering Research Institute (Institutul de Cercetari Hidrotehnice—I.C.H.)

In the hydraulic laboratory of the Hydraulic Engineering Research Institute of Bucharest (Institutul de Cercetari Hidrotehnice—I.C.H.) a new device, made of fixed parts and allowing non-symmetrical hydraulic resistance, was designed and tested. This device presents the following advantages:

- ensures relatively large ratios of non-symmetry;
- has a simple design.

Description of the device

The device is made of *n* identical frusta of cone-like parts and a double frustum of a cone-like part placed in a certain determined position inside a duct [12, 18].

Figure 3.9 shows the scheme of the device fitted on a duct denoted by 1. The parts of the device are the following: (2) attachment flange; (3) tubing length; (4) part of frustum-of-a-cone shape; (5) part of double frustum-of-a-cone shape; (6) tubing length (it may be missing). Their relative diameters with respect to the diameter *D* of the pipe are shown in the layout. For the frustum-of-a-cone shape no 4, the ratio of the diameters of the two bases is $\alpha = d/D = 0.4 - 0.6$ and the length ℓ is equal to the diameter *D* of the main duct. Figure 3.10 shows the laboratory setup for testing this device.

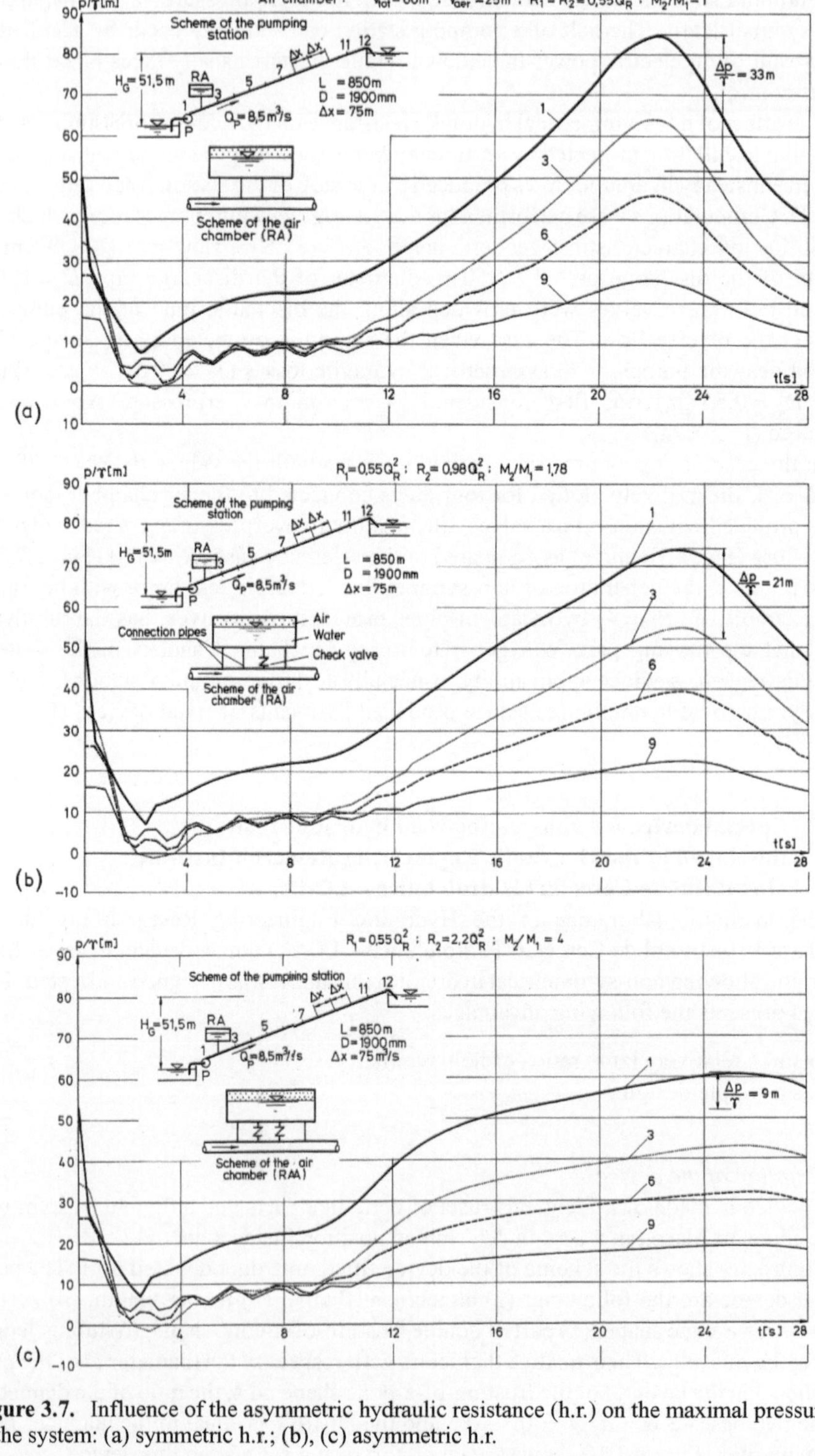

Figure 3.7. Influence of the asymmetric hydraulic resistance (h.r.) on the maximal pressures in the system: (a) symmetric h.r.; (b), (c) asymmetric h.r.

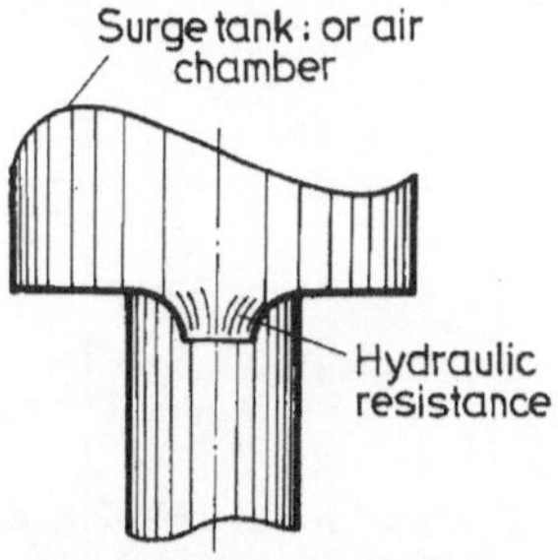

Figure 3.8. Device for simple asymmetric hydraulic resistance.

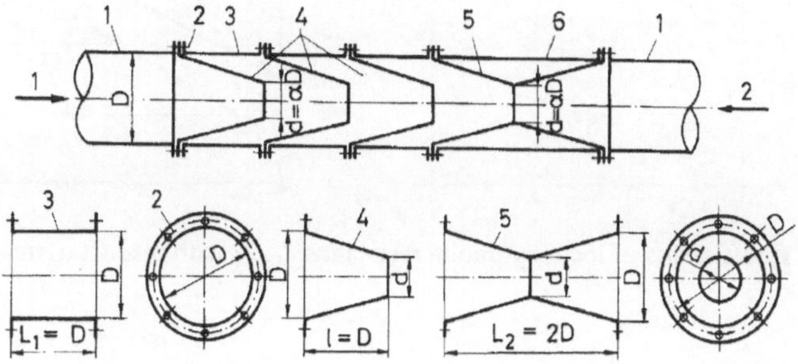

Figure 3.9. Mounting scheme of the device of asymmetric hydraulic resistance of the ICH type.

Figure 3.10. Laboratory experimental installation for the determination of the hydraulic characteristics of the device.

The hydraulic characteristics of the device

The coefficients of local hydraulic resistance of the device for the two flow directions were experimentally determined with the aid of a laboratory installation. With the subscripts 1 and 2 for the two flow directions along the pipe, the ratio of non-symmetry

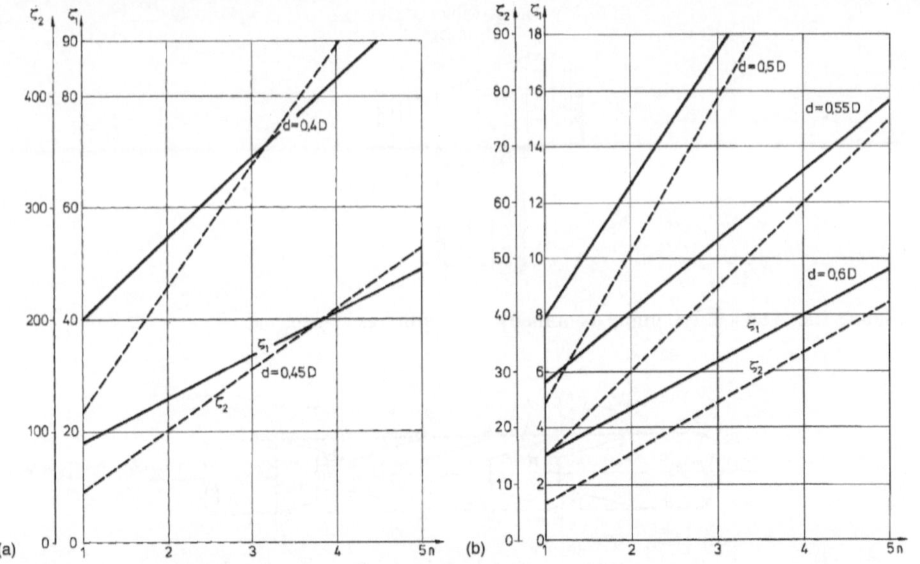

Figure 3.11. Coefficients of local hydraulic resistance ζ_1, ζ_2 with respect to the number n of truncated pieces.

of the head losses is given by $r = \zeta_2/\zeta_1$, when $\zeta_2 > \zeta_1$, and ζ is the coefficient of local head losses ($\Delta h = \zeta V^2/2g$).

In Fig. 3.11 there are plots of the coefficients ζ_1 and ζ_2 versus the number n of frustum-of-a-cone parts and versus the ratio $\alpha = d/D$ (0.4, 0.6). Figure 3.12 shows values of the ratio ζ_2/ζ_1 versus the same parameters [12]. Usually, the simple design devices made with fixed parts exhibit hydraulic non-symmetry ratios within the range 1.5–2 (Fig. 3.8). It is easy to notice that this device has ratios of non-symmetry within the range 4–5 for the versions with two or three frustum-of-a-cone parts.

Literature in the field mentions a device with non-symmetrical resistance made many decades ago [23]. This device was thereafter improved when applied to a surge tank of a hydroelectric power plant [19]. This device presents large ratios of non-symmetry, but can hardly be used in practice because of its large gauge [3].

3.4.3. Theoretical, experimental and applied results

Usually, protection devices like air chambers placed in the vicinity of pumping stations limit—for unsteady hydraulic flows—the maximum overpressures in the discharge pipe to values around 20–30% of the steady pressure [6, 21].

At the Hydraulic Engineering Research Institute (I.C.H.) an efficient system to protect the pumping stations of the type "air chamber provided with special device for non-symmetrical l head losses" was designed. It is able to significantly reduce or even to remove in certain conditions the overpressures generated by the unsteady hydraulic flows in the given hydraulic system [13, 14]. This new protection system was applied in the design of more than 20 pumping stations in Romania, most of them already in operation. The effect of this protection system on the hydrodynamic pressures inside the discharge pipe is illustrated in this book by hydraulic computations (see Section 3.5.4) and laboratory experiments (see Section 7.1.1).

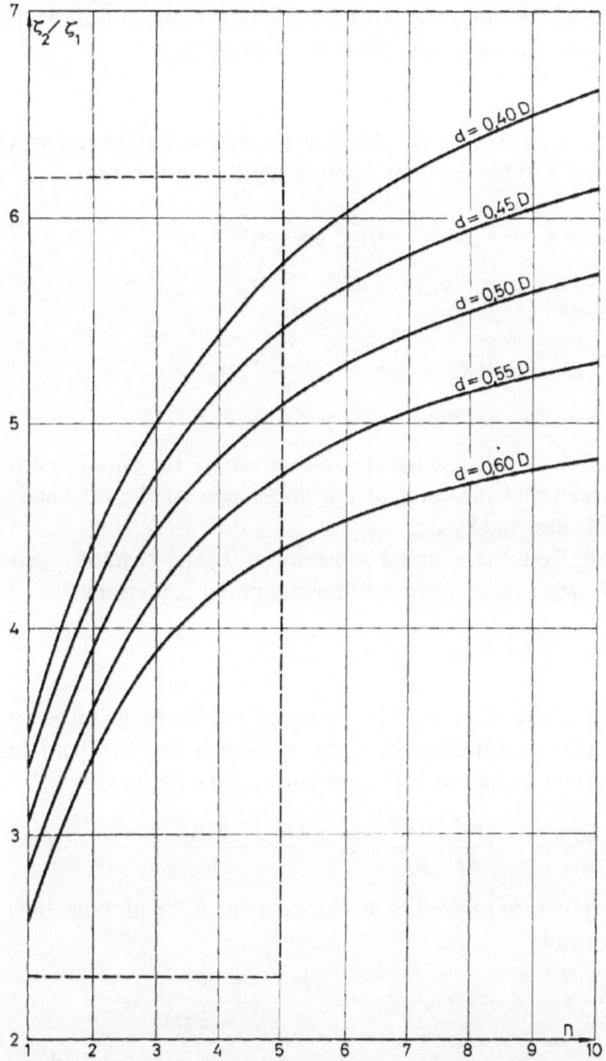

Figure 3.12. Curves of variation of the ratio ζ_2/ζ_1 with respect to the number n of truncated pieces.

A similar protection system was obtained by fitting the ICH type of device for non-symmetrical head losses to the surge tank. In this case, its efficiency consists in reducing the amplitude of the maximum levels inside the tank.

3.4.4. Computation of the optimum hydraulic resistance for air chambers

Usually, the hydraulic computations of the unsteady flow in pumping stations—which imply the choice and the adequate dimensioning of protection systems against waterhammer—are performed by numerical methods and generally involve the analysis of about 10–15 variants, out of which the best is chosen for the operation of the respective protection system. A direct hydraulic computation method to optimally dimension the protection systems of the type "air chamber provided with non-symmetrical head losses" is presented here together with the practical application, to a pumping station previously designed by numerical methods [5, 10, 15].

The mathematical model

For the pumping station with air chamber located in its lower part (Fig. 3.1a), the general equations (3.3) for the water motion, written in the form

$$\frac{L}{gf}\frac{\mathrm{d}Q}{\mathrm{d}t} - (H+Z) + PQ\,|Q| - MQ_R\,|Q_R| = 0$$

$$Q_P = F\frac{\mathrm{d}Z}{\mathrm{d}t} + Q \tag{3.6}$$

$$(H+h_a)V = (H+h_a)(Z^* - Z)F = (H_o + h_a)V_o = C$$

$$Q_P = Q_P(t)$$

are used, where the positive sense of Q_R is reversed, the simplified notation $P' = P$, $M' = M$ is adopted and the level of the upper part of the air chamber against the reference plane is denoted by Z^*.

For the steady flow the adopted subscript is 0. The case of a sudden halt of the station ($Q_P = 0$), which assumes the following initial conditions

$$Z(t)|_{t=o} = Z_o \qquad \left.\frac{\mathrm{d}Z}{\mathrm{d}t}\right|_{t=o} = -\frac{Q_o}{F} = -\frac{f}{F}W_o \tag{3.7}$$

will be analysed, where W is the velocity inside the discharge pipe.

In this case the system of eqns. (3.6), written in the $Z(t)$ variable, leads to the following differential equation of the oscillatory flow in the system:

$$\frac{LF}{gf}\frac{\mathrm{d}^2Z}{\mathrm{d}t^2} + F^2(P+M)\frac{\mathrm{d}Z}{\mathrm{d}t}\left|\frac{\mathrm{d}Z}{\mathrm{d}t}\right| + Z + \frac{(H_o+h_a)(Z^*-Z_o)}{Z^*-Z} - h_a = 0 \tag{3.8}$$

while the water pressure inside the discharge pipe, at the air tank section, is given by the following formula

$$\frac{p_A(t)}{\gamma} = H + Y + Z + MQ_R\,|Q_R| = \frac{(H_o+h_a)(Z^*-Z_o)}{Z^*-Z(t)} + Y + Z(t) + MF^2\frac{\mathrm{d}Z}{\mathrm{d}t}\left|\frac{\mathrm{d}Z}{\mathrm{d}t}\right| - h_a \tag{3.9}$$

The extreme (maximum and minimum) values of this pressure depend on the extreme value of Z according to the relation

$$\frac{p_A}{\gamma} = \frac{(H_o+h_a)(Z^*-Z_o)}{Z^*-Z} + Z + Y - h_a \tag{3.10}$$

The theoretical bases of the computation method

In order to obtain the maximum value of the water level Z in the tank, the following two initial stages of the flow are considered:

- the first stage, where the water level in the tank decreases within the limits Z_o and Z_{min};
- the second stage, where the water level inside the tank increases within the values Z_{min} and Z_{max}.

As in both stages $\mathrm{d}Z/\mathrm{d}t \neq 0$, the function $t \to Z(t)$ can be inverted. We denote by τ the inverse function and by α the function defined by

$$\alpha(x) = \frac{1}{2}\left[\frac{\mathrm{d}Z}{\mathrm{d}t}\Big|_{t=\tau(x)}\right]^2 \tag{3.11}$$

Remark that this function represents the kinetic energy of the mass unit oscillations in the system. The change to variable α for the study of the water level oscillations inside the air tank is classical.

With these variable changes, the following equations are obtained:

(a) for the first stage

$$\frac{\mathrm{d}\alpha}{\mathrm{d}x} = \frac{2gfF}{L}(P+M)\alpha(x) - \frac{fg}{LF}\left[x + \frac{(H_o + h_a)(Z^* - Z_o)}{Z^* - x} - h_a\right] \tag{3.12}$$

(b) for the second stage

$$\frac{\mathrm{d}\alpha}{\mathrm{d}x} = \frac{2gfF}{L}(P+M)\alpha(x) - \frac{fg}{LF}\left[x + \frac{(H_o + h_a)(Z^* - Z_o)}{Z^* - x} - h_a\right] \tag{3.13}$$

with the initial conditions

$$\alpha(Z_o) = \frac{f^2}{2F^2}W_o^2 \tag{3.14}$$

for the first stage and

$$\alpha(Z_{\min}) = 0 \tag{3.15}$$

for the second stage, where $Z_{\min}$ is the minimum value of the water level obtained in the first stage.

During the first stage the equation is studied for $x < Z_o$, while during the second stage, for $x > Z_{\min}$. The value $Z_{\min}$ is computed from the equation $\alpha(Z_{\min}) = 0$, α being the solution of eqn. (3.12) corresponding to the first stage.

Introducing the notation

$$\rho = \frac{2gfF}{L}(P+M) \tag{3.16}$$

for $Z_{\min}$, the equation

$$\Phi(\rho, Z) = 0$$

is obtained, where

$$\Phi \overset{\text{def}}{=} \int \mathrm{e}^{\rho(Z_o - \xi)}\left[\xi + \frac{(H_o + h_a)(Z^* - Z_o)}{Z^* - \xi} - h_a\right]\mathrm{d}\xi - \frac{fL}{2gF}W_o^2 \tag{3.17}$$

For the second stage, the value $Z_{\max}$ is computed from equation

$$\Psi_{Z_{\min}}(\rho, Z) = 0$$

where

$$\Psi_{Z_{\min}} \overset{\text{def}}{=} \int_{Z_{\min}} \mathrm{e}^{\rho\xi}\left[\xi + \frac{(H_o + h_a)(Z^* - Z_o)}{Z^* - \xi}\right]\mathrm{d}\xi \tag{3.18}$$

with $Z_{\min}$ obtained from the first stage.

The computation procedure

The following computation procedure is recommended:

(a) A first ρ value is chosen, hence a first value for the hydraulic resistance coefficient M of the hydraulic connection.

The non-linear second order differential eqn. (3.8) is solved numerically. With the notation (3.16) it takes the form

$$\frac{LF}{gf}\frac{\mathrm{d}^2Z}{\mathrm{d}t^2}+\frac{FL\hat{\rho}}{2gf}\frac{\mathrm{d}Z}{\mathrm{d}t}\left|\frac{\mathrm{d}Z}{\mathrm{d}t}\right|+Z+\frac{(H_\mathrm{o}+h_\mathrm{a})(Z^*-Z_\mathrm{o})}{Z^*-Z}-h_\mathrm{a}=0 \tag{3.19}$$

while the initial conditions become

$$Z(t)|_{t=\mathrm{o}}=Z_\mathrm{o} \quad \text{and} \quad \left.\frac{\mathrm{d}Z}{\mathrm{d}t}\right|_{t=\mathrm{o}}=-\frac{Q_\mathrm{o}}{F}=-\frac{f}{F}W_\mathrm{o} \tag{3.20}$$

and the minimum value of Z is thus determined for the first oscillation stage; this value is denoted by $\hat{Z}$.

(b) The following differential equation

$$\frac{\mathrm{d}\rho}{\mathrm{d}Z}=-\frac{\dfrac{\partial\Phi}{\partial Z}(\rho,Z)}{\dfrac{\partial\Phi}{\partial\rho}(\rho,Z)} \tag{3.21}$$

is formed with function $\Phi(\rho, Z)$ defined by eqn. (3.17) and is solved for the initial condition

$$\rho(\hat{Z})=\hat{\rho}$$

If a certain value for the minimum pressure in the pipe $(p_\mathrm{A}/\gamma)_{\mathrm{min}}$ is adopted, and hence a Z_{min}, the integration from $\hat{Z}$ to Z_{min} is performed and the corresponding value $\rho(Z_{\mathrm{min}})$ is stored; from eqn. (3.16), the value M_1 of the head losses coefficient of the hydraulic connection is obtained for the tank discharge pipe flow direction.

Alternatively, the differential equation

$$\frac{\mathrm{d}Z}{\mathrm{d}\rho}=-\frac{\dfrac{\partial\Phi}{\partial\rho}(\rho,Z)}{\dfrac{\partial\Phi}{\partial Z}(\rho,Z)} \tag{3.22}$$

can be computed with the initial condition $Z(\hat{\rho})=\hat{Z}$ and by performing the computations for a range of ρ values, one can choose ρ values which lead to acceptable values of Z_{min} (hence to acceptable minimum pressures in the discharge pipe).

Mention should be made that this last equation reads

$$\frac{\mathrm{d}Z}{\mathrm{d}\rho}=-\frac{\displaystyle\int_{Z_\mathrm{o}}^{Z}\mathrm{e}^{\rho(Z-\xi)}(Z_\mathrm{o}-\xi)\left[\xi+\frac{(H_\mathrm{o}+h_a)(Z^*-Z_\mathrm{o})}{Z^*-\xi}-h_\mathrm{a}\right]\mathrm{d}\xi}{Z+\dfrac{(H_\mathrm{o}+h_\mathrm{a})(Z^*-Z_\mathrm{o})}{Z^*-Z}-h_\mathrm{a}} \tag{3.23}$$

(c) The value of ρ, denoted by $\tilde{\rho}$, convenient for the first stage of the flow, is retained and with this value, instead of $\hat{\rho}$, eqn. (3.19) with the initial conditions (3.20) is solved for both stages of the flow in the system in order to calculate the maximum value of Z, denoted by Z.

(d) The differential equation

$$\frac{\mathrm{d}Z}{\mathrm{d}\rho} = -\frac{\dfrac{\partial \Psi_{Z_{\min}}(\rho, Z)}{\partial \rho}}{\dfrac{\partial \Psi_{Z_{\min}}(\rho, Z)}{\partial Z}} \tag{3.24}$$

is written explicitly in the form

$$\frac{\mathrm{d}Z}{\mathrm{d}\rho} = -\int_{Z_{\min}}^{Z} \frac{\mathrm{e}^{\rho(\xi - Z)}\xi\left[\xi + \dfrac{(H_o + h_a)(Z^* - Z_o)}{Z^* - \xi} - h_a\right]}{Z + \dfrac{(H_o + h_a)(Z^* - Z_o)}{Z^* - Z} - h_a}\,\mathrm{d}\xi \tag{3.25}$$

and is solved for

$$Z(\tilde{\rho}) = \tilde{Z}$$

A value for ρ is chosen, that is a value for the M_2 coefficient of the head losses of the hydraulic connection in the discharge pipe–tank direction, which ensures acceptable values for the maximum pressure $(p_A/\gamma)_{\max}$ in the pipe for $Z_{\max}$, respectively.

In this way, the values for the local hydraulic resistance of the hydraulic connection in both directions are obtained with respect to the acceptable values of the minimum and maximum pressures in the discharge pipe. The characteristics of the non-symmetrical hydraulic resistance of the hydraulic connection (M_1, M_2, M_1/M_2) are computed.

Note that the computation procedure for the solution of implicit functions problems by reducing them to Cauchy problems for differential equations, adopted here, is due to Davidenko and was used in different applications by several authors [5]. Finally, the method presented here can be also applied to pumping stations with surge tanks provided with a device for non-symmetrical hydraulic resistance (Fig. 3.1b). In this case, the computations are much more simple.

Application

The method was applied and checked at several pumping stations. Here, as an example we have chosen the SPA Oprisani basic pumping station from the hydraulic system Prut-Birlad, which will be deeply studied in Section 3.5.4.

The pumping station was suitable for a protection solution with air chamber equipped with a device for non-symmetrical hydraulic resistance. Indeed, as the steady pressure is rather high, about 10 atm, in order to ensure an optimum hydraulic operation under unsteady flow it is necessary that the overpressures in the system be as low as possible.

The application of this method to the design of the air chamber required several computations being simultaneously made for some acceptable values of the air cushion in steady flow. (For example, the volume of the air chamber is taken about 2% of the volume of the discharge pipe, and the volume of the air cushion in steady flow around 0.3–0.5 of the volume of the air tank.)

In this case, all the computations required by the method were performed for two volumes of the air cushion in steady flow

- the first variant: $V_o = 50\,\mathrm{m}^3$, and
- the second variant: $V_o = 70\,\mathrm{m}^3$.

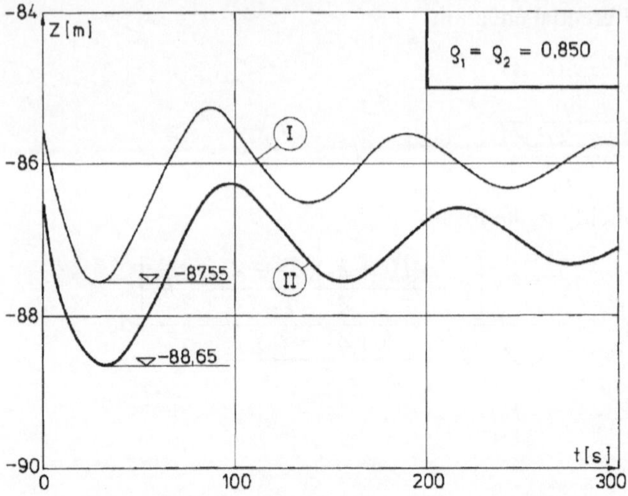

Figure 3.13. Curves of time variation of the water level in air chamber at the manoeuvre of sudden stop of the pumping station for a certain hydraulic resistance of the branch pipe $\rho_1 = \rho_2 = 0.850$.

The most dangerous operation—the halt of the pumping station while working at the maximum flow rate $Q_p = 3\,\mathrm{m}^3/\mathrm{s}$—was taken into account.

In preliminary computations, a hydraulic resistance coefficient of the hydraulic connection $M_1 = M_2 = 8$ was chosen, which led to $\rho_1 = \rho_2 = 0.85$ (eqn. (3.16)). The numerical integration of eqn. (3.19) with initial conditions of type (3.20) was performed. The following results were obtained:

- for the first variant

$$Z(t)\big|_{t=0} = -85.6\,\mathrm{m}; \qquad \left.\frac{dZ}{dt}\right|_{t=0} = -0.15\,\mathrm{m/s}$$

- for the second variant

$$Z(t)\big|_{t=0} = -86.5\,\mathrm{m}; \qquad \left.\frac{dZ}{dt}\right|_{t=0} = -0.15\,\mathrm{m/s}$$

In this computation the values $Z^* = -83\,\mathrm{m}$ and $F = 20\,\mathrm{m}^2$ were also involved. The computation results are plotted in Fig. 3.13, the minimum encountered values for level Z being

- for the first variant: $Z_{min} = -87.55\,\mathrm{m}$;
- for the second variant: $Z_{min} = -88.65\,\mathrm{m}$.

With these results assumed as initial conditions $Z(\rho_1) = Z_{min}$ the differential equation (3.23) is numerically integrated for both variants, wherefrom the plots $Z_{min}(\rho_1)$ and $p_A/\gamma(\rho_1)$ are obtained and presented in Figs. 3.14 and 3.15.

Then, in order to choose the optimum values for the hydraulic resistance in the hydraulic connection, when using, for example, the second variant ($V_o = 70\,\mathrm{m}^3$), we

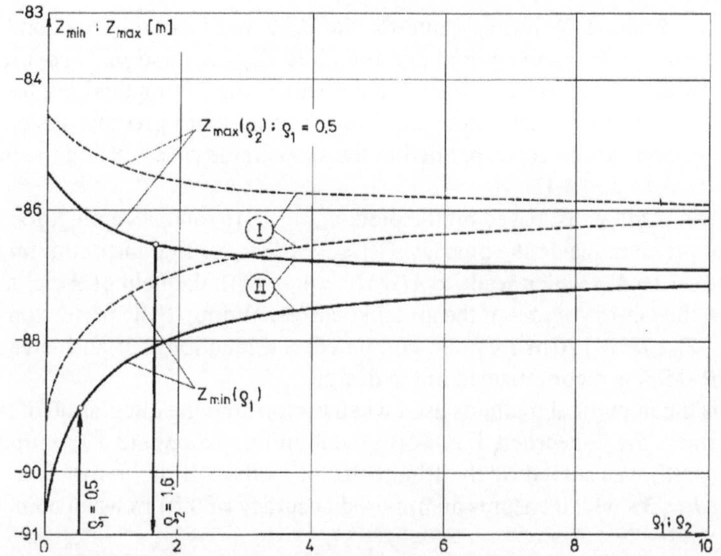

Figure 3.14. Plots of the minimal and maximal water levels in air chamber as functions of the hydraulic resistances of the branch pipe in the two flow directions ρ_1, ρ_2.

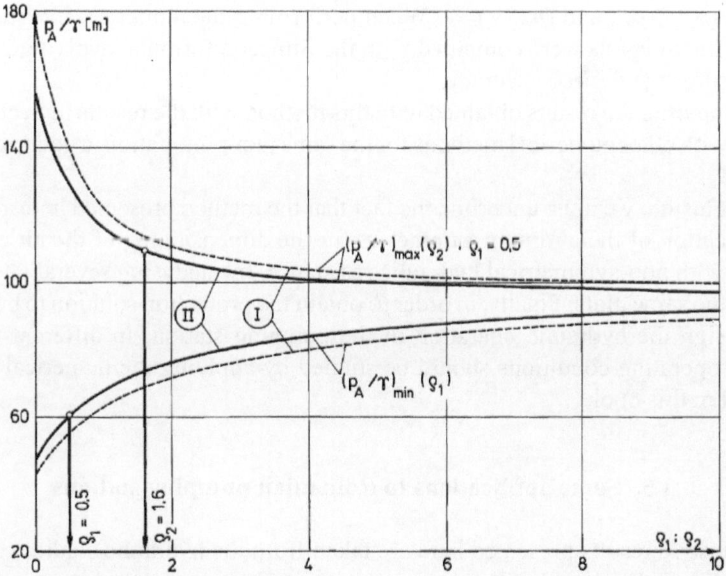

Figure 3.15. Plots of the minimal and maximal values of the water pressure in the pipe at the section of the air chamber in terms of ρ_1 and ρ_2 respectively.

take the minimum pressure in the pipe $(p_A/\gamma)_{min} = 60$ m, from the plots of Fig. 3.15 and obtain the value $\rho_1 = 0.5$ ($M_1 = 0.405$).

In order to find the second coefficient of hydraulic resistance of the hydraulic connection M_2, eqn. (3.19) with boundary conditions (3.20) and $\rho_1 = \rho_2 = \hat{\rho} = 0.5$ is numerically integrated. From the plots of $Z(t)$ the maximum value Z_{max} after the first oscillation is obtained. With this value for the Z_{max}, used as an initial condition, the differential eqn. (3.25) is integrated and the plots $Z_{max}(\rho_2)$ and $p_A/\gamma(\rho_2)$, presented in Figs. 3.14 and 3.15, are obtained. At the same time, computations for the first variant $V_o = 50\,m^3$ were performed, choosing the minimum pressure inside the pipe $(p_A/\gamma)_{min} = 55$ m, which corresponded to the same value $\rho_1 = 0.5$ and the plots were those in Figs. 3.14 and 3.15.

These results allow us, based on the plots in Fig. 3.15, and accepting, for example, a minimum pressure inside the pipe $(p_A/\gamma)_{min} = 108$ m, to find the optimum value for $\rho_2 = 1.6(M_2 = 16.43)$ which leads to $M_2/M_1 = 4.1$. With the help of the plot $Z_{min}(\rho_1)$ in Fig. 3.14 the entire volume of the air tank, corresponding to the protection solution: $V_t = (Z^* - Z_{min})F = 120\,m^2$, can be computed; we mention that an increase in this valve by 10–15% is recommended in the design.

As far as the numerical methods used when performing the integration of the differential equations are concerned, it is worth mentioning that eqn. (3.19) with the initial condition (3.20) was solved by the Runge–Kutta method of the fourth order with the time step $\Delta t = 3$ s which ensures an imposed accuracy of 0.01 m when computing the level Z.

For the numerical solution of the differential eqns. (3.23) and (3.25) with the corresponding initial conditions, the same Runge–Kutta method was used to obtain the values $Z_{min}(\rho)$ and $Z_{max}(\rho)$ with the steps $\Delta\rho = 0.2$ in the direct sense and $\Delta\rho = 0.05$ in the reverse sense, an accuracy of 0.1 m being achieved in computing the extreme pressures $(p_A/\gamma)_{min}$ and $(p_A/\gamma)_{max}$. When performing the numerical integration, the values of the integrals were computed with the Simpson formula involving 25 points, which ensures the desired accuracy.

By comparing the results obtained with this method with the results in Section 3.5.4 obtained with other numerical methods for the same pumping station, a good agreement is noticed.

In conclusion, we must underline the fact that the method presented here allows the direct selection of the optimum parameters for the dimensioning of the air chambers provided with non-symmetrical hydraulic resistance, by analysing several alternatives based on the same plots. Finally, in order to obtain the protection solution to be adapted in the design the hydraulic operation of the pumping station, in different unsteady flow and operating conditions should be studied by applying the numerical methods presented in this book.

3.5. Some applications to Romanian pumping stations

The applicative results presented here are taken from the hydraulic studies conducted at the Hydraulic Engineering Research Institute of Bucharest to design and dimension the protection systems for over one hundred pumping stations already in operation in Romania [7, 14, 17].

3.5.1. Pumping stations provided with surge tanks

The SP-2 Sarichioi pumping station has a flow-rate of $3\,m^3/s$ for a geodetic height of 21 m. The discharge is made through one 420 m long metallic pipe of 1.25 m in diameter. The station is fitted with five type 18 NDS-C pumps, with the characteristics $Q_o = 0.71\,m^3/s$ and $H_o = 22$ m driven by electric motors of the type MAB-355 with

Figure 3.16. Pumping station SP-2 Sarichioi equipped with a tank at the base; overall view.

power $N = 200$ kW and speed $n = 750$ rpm. Along the discharge pipes of the pumps (0.6 m in diameter) valves and non-return valves are fitted. The pipe is provided with air valves in the area of discharge tank ensuring the breakdown of the water column for the return flow when the station is stopped. The protection of the station is ensured by a low-level surge tank with a height of $H = 17$ m and $D = 1.9$ m in diameter, connected via a pipe of diameter $d = 1$ m to the system. This solution was chosen by technical and economical reasons from many design alternatives. The station was commissioned in 1974 (Fig. 3.16).

The plots of the pressure variation with time in the discharge pipe, as well as of the level in the surge tank, when the station is stopped while fully operating as well as the starting-up of the fifth pump, are presented in Fig. 3.17.

The SP-1 Babadag pumping station has a flow-rate of 11 m^3/s for a geodetic height of 15 m. The discharge is made via two independent steel pipes, with a length of 1220 m each and a diameter of 1.6 m. The station is provided with eight pumps, type Siret-900-a, with the hydraulic characteristics $Q_o = 1.38\ m^3/s$ and $H = 22$ m, driven by asynchronous electric motors, type MAB-560, of power $N = 500$ kW and speed $n = 590$ rpm. Along the discharge pipes (0.7 m in diameter) from the pump to the two discharge pipes of the station electrically operated and butterfly valves are fitted. The discharge pipes are fitted with air valves in the proximity of the discharge tank, ensuring the water-column breakdown at reverse flow when the station is stopped.

To protect the station, the solution with single surge tank for both discharges pipes was adopted with the characteristics $H = 22$ m, $D = 3.5$ m and diameter of connecting pipes $d = 0.8$ m. The station was commissioned in 1975 (Fig. 3.18).

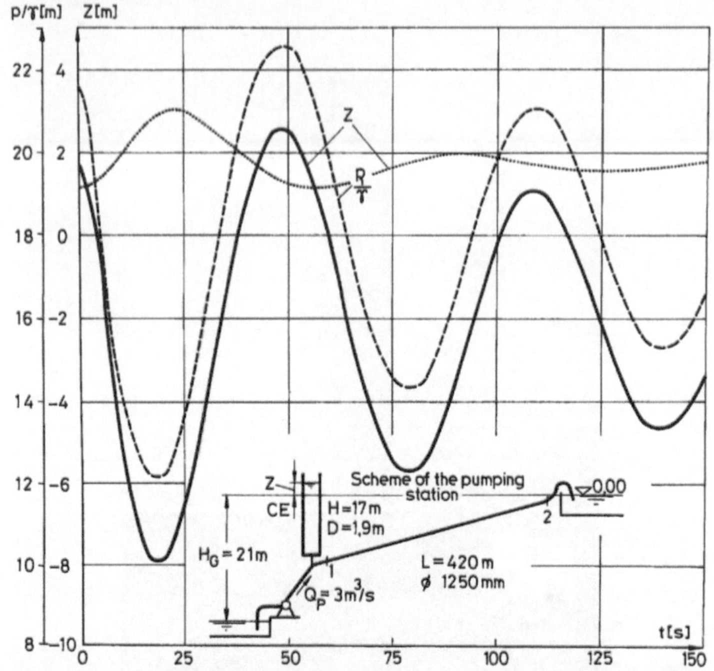

Figure 3.17. SP-2 Sarichioi pumping station. Operation representations at transient hydraulic regimes.

Figure 3.18. SP-1 Babadag pumping station with a surge tank at the overall view.

The unsteady hydraulic flow caused by the total halt of the station when fully operating as well as the simultaneous starting-up of two pumps are shown in Fig. 3.19, as obtained by numerical simulation. The protection solution was adopted after a cost–efficiency analysis of several variants with either air chambers or surge tanks.

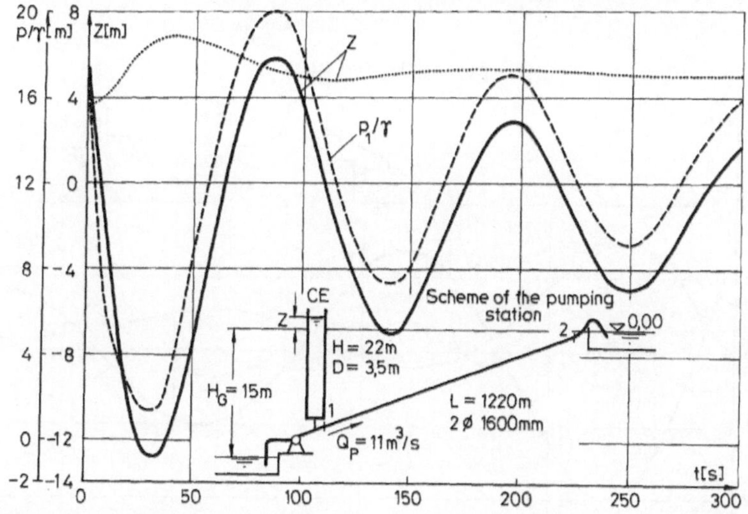

Figure 3.19. SP-1 Babadag pumping station. Operation representations at transient hydraulic regimes.

The SPP Pecineaga floating pumping station has a steady flow rate of 8.5 m^3/s for a geodetic height of 7 m, the pumping height varying in terms of the water level in the Danube.

The discharge is made from two identical independent metallic pipes, 162 m long and 1.25 m in diameter. The station is provided with eight Brates-600 pumps (four pumps for each discharge pipe), driven by asynchronous electric MIP-335 motors, each of power $N = 132$ kW and speed $n = 600$ rpm. The pipes are provided with blade valves in area of the pressure basin to avoid reverse flow when the station is shut down.

The initial design of the station expected that the discharge pipelines be drawn according to variant A (Fig. 3.20), practically following the terrain configuration and crossing the Danube dyke in that region.

By analysing various protection alternatives of the station against waterhammer and by taking into account the long run of discharge pipes specific to this kind of pumping station, a variant which entirely changed the hydraulic layout of the station was reached.

The adopted design solution showed the need to provide a water *storage basin* along the discharge pipe-run, from where the water is driven in the pressure basin *by gravitation*, i.e. the variant B of the discharge pipes in Fig. 3.20. The analysis of hydraulic unsteady flow of the station makes it obvious that the pressure basin is acting like a surge tank, being calibrated in this way.

The hydraulic computations were performed for one discharge pipe-run in the following hypotheses: the sudden halt of four pumps, the simultaneous start of two pumps and the start of the fourth pump. The results are shown in Fig. 3.20 for the adopted design alternative, with a pressure basin of the following dimensions: length $L = 20$ m width $b = 2.5$ m, height $h = 4.5$ m.

As can easily be seen, the dynamic pressure flow was actually eliminated from the system. That is why the discharge pipes were redesigned once more to take into account the smaller maximum pressures obtained. At the same time, this study is an

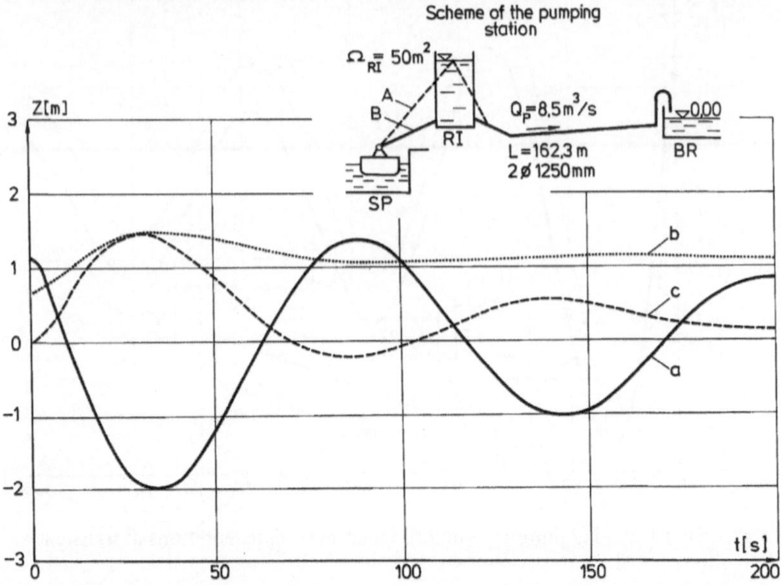

Figure 3.20. SPP Pecineaga floating pumping station. Curves of time variation of the water level in chamber at the manoeuvres of the total stop of the station (a), start with two pumps (b) and start of a pump (c).

example that the computations for protection against waterhammer can fully reshape the hydraulic system, leading to new, more economical and operation-safe constructive solutions instead of simply enabling to establish and dimension a protection means against waterhammer.

The SPA-Bordusani pumping station has a flow-rate of $17.2\,m^3/s$ for a geodetic height of 12.5 m and is equipped with seven pumps, type MV-1001/III, driven by asynchronous electric motors, type MAB-V, of power $N = 630$ kW and speed $n = 500$ rpm. The discharge is ensured by two independent metallic pipes, the first of diameter $D_1 = 2$ m to which three pumps are connected and the second of diameter $D_2 = 2.2$ m to which four pumps are connected. The pipes have the same length $L = 270$ m (Fig. 3.21).

The protection solution of the station against waterhammer is presented in two variants: with single surge tank for both discharge tanks and with two separate surge tanks, one for each pipe, with the following characteristics:

(a) *The variant of protection with a single surge tank*:
- the distance from the station: 10–15 m;
- the height of the surge tank with respect to the ground level: $H = 11$ m;
- the diameter of the surge tank: $D = 3$ m;
- the diameter of the connection pipes between the surge tank and the discharge pipes: $d_1 = 0.5$ m for the discharge pipe of diameter $D_1 = 2$ m and $d_2 = 0.6$ m for the discharge pipe of diameter $D_2 = 2.2$ m.

(b) *The variant with separate surge tanks*:
- the distance from the station: 10–15 m;
- the height of the surge tank with respect to the ground level: $H = 11$ m;

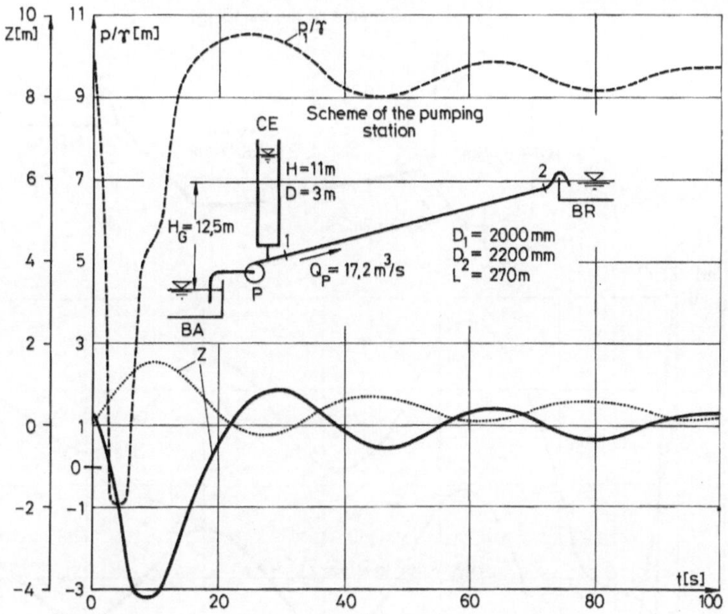

Figure 3.21. SPA-Bordusani pumping station. Curves of time variation of the water level in the surge tank (Z) and of the water pressure in pipe (p/γ) at the manoeuvres of total stop of the station and of start of a pump.

- the diameters of the surge tanks
  $D_1 = 2$ m for the discharge pipe of diameter $D_1 = 2$ m, and
  $D_2 = 2.2$ m for the discharge pipe of diameter $D_2 = 2.2$ m;
- the diameters of the pipes which connect the surge tank with the discharge pipes $d_1 = 0.5$ m for the 2 m diameter pipe and $d_2 = 0.6$ m for the 2.2 m diameter discharge pipe [11].

Among the results of the performed hydraulic computations, in Fig. 3.21 is exhibited, for example, the variation in time of the water level inside the surge tank and of the water pressure inside the discharge pipes in the case of variant (a).

The SRP-3 Bordusani pumping station. This station has a flow rate of 4 m^3/s for a geodetic height of 6.4 m and is provided with four type Brates 600-a pumps with the characteristics: $Q_o = 1$ m^3/s and $H_o = 13$ m, driven by electric motors of power $N = 160$ kW and speed $n = 600$ rpm.

The discharge is made from two pipes of reinforced concrete, type PREMO, each being 620 m long and 1 m in diameter. Along the discharge pipes (diameter $d = 0.6$ m) of the pumps retention valves are fitted. Near the outflow basin the two discharge pipes are fitted with blade valves to prevent reverse flow when the station is stopped.

The protection solution against waterhammer consists of a single surge tank for both discharge pipes located near the station, with the following characteristics:

- the distance from the station: 10–15 m;
- the height of the surge tank with respect to the ground level: $H = 8$ m;
- the diameter of the surge tank: $D = 3$ m;
- the diameter of the connecting pipes: $d_1 = d_2 = 0.45$ m.

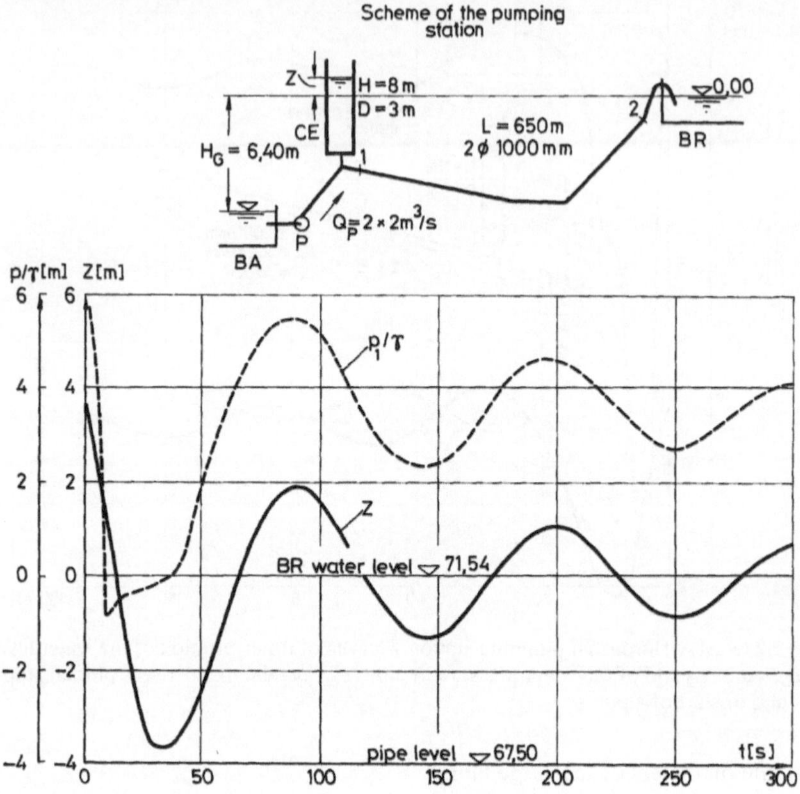

Figure 3.22. SRP-3 Bordusani pumping station. Curves $Z(t)$ and $p/\gamma(t)$ at the manoeuvre of the total stop of the station.

The relatively small height of the surge tank was also the result of a favourable ground configuration for the pipeline, exhibiting a small hill near station building (Fig. 3.22).

The results of the hydraulic computations presented like plots of pressure versus time inside the discharge pipes and of water levels inside the surge tank, at the station shut-down in full operation are presented in Fig. 3.22.

3.5.2. Pumping stations provided with air chambers

The SRP-2 Daeni pumping station has a flow rate of 16 m^3/s for a geodetic height of 59 m and it is provided with 16 type 24 NDS pumps, driven by asynchronous electric motors, MAB-2-630, of power $N = 1250$ kW and speed $n = 740$ rpm. The discharge consists of two identical ducts 715 m long and 2 m in diameter, equipped with electrically operated valves in the area of the outflow basin to let air inside the pipes when stopping the station, thus allowing the water column breakdown at reverse flow (the valves do not operate for accidental halts due to electric power breakdown).

In the hypothesis in which the station is operated without protection devices, hydraulic computations identified the presence in the system of unsteady hydraulic flows like those presented in Fig. 3.23, with sudden variation and unacceptable extreme values for the pressure $p_{max} = 12$ atm, $p_{min} = -1$ atm (cavitation). Among the computed

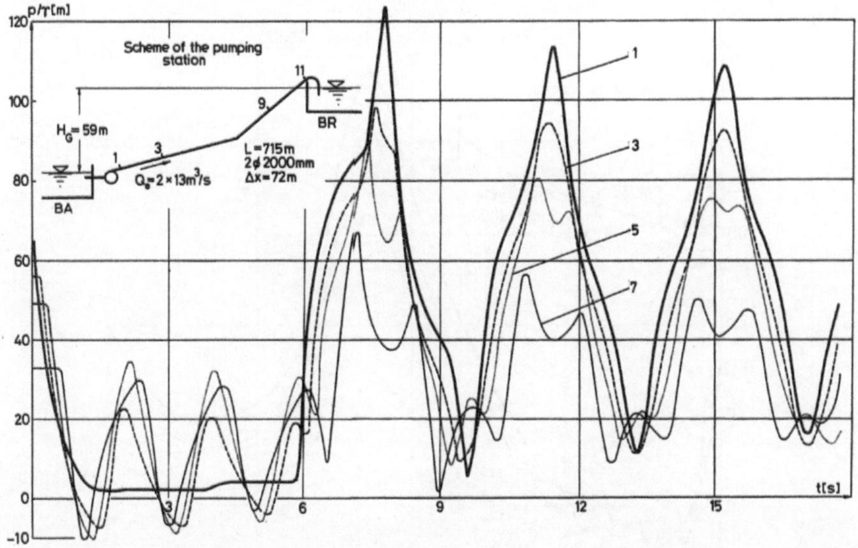

Figure 3.23. SRP-2 Daeni pumping station; hydrodynamical pressures calculated in the pipe in the absence of the protection devices.

and analysed protection solutions against waterhammer the following was chosen (for a single discharge pipe):

- the number of air chambers: 2 of 30 m^3;
- the volume of the air cushion in steady flow: 15 m^3 for each chamber;
- the connection distance from the station: 10–15 m;
- the diameter of the connecting pipe: $d = 0.35$ m for each chamber.

The numerical simulation results of the unsteady flow produced by the complete halt of the station while fully operating are graphically presented in Fig. 3.24, both for the design solution and for a comparative variant.

The total volume of an air chamber of a pumping station is computed by increasing for safety reasons by about 10–15% the maximum volume reached by the air cushion during the phenomenon of mass oscillation inside the system. This mass oscillation is the result of the most unfavourable manoeuvre encountered or possible in the operation of the perspective pumping station.

The SRP-3 Daeni pumping station. The station has a maximum flow rate of 1.8 m^3/s for a geodetic height of 37 m and is provided with six vertical pumps, type MV-403-a, driven by electric motors, type MAB-2-NV of power $N = 315$ kW and speed $n = 1{,}485$ rpm. Along the discharge pipes of each pump (0.4 m in diameter) non-return valves are fitted. The station has reinforced concrete pipes, type PREMO, with the length of 2,400 m and diameter of 0.8 m, provided with an air valve in area of the pressure basin.

The protection solution consists of an air chamber located near the station, with the following geometric and hydraulic characteristics:

- the total volume of the tank: 30 m^3;
- the volume of this air cushion in steady flow: 15 m^3;

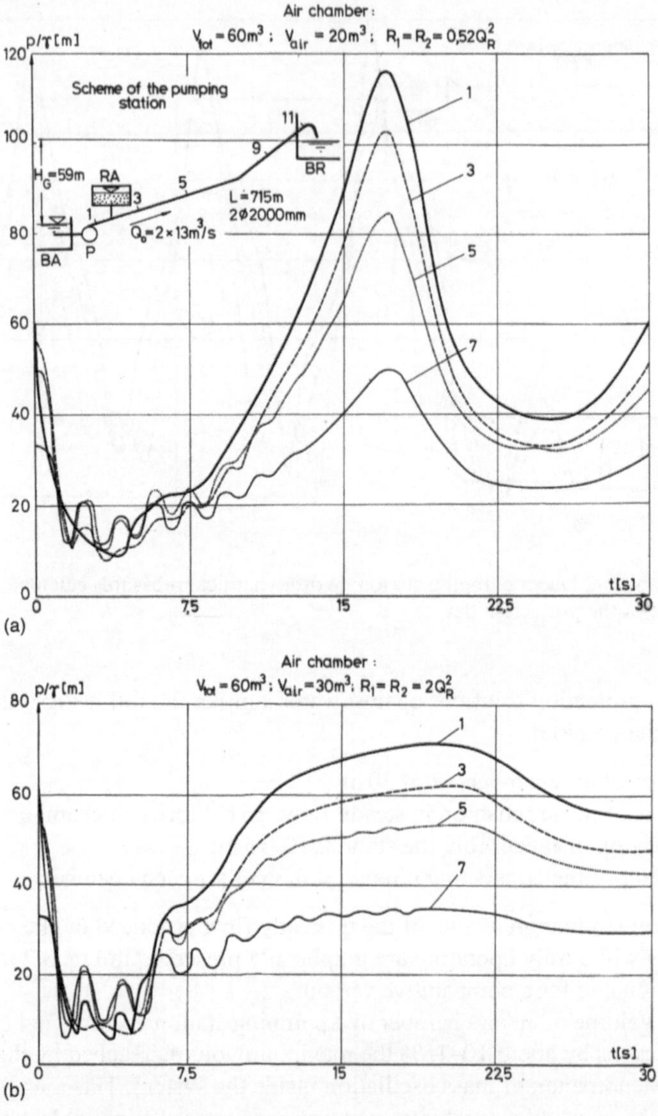

Figure 3.24. SRP-2 Daeni pumping station: pressure calculated in the pipe; (a) comparison variant, (b) design solution.

- the connection distance from the station: 10 m;
- the diameter of the pipe which connects the tank to the discharge pipe: $d = 0.35$ m ($R_1 = R_2 = 8.25Q_R^2$).

The results of hydraulic computations prove the efficiency of the protection solution adopted which generates in the system—for unsteady flows—a slowly varied flow with acceptable maximum and minimum pressure values (Fig. 3.26).

The SRP-2 Seimeni pumping station (first stage) This station has a maximum flow rate of 1.53 m³/s for a geodetic height of 35 m and is equipped with three pumps 18 NDS-600 with the steady characteristics $Q_o = 0.5$ m³/s and $H_o = 49$ m, driven by electric

Figure 3.25. Battery of air chambers of a pumping station from the Daeni irrigation system.

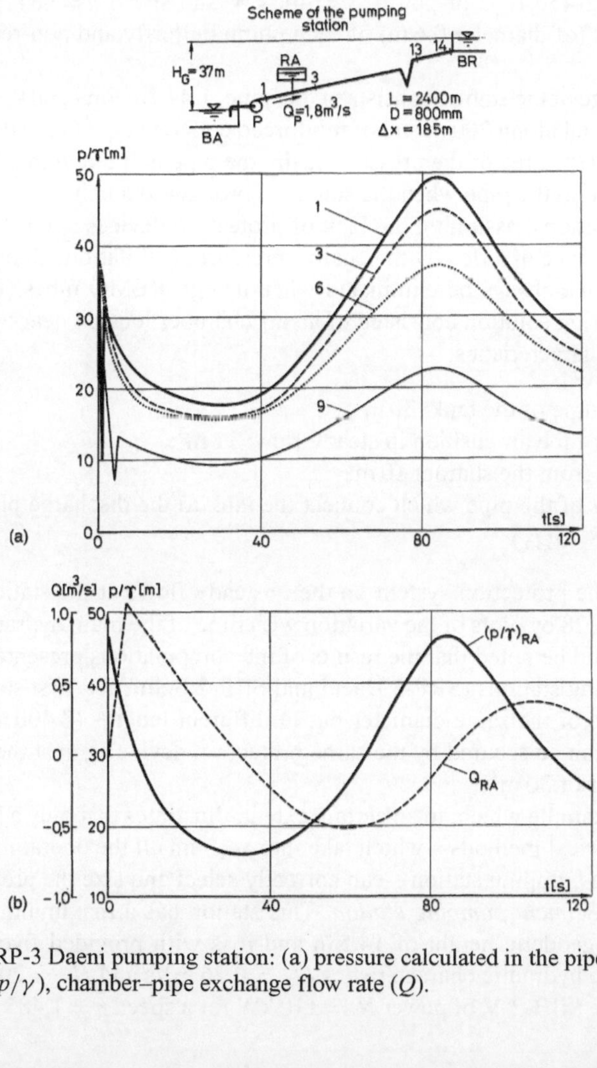

Figure 3.26. SRP-3 Daeni pumping station: (a) pressure calculated in the pipe; (b) pressure of the air cushion (p/γ), chamber–pipe exchange flow rate (Q).

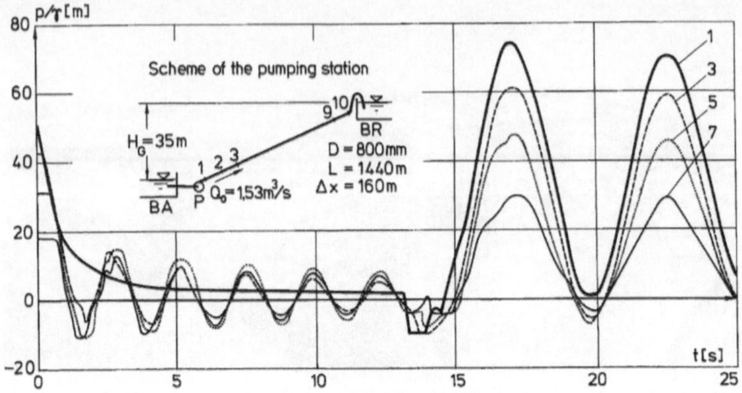

Figure 3.27. SRP-2 Seimeni pumping station (step I): hydrodynamical pressures calculated in the pipe.

motors, MAB-2-450 type of power $N = 400$ kW and speed $n = 984$ rpm. Along the discharge pipes (of diameter 0.6 m) of each pump butterfly and non-return valves are fitted.

The discharge of the station consists of one pipe, 1,440 m long and 0.8 m in diameter and made of metal along 200 m and of reinforced concrete type PREMO for the rest of the pipe-run. In the area of the pressure basin, the pipe is fitted with a 0.4 m diameter valve to let air into the pipe when the station is brought to a halt.

The computations, assuming the lack of protection devices against waterhammer, exhibit the existence of a flow with negative pressures (cavitation) along the pipe. This kind of load must always be eliminated when using PREMO tubes (Fig. 3.27). The adopted protection solution consisted of an air chamber located near the station, with the following characteristics:

- the total volume of the tank: 30 m^3;
- the volume of the air cushion in steady flow: 12 m^3;
- the distance from the station: 10 m;
- the diameter of the pipe which connect the tank to the discharge pipe: $d = 0.35$ m ($R_1 = R_2 = 8.25 Q_R^2$).

The effect of the protection system on the unsteady flows at the station shutdown is shown in Fig. 3.28 by plots of the variation with time of the main hydraulic parameters of flow. It should be noted that the results of the computations presented above reveal that two pumping stations (SRP-3 Daeni and SRP-2 Seimeni—first stage) fitted with discharge pipes of the same diameter but of different lengths (2,400 m, and 1,435 m respectively), can be secured by the same protection device against the waterhammer (an air chamber of 30 m^3).

This is an example which, to an certain extent, illustrates that only a hydraulic study based on numerical methods—which take into account all the operating and technical parameters of a pumping station—can correctly select and size the protection system.

The SRP-4 Seimeni pumping station. This station has a maximum rate of flow of 1.3 m^3/s for a geodetic height of 14.5 m and it is with provided five type MV 303 pumps with the hydraulic characteristics $Q_o = 0.26$ m^3/s and $H_o = 20$ m, driven by a electric motors, MIB-2-V, of power $N = 110$ kW for a speed $n = 1{,}485$ rpm. Along the

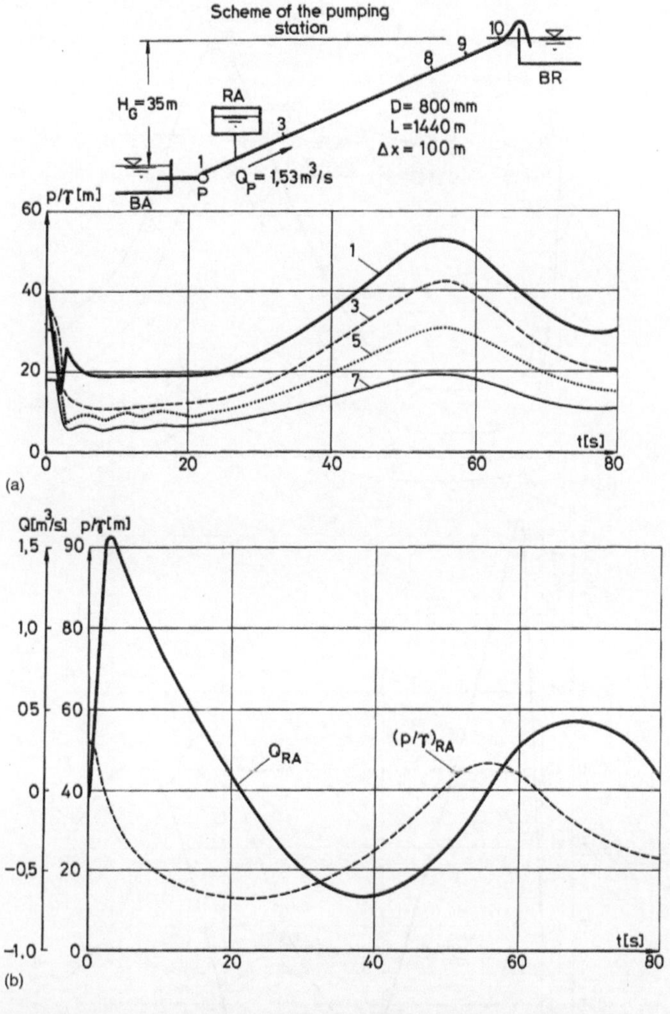

Figure 3.28. SRP-2 Seimeni pumping station (step I): (a) pressure calculated in the pipe; (b) pressure of the air cushion (p/γ), chamber–pipe exchange flow rate (Q).

discharge pipes, butterfly and non-return valves are fitted. The station uses discharge pipes with a length of 550 m and a diameter of 0.8 m, made of PREMO tubes. Near the pressure basin, the discharge pipe is fitted with a DAD-3 valve for air inflow/outflow when stopping and starting the station. The adopted protection solution consists of two air tanks, disposed in battery, near the station, with the following geometric and hydraulic characteristics:

- the volume of the air chamber: 10 m^3;
- the volume of the air cushion is steady flow: 4 m^3;
- the connection distance from the station: 10–15 m;
- the diameter of the connecting pipe to the air chamber: $d = 0.35$ m.

The results of the numerical simulation of the unsteady hydraulic flow produced by the complete halt of the station in full operation are plotted in Fig. 3.29 and prove

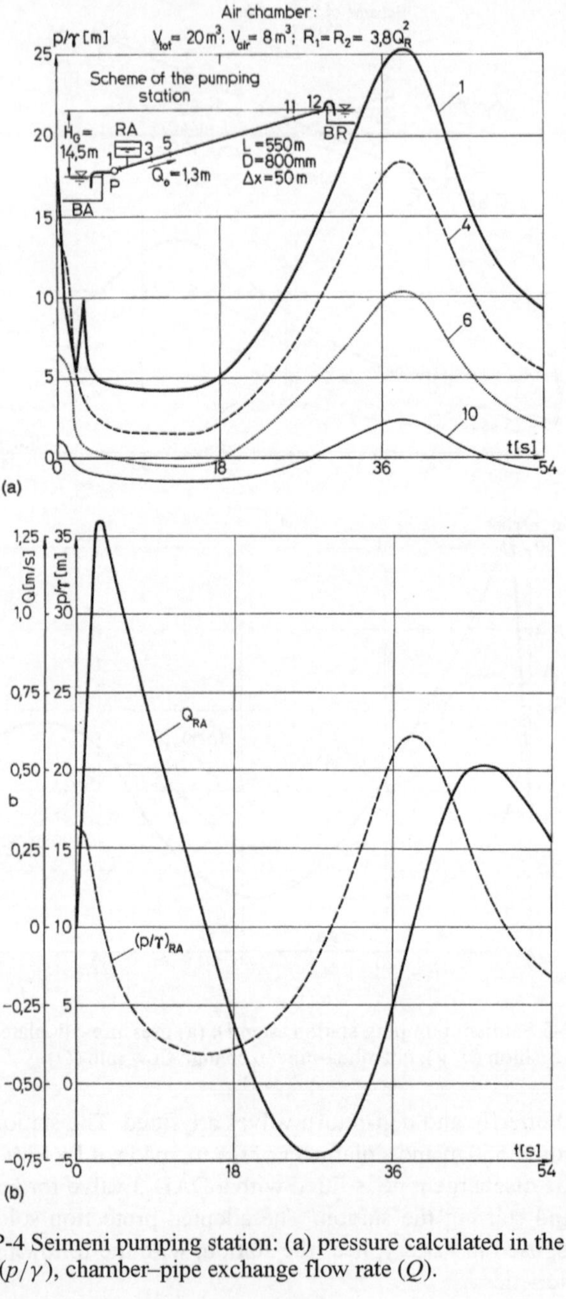

Figure 3.29. SRP-4 Seimeni pumping station: (a) pressure calculated in the pipe; (b) pressure of the air cushion (p/γ), chamber–pipe exchange flow rate (Q).

the efficiency of the adopted protection solution, which generates inside the system a slowly varied flow with acceptable maximum and minimum pressure values.

This example can raise questions with respect to the previously adopted solutions: the air chamber type protection was now adopted for a station with small geodetic height, $H_G = 14.5$ m, while usually a surge tank solution is recommended. The explanation is quite simple. The protection solution against waterhammer of the station (of air chamber type) proved to be more eligible in this case from the investment point of view, as compared to the solution with surge tank. The more general conclusion that can be drawn from the few examples presented here is that only a correct analysis based upon hydraulic computations and investment assessments can decide on the type and the optimum protection solution in a given case.

The SRP-I Seimeni pumping station. The station has a maximum steady flow rate of 16 m^3/s for a geodetic height of 20 m and it is fitted with 10 SIRET-900 type pumps, driven by electric motors, MAB-2-500 V, each of power $N = 630$ kW and speed $n = 750$ rpm. The station has two identical discharge pipes of 528 m and diameter 1.6 m, made of reinforced concrete, type SENTAR. Near the pressure basin, the pipes are fitted with electrically operated valves (of 0.5 m diameter) to allow the air in the system when the station is stopped, thus ensuring the water column breakdown at reverse flow. The valves do not operate in accidental halts of the station due to electric power breakdown.

The protection solution, for one discharge pipeline flow, consists of an air chamber located near the station, with the following geometric and hydraulic characteristics:

- the volume of the air chamber: 30 m^3;
- the volume of the air cushion in steady flow: 15 m^3;
- the connection distance from the station: 10 m;
- the diameter of the connecting pipe: $d = 0.45$ m ($R_1 = R_2 = 3.08Q_R^2$).

The hydraulic computations results presented under graphical form in Fig. 3.30, exhibit the pressure distribution along the discharge pipe while working in unsteady flow, after the total halt of the station when fully operating and draw the attention on some classical solutions of the air tank type which in certain cases can significantly reduce the overpressures in the system.

3.5.3. Pumping stations provided with air chambers and surge tanks

The SRP-7 Daeni pumping station. The station has a flow rate of 1.4 m^3/s for a geodetic height of 67.5 m and is equipped with six vertical pumps, MV-303 × 2, with the characteristics in steady flow: $Q_o = 0.23$ m^3/s and $H_o = 73$ m, driven by electric motors of power $N = 250$ kW and speed $n = 1{,}500$ rpm. The discharge pipe has a length of 1,580 m, a diameter of 0.8 m and is made of reinforced PREMO concrete. The adopted protection system contains an air chamber near the station and a surge tank along the pipeline (Fig. 3.31), with the following characteristics:

- *the air chamber*:
  $V_{tot} = 10$ m^3—the total volume of the tank;
  $V_{air} = 5$ m^3—the volume of the air cushion in steady flow;
  $\ell_1 = 10$ m—the distance from the station to the attachment section;
  $d_1 = 0.3$ m—the diameter of the pipe which connects the tank to the discharge pipe ($R_1 = R_2 = 15.4Q_R^2$),
- *the surge tank*:
  $H = 20$ m—the height of the tank;
  $D = 1.5$ m—the diameter of the tank;
  $\ell_2 = 810$ m—the distance from the station to the attachment section of the tank;
  $d_2 = 0.45$ m—the diameter of the connection pipe.

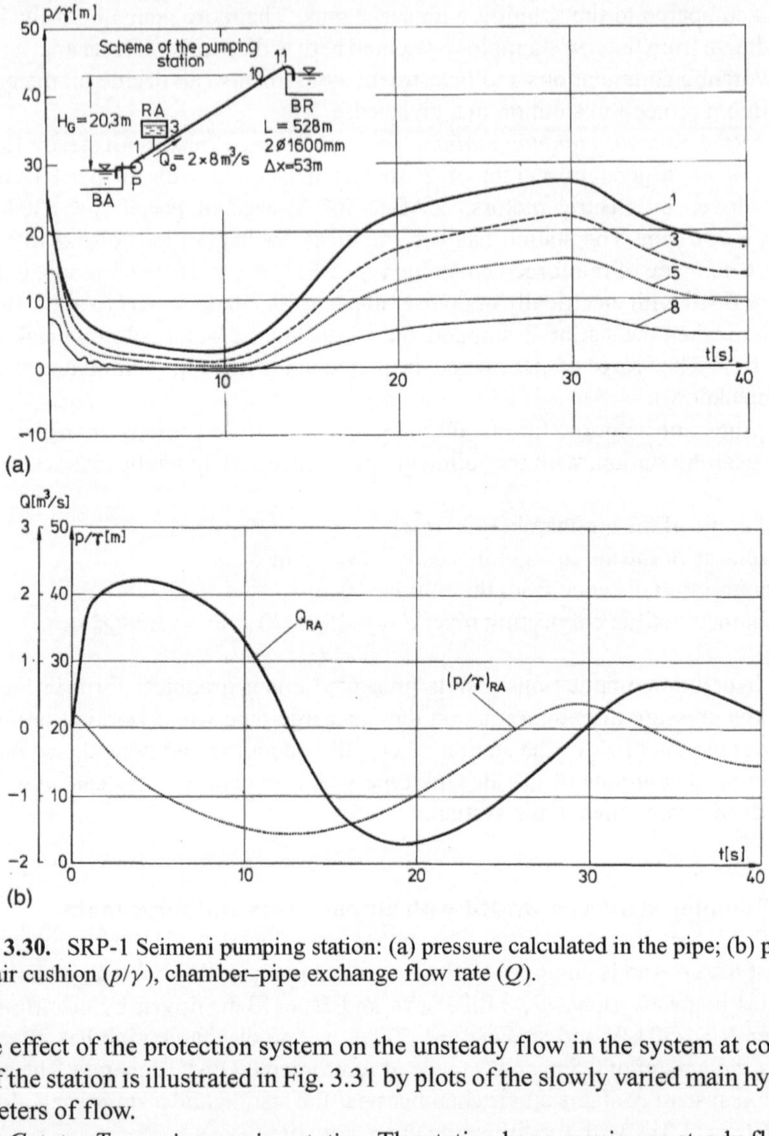

Figure 3.30. SRP-1 Seimeni pumping station: (a) pressure calculated in the pipe; (b) pressure of the air cushion (p/γ), chamber–pipe exchange flow rate (Q).

The effect of the protection system on the unsteady flow in the system at complete stop of the station is illustrated in Fig. 3.31 by plots of the slowly varied main hydraulic parameters of flow.

The Cetatea Troesmis pumping station. The station has a maximum steady flow rate of 6.5 m^3/s for a level difference of 36 m, along a discharge pipe of 830 m with the diameter of 1.9 m. In its upper part, the discharge pipe exceeds the level of the outflow basin by several meters (Fig. 3.32).

The adopted protection system consists in a system made of an air chamber, a surge tank and air valves with the following characteristics:

- *the air chamber*:
  $V_{tot} = 30\,m^3$; $V_{air} = 15\,m^3$; $\ell_1 = 10\,m$; $d_1 = 0,5\,m$;
- *the surge tank*:
  $H = 10\,m$; $D = 1.9\,m$; $\ell_2 = 180\,m$; $d_2 = 0.8\,m$;
- *the air valves*: of type DAD-6, in number of three, are fitted in the upper part of the discharge pipe.

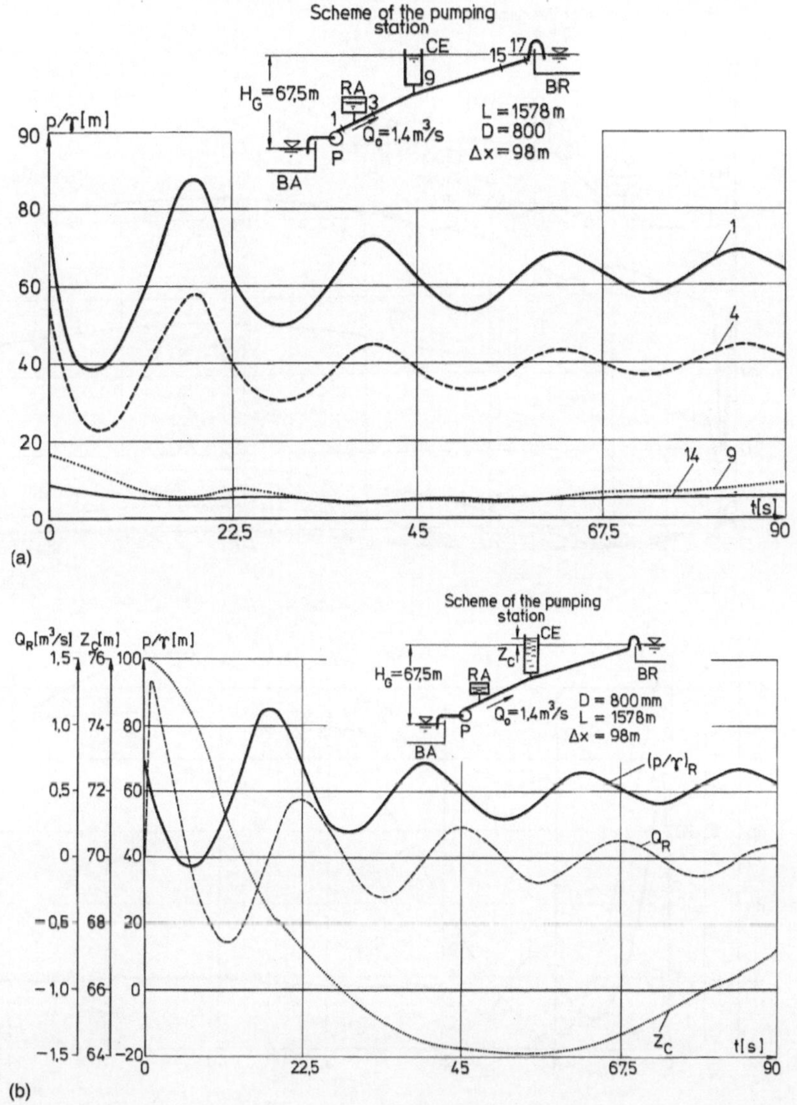

Figure 3.31. SRP-7 Daeni pumping station: (a) pressure calculated in the pipe; (b) pressure of the air cushion (p/γ), chamber–pipe exchange flow rate (Q_R), the water level (Z) in the surge tank.

The hydraulic operation of the station—at a total halt—is illustrated in the plots of Fig. 3.32, which show the possible appearance of negative pressures in the upper part of the discharge pipe, justifying the supplementary use of air valves.

The SRP-8 Daeni pumping station. The station is designed on a complex hydraulic scheme which ensures a steady flow rate of 15.2 m^3/s for two levels of height, namely: the first stage with $H_G = 39$ m and $Q_o = 14.5$ m^3/s and the second stage with $H_G = 54$ m and $Q_o = 0.7$ m^3/s.

The suction, which is common to both stages, is made of two identical pipes (CA), 468 m long and 2 m in diameter; the discharge pipes (CR) of the station have

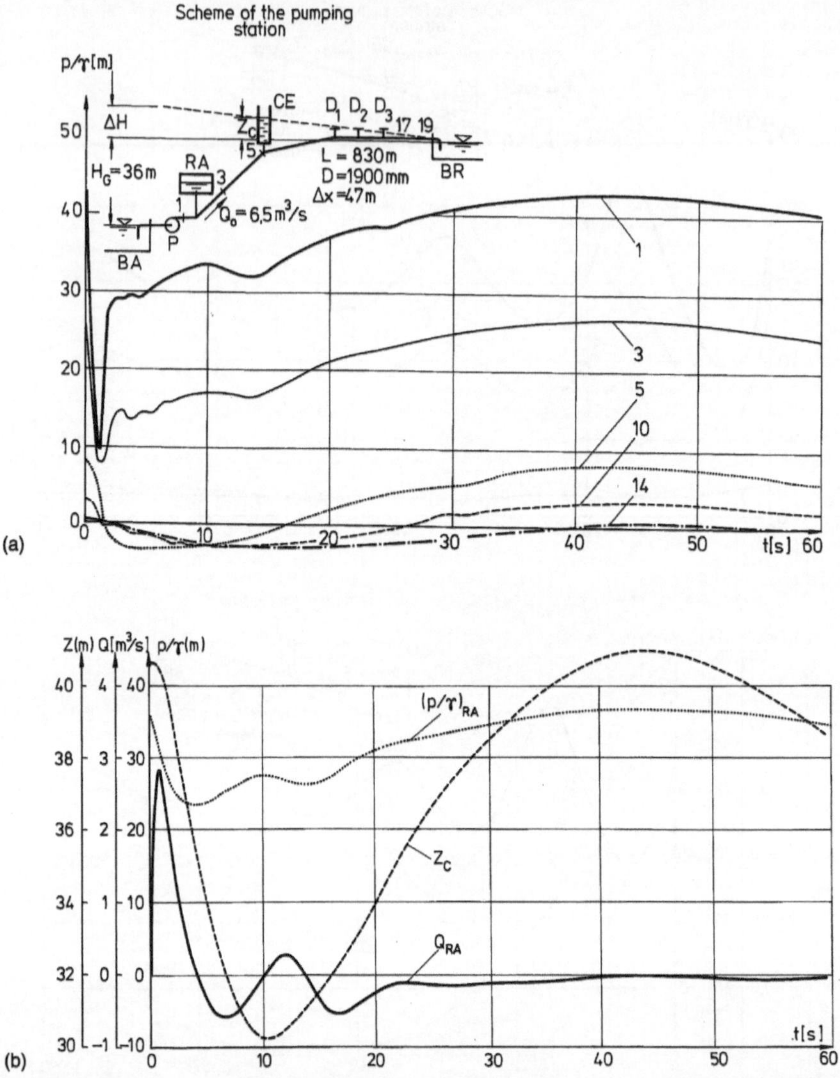

Figure 3.32. SPA Cetate Troesmis pumping station: (a) pressure calculated in the pipe; (b) pressure of the air cushion (p/γ), chamber–pipe exchange flow rate (Q_{RA}), the water level (Z) in the surge tank.

the following characteristics: the first stage $L = 493$ m, $D = 1.6$ m; the second stage $L = 740$ m, $D = 0.5$ m (Fig. 3.33). The adopted protection system for both stages of the pumping station for suction (CA) and discharge (CR) is the following:

- *the discharge of the first stage* (for one pipeline): an air chamber with the following characteristics: $V_{tot} = 30\,m^3$; $V_{air} = 15\,m^3$; $\ell_1 = 10$ m; $d_1 = 0.45$;
- *the discharge of the second stage*: an air chamber with the following characteristics: $V_{tot} = 10\,m^3$; $V_{air} = 5\,m^3$; $\ell_2 = 10$ m; $d_2 = 0.25$ m;
- *the suction of the first and second stages* (for one pipeline only): surge tank with the following characteristics: $H = 10$ m; $D = 1.5$ m; $\ell = 10$ m; $d = 0.35$ m.

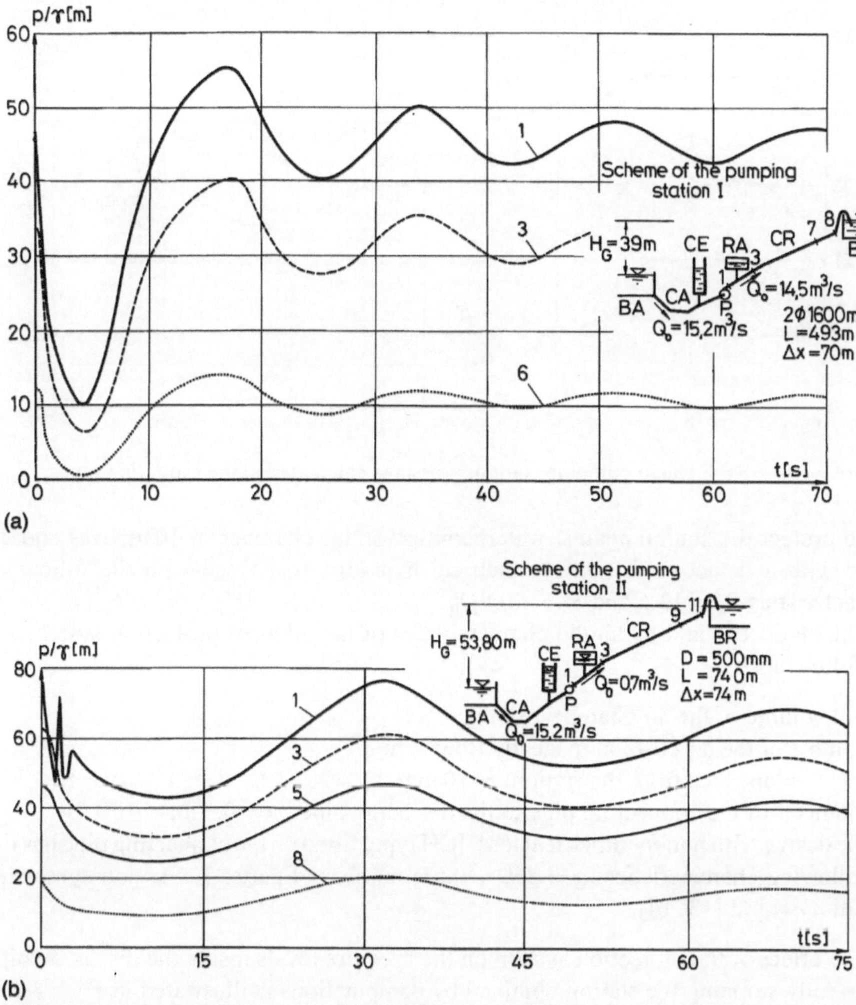

Figure 3.33. SRP-8 Daeni pumping station: pressures calculated in the pipes from step I (a) and step II (b).

The results of numerical simulation of the unsteady flow in the system, generated at a total halt of the station in full operation, are presented in the graphs, separately for the discharge pipes (Fig. 3.33) and for the suction pipes (Fig. 3.34).

3.5.4. Pumping station provided with ICH type air chamber

The SRP-2 Pecineaga-Turcoaia pumping station. The station has a maximum flow rate of 2 m³/s for a geodetic height of 44 m and is equipped with six pumps, type 12 NDS-430, with the steady flow characteristics: $Q_o = 0.334\,\text{m}^3/\text{s}$ and $H_o = 48$ m, driven by MAB-315 type electric motors with power $N = 200$ kW and speed $n = 1{,}500$ rpm. The 708 m long and 1 m diameter discharge pipe is made of PREMO tubes of reinforced concrete and is fitted near the pressure basin with air valves to break down the water column at reverse flow, when the station is stopped. Along the 0.4 m diameter discharge pipes, of each pump, non-return valves and hand-operating valves are fitted.

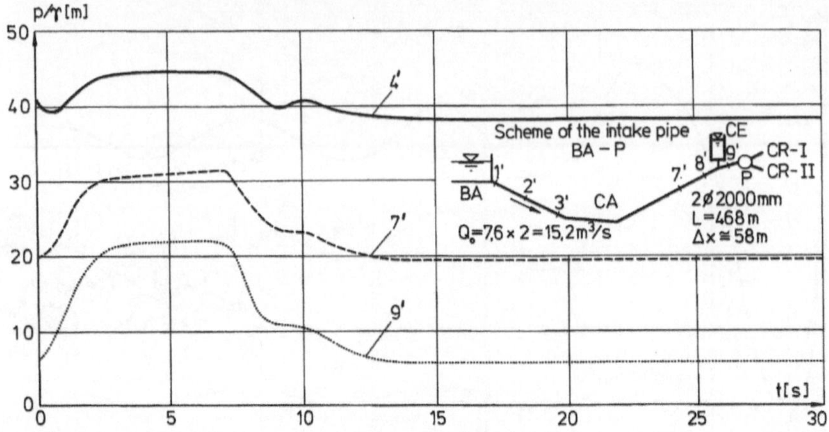

Figure 3.34. SRP-8 Daeni pumping station: pressures calculated in the aspiration pipes.

To protect the station against waterhammer, an air chamber of $10\,\text{m}^3$ was chosen fitted with a device with non-symmetrical hydraulic resistance with the following characteristics $R_1 = 13.6$ and $R_2 = 40.8Q_R^2$.

The geometric and hydraulic characteristics of the adopted protection system are the following:

- full volume of the air chamber: $10\,\text{m}^3$;
- volume of the air cushion in steady flow: $5\,\text{m}^3$;
- location distance from the station: 5–10 m;
- diameter of the connecting pipe to the discharge pipe: $D = 0.5$ m;
- the device with non-symmetrical loss, ICH type, fitted on the connecting pipe has the following characteristics: $\alpha = d/D = 0.55$; number of parts: $3 + 1$; non-symmetry ratio: $r = 4.2$ [13, 14].

The effect of the protection system on the pressure levels inside the discharge pipe when fully stopping the station obtained by computations is illustrated in Figs. 3.35b and 3.36b, compared, for example, to the effect of an ordinary protection system with an air chamber of the same volume, but connected to a simple duct, $R = 8.25Q_R^2$, in Figs. 3.35a and 3.36a.

The SRP-2 Seimeni pumping station (stage II). The station has a steady flow of $10.7\,\text{m}^3/\text{s}$ for a geodetic height of 69 m and is provided with 8 pumps, type 24 NDS-955, with the characteristics: $Q_o = 1.35\,\text{m}^3/\text{s}$ and $H_o = 76$ m, driven by electric motors MAB-2-630 of power $N = 1{,}400$ kW and speed $n = 740$ rpm. Along the discharge pipe of each pump (0.6 m in diameter) non-return valves and butterfly valves are fitted. The 2,340 m long and 2.2 m diameter discharge pipe, is made of metal and provided, in the proximity the outflow basin, with an electrically driven valve (0.5 m in diameter) for aeration of the pipe when the station is stopped.

The adopted protection system consists of four identical air chambers with the following geometrical and hydraulic characteristics:

- total volume of each chamber: $45\,\text{m}^3$;
- volume of the air cushion for steady flow of each chamber: $20\,\text{m}^3$;
- location distance from the station: 10–25 m;

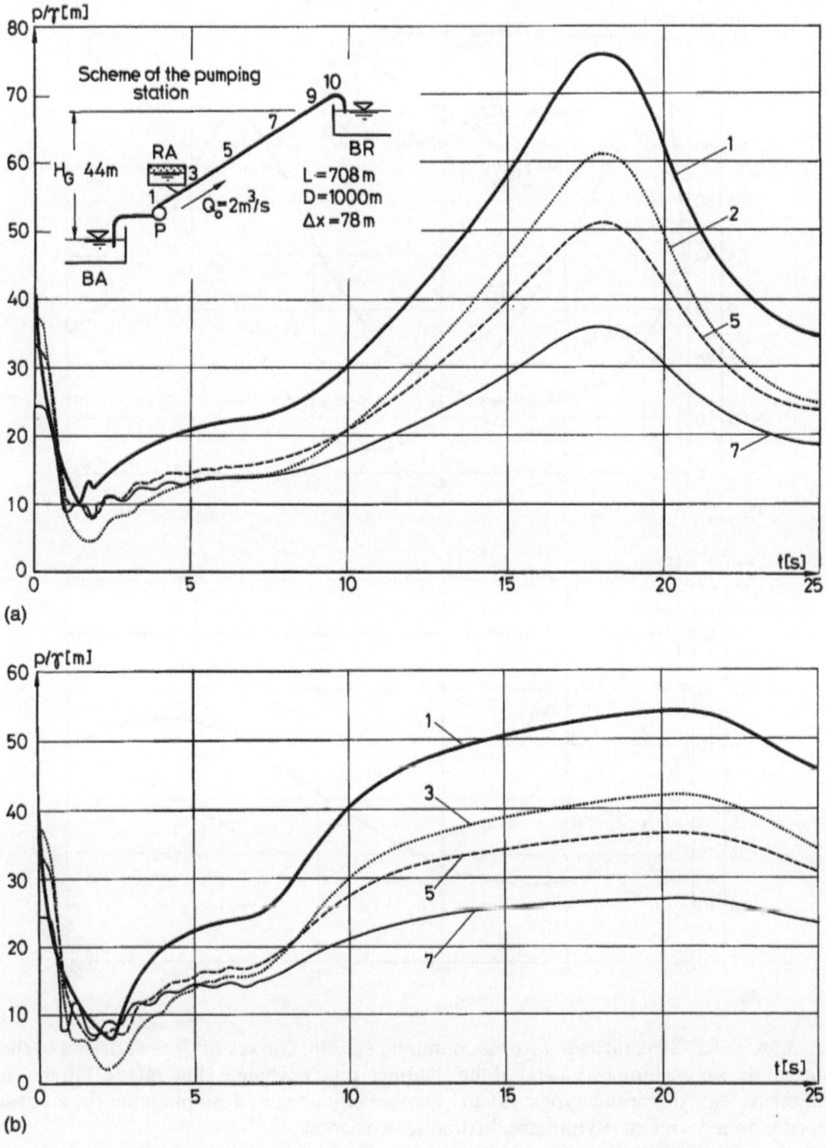

Figure 3.35. SRP-2 Pecineaga–Turcoaia pumping station. Curves of time variation of water pressure in the pipe in the cases of various types of protection systems: (a) air chamber branched with simple pipe; (b) air chamber equipped with a device of asymmetric hydraulic resistance.

- number of connecting pipes ($D = 0.5$ m) for each chamber: 2;
- the device of head losses, type ICH, fitted along the connecting pipe has the following characteristics: $\alpha = d/D = 0.5$; number of parts: $2 + 1$; non-symmetry ratio $r = 3.90$ [13, 14].

The effect of the protection system on the pressure regime in the discharge pipe, produced by a full halt of the station while fully operating, is illustrated in Fig. 3.37 by

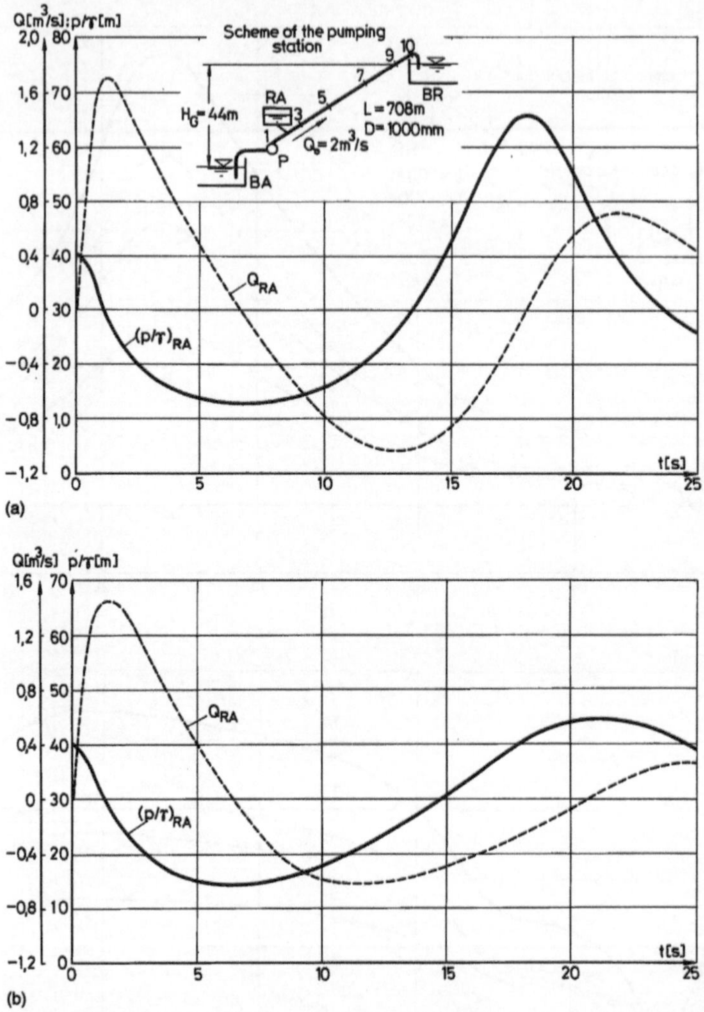

Figure 3.36. SRP-2 Pecineaga–Turcoaia pumping station. Curves of time variation of the pressure of the air cushion (p/γ) and of the chamber–pipe exchange flow rate (Q) in the case of protection systems of various types: (a) air chamber branched to a simple pipe; (b) air chamber equipped with a device of asymmetric hydraulic resistance.

plots, obtained by hydraulic computations, both for the adopted solution (Fig. 3.37b), and for a classical protection solution to enable comparison (Fig. 3.37a).

The Prut–Birlad dirivation

The term of dirivat is used for a watershed project in the North-East of Romania, which ensures the transfer by a two-stage system, of a steady flow of 3 m³/s from the river Prut in the river Vasluiet for a total geodetic height of 166 m. The water is used for various purposes in the villages and towns in area of the Birlad River watershed.

Thus, among the most important uses is the water supply of the city of Vaslui, from the Golesti water storage, located at about 30 km downstream of the entrance point

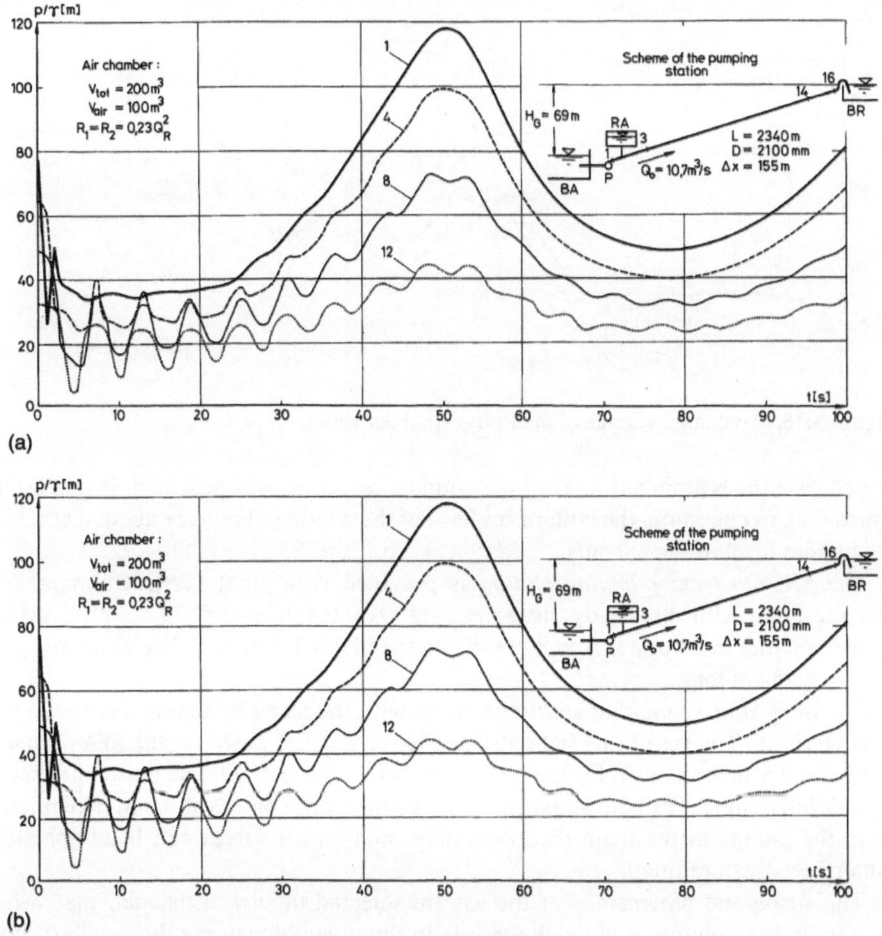

Figure 3.37. SRP-2 Seimeni pumping station (step II). Curves of time variation of the pressure of the water in the pipe for: (a) protection solution of classical type; (b) solution of protection with ICH type air chamber.

of the diversion in the Vasluiet river. The Prut–Birlad dirivation is the first project of this kind in Romania, characterized by the fact that among all studied versions of gravitational or pumping schemes [16], its hydraulic pumping scheme proved to be the most rational from the point of view of investment. The hydraulic scheme of the diversion, presented in Fig. 3.38, exhibits a metallic pipe, 11,720 m long and 1.4 m in diameter, with two pumping stations with similar characteristics, placed along the pipe. Relatively small lengths of this pipe are made of reinforced concrete tubes, type SENTAB [7, 24].

The water intake of the Prut River is situated near Opriseni village and is built as a pumping station—SPA Opriseni—of type caisson with wet vat, placed in the riverbank.

The water enters the station vat through windows cut out in the caisson's body the bottom sills of which are below the minimum water level in the river. To allow an optimum water caption rate, a special gate with detachable elements, that can increase

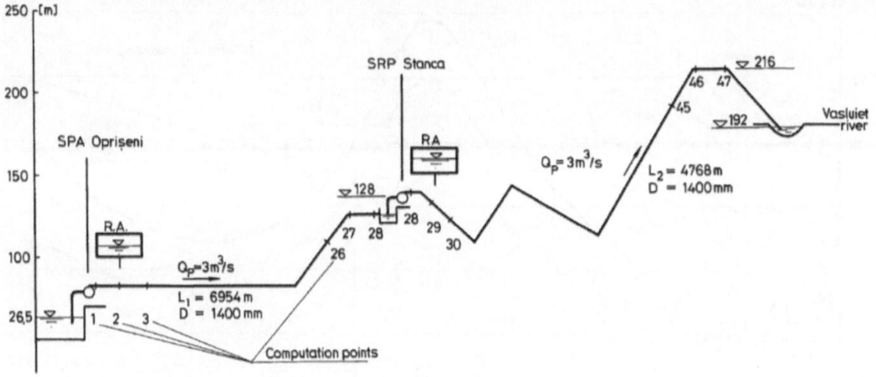

Figure 3.38. Hydraulic scheme of the Prut–Birlad derivation.

or decrease the bottom sill level of the window by 0.5 m, was provided. It was established that, in operation, the bottom sill level of the window should be about 0.6 m over the bottom level of the channel.

The SPA-Opriseni pumping station is provided with eight vertical pumps, NV 403 × 2 type, with the steady characteristics $Q_o = 0.4\,\text{m}^3/\text{s}$ and $H_o = 100\,\text{m}$, driven by electric motors of power $N = 800\,\text{kW}$ and speed $n = 1{,}500\,\text{rpm}$. The discharge pipe is $L_1 = 6{,}954\,\text{m}$ long.

The SRP-Stinca pumping station is fitted with the same hydraulic devices as the SPA Opriseni. The vat of this station stands for an outflow basin for the SPA Opriseni station and a suction basin for the SRP-Stinca at the same time. The discharge pipe is $L_2 = 4{,}768\,\text{m}$ long. For both stations, along the discharge pipes (of 0.5 m in diameter) from the pumps to the main discharge pipe, non-return valves and hand-operating butterfly valves are fitted.

The shape and dimensions of the vat are selected in such a manner, that, when operating, the volume of alluvial deposits in the pumped water is the smallest. Possible alluvial deposits accumulated when the pumping station is not in operation, are evacuated from the suction basins either by hydro-elevators in the SPA Opriseni, or by opening the bottom water outlet in the SRP Stinca.

In order to ensure the serial operation of the two pumps the following solutions have been adopted:

- the stations were provided with the same hydraulic and mechanical devices (same type and same number of pipes);
- the flow rate of the SPA Opriseni station is automatically adjusted by a floating valve which opens or closes function of the water level in the vat of the SPR-Stinca station;
- there exists a common system of optimal signals for the operation of each pump at both pumping stations;
- the pumping stations were located along the hydraulic path in such a manner as to obtain nearly the same pumping pressures;
- a direct telephone line links the two stations.

All significant operation cases of the two pumping stations in unsteady flow with and without waterhammer protection systems have been analysed by computations and

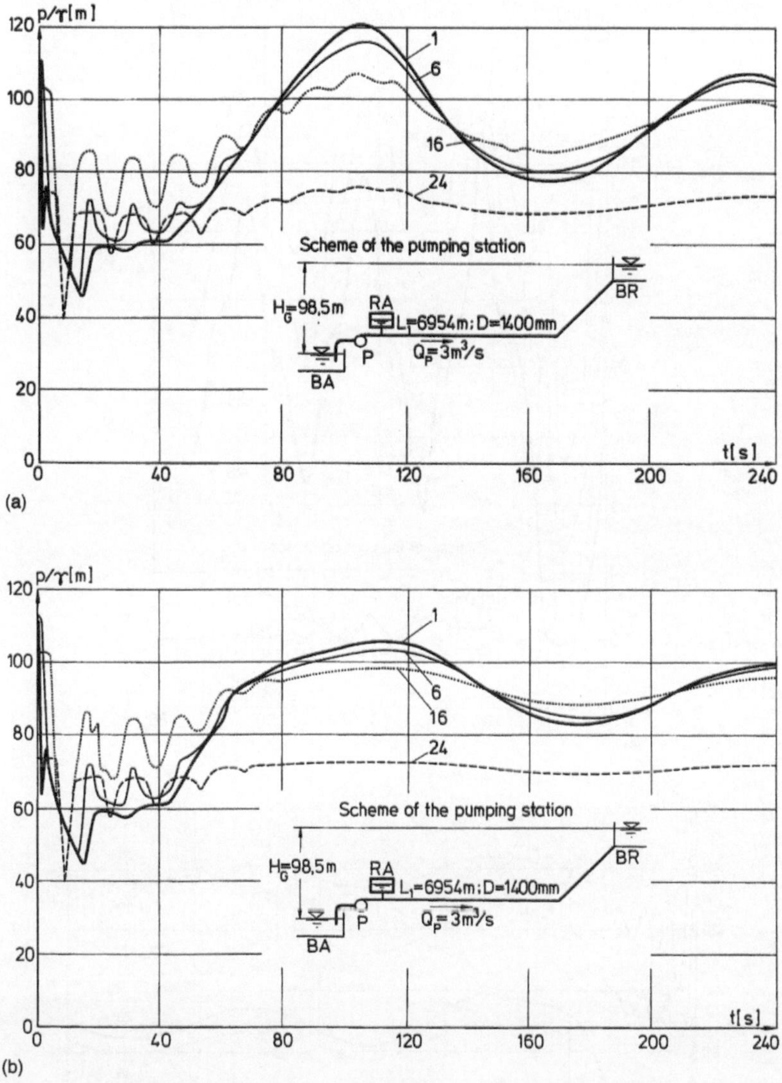

Figure 3.39. SPA Opriseni pumping station. Time variation curves of the water pressure in the pipe for: (a) protection solution of classical type; (b) solution of protection with ICH type air chamber.

special attention was paid to the most unfavourable operation situation: the sudden stop of all pumps at one station produced by an electric power breakdown.

From the large amount of hydraulic computations performed for this project, only some significant results are graphically presented here. Taking into account the fact that the operating pressures at the two stations are relatively high (10 atm), consideration was given to waterhammer protection solutions, involving as small overpressures as possible.

The adopted protection alternative for the two pumping stations consists of two special air chambers type ICH (of 30 m^3 volume), located in batteries near the stations,

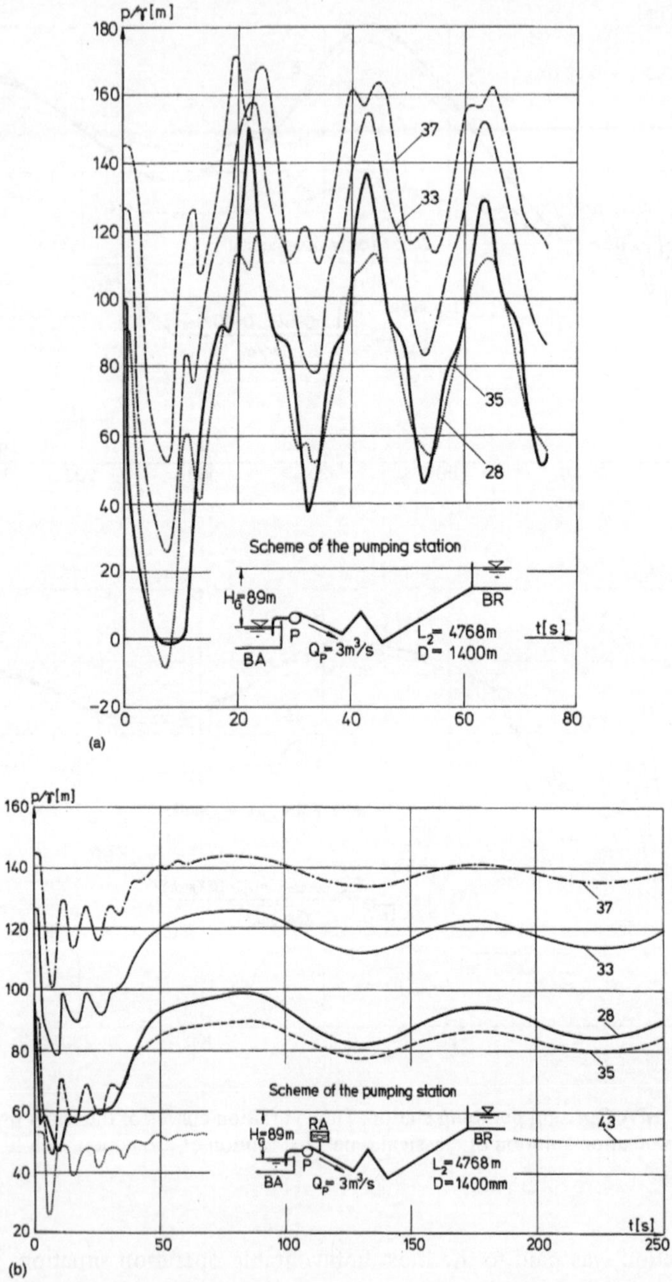

Figure 3.40. SRP Stinca pumping station. Time variation curves of the water pressure in the pipe: (a) in the absence of the protection devices; (b) with protection ICH type air chambers.

with the following geometric and hydraulic characteristics per air chamber:

- volume of the air cushion in steady flow: 15 m^3;
- location distance from the pumping station: 10–20 m;

- diameter of the connection pipe to the discharge pipe: $D = 0.3$ m;
- the device for non-symmetrical hydraulic losses fitted on the connection pipe has the following characteristics: $\alpha = d/D = 0.5$, the number of parts: $2 + 1$, the ratio of non-symmetry $r = 3.5$ [13, 14].

The batteries of air chambers are made of five tanks at SPA Opriseni and of four tanks at SPR Stinca, respectively. To ensure preservation of the air cushion in the air chambers under steady flow, the pumping stations are fitted with the required compressors for air supply.

The results of hydraulic computations, presented in Figs. 3.39 and 3.40, show that the adopted protection solutions actually eliminate the overpressure inside the discharge pipe of the two pumping stations after an emergency stop of the station in full operation.

References

1. Cioc, D., Tatu, G., Anton, A.: Practical Results Obtained in Utilisation of a Complex Hydrophore for the Protection against Waterhammer. *Hidrotehnica*, 10 (1982) (in Romanian).
2. Fok, A.: Design Charts for Surge Tanks on Pump Discharge Lines. *Proceedings of the 3rd International Conference on Pressure Surge*, Canterbury, England, 1980.
3. Hancu, S., Popescu, M., Zaharescu, E.: *Applied Hydraulics*. Ed. Tehnica, Bucharest, 1985.
4. Jones, S., Wood, D.: Prediction and Control of Pressure Surges due to Valve Closures. *Proceeding of the 1st International Conference on Pressure Surges*, Canterbury, England, 1972.
5. Jung, G., Schneider, E.: Design and Optimisation of Seawater Pumping Stations for Industrial Plants. *International Conference on the Hydraulics of Pumping Stations*, Manchester, England, 17–19 September, 1985.
6. Parmakian, J.: *Waterhammer Analysis*. Longmans, London, 1955.
7. Popescu, M.: *Hydraulic Studies Concerning the Design of Protection Devices against Waterhammer at Pumping Stations in Romania*. Internal Reports I.C.H., Bucharest, 1970–1980.
8. Popescu, M.: Pumping Station Provided with Surge Tank. *Hidrotehnica*, 6 (1972), 308–311.
9. Popescu, M., Vlase, P.: *Robinet Antiretur cu Clepra Inclinata de Forme Hidrodinamica* (Check-Valve with Hydrodynamic Flap). Invention, Patent O.S.I.M., Romania, nr. 98592, 1990.
10. Popescu, M., Halanay, A.: Le Calcul de Resistences Hydraulique Optimales des Systemes Speciaux de Protection des Stations de Pompage contre le Coup de Belier. *Proceedings of the XXIV IAHR Congress*, Vol. 4, Madrid, 1991, pp. 41–53.
11. Popescu, M., Stere, C., Jelev, I.: *Waterhammer Protection Device for Pumping Stations*. Invention, Patent O.S.I.M., Romania, nr. 71287/1979.
12. Popescu, M.: Recherches sur Certaines Dispositifs Nouveaux Fournissant des Résistances Hydrauliques Asymétriques dans les Systèmes sous Pression. *XVIII Convegno di Idraulica e Construzioni Idrauliche*, Bologna, Italia, 1982.
13. Popescu, M.: Système de Protection des Stations de Pompage contre le Coup de Belier a Effet de Performance. *XIX Convegno di Idraulica e Construzioni Idrauliche*, Pavia, Italia, September 6–8, 1984.
14. Popescu, M.: Highly Efficient Protection Systems Against Waterhammer for Pumping Stations. *International Conference on the Hydraulics of Pumping Stations*, Manchester, Anglia, September 17–19, 1984.
15. Popescu, M., Halanay, A., Ghinda, T.: Calculul Rezistentelor Hidraulice Optime la Sisteme Speciale de Protectie la Soc Hidraulic a Statiilor de Pompare (The calculation of the optimal hydraulic resistances for special systems for waterhammer protection of the pumping stations). *Studii si Cercetari de Mecanica Aplicata*, 42 (1990), 207–217.

16. Popescu, M., Florescu, I.: Application of a Romanian Invention at Pumping Stations on Prut–Birland Derivation. *Hidrotehnica*, 10 (1981).
17. Popescu, M., Arsenie, D.: Problemes Dynamiques dans le Projet et le Fonctionnement des Usines Hydroelectriques et des Stations de Pompage. *Proceedings of the XXVI IAHR Congress*, Vol. 1, London, 1995, pp. 613–615.
18. Popescu, M., Constantinescu, G., Vlase, P., Jelev, I.: *Hydraulic Resistance Module and Fixed Modular Elements Battery*. Invention, Patent O.S.I.M., Romania, nr. 71286/1979.
19. Seeber, G.: Das Wasserschloss des Kauntertal-Kraftwerkes der TIWAG—Ein neuer Typ eines ruckstromgedrossten Kammerwasser-schlosses. *Schweizerische Bauzeitung*, 1 (1970).
20. Stere, C.: *Contributions to Numerical Computing of Waterhammer in Unifilar Pressure Pipes*. Ph. D. Thesis, Institutul de Constructii, Bucharest, 1977.
21. Streeter, V., Wylie, E.: *Hydraulic Transients*. McGraw-Hill, New York, San Francisco, London, 1967.
22. Vlase, P.: *Contributions to the Study of the Hydraulic Systems with Automatic Regulators*. Ph.D. Thesis, University of Constantza, 1999.
23. Zobel, R.: Versuche an der hydraulischen Ruckstromdrossel. *Mitteilungen HITH Munchen*, 8 (1936).
24. Bartha, I., Cismaru, C., Luca, M.: *Hydraulic and Operating Characteristics of Laterals for Sprinkling Irrigation Made up Using Zinc Plated Steel Pipes*. Ovidius University Annals of Constantze, Year III–IV, Series Constructions, 2002, pp. 465–470.

Chapter 4

Hydraulic Resonance in Hydroelectric Power Plants and Pumping Stations

Theoretical and experimental studies of the phenomenon of hydraulic resonance in hydraulic surge systems have been performed since the beginning of our century, but started to have a systematic character since very dangerous accidents occurred in hydroelectric power plants, mainly caused by the appearance and development of some phenomena of hydraulic resonance with violent flow inside the system.

It is generally accepted that the occurrence of hydraulic resonance inside a pressure system is due to the vibrations in the hydraulic circuit induced by some hydraulic devices, such as valves, regulators, etc. The induced vibrations should be understood as the vibrations introduced by these devices in the fluid of hydraulic systems, which act like perturbations, and not the mechanical vibrations of these devices which have usually much larger frequencies than the frequency of hydraulic resonance. Resonance in a hydraulic circuit can cause many problems, from the annoyance introduced by noises and vibrations, to the break of pipes and galleries due to negative pressures (cavitation) or to large amplitude positive pressures or to the total failure of the hydraulic system operation due to large variations in the hydraulic parameters of flow.

The lack of detailed information about the damage produced in hydroelectric power plants due to the appearance of resonance phenomena, is one of the main reasons that delayed the development of the theory of hydraulic resonance.

Among the many accidents and cases of hydraulic resonance produced at hydroelectric power plants, reported by the literature in the field, we shall mention the most significant [15, 20]:

The water project Lac Blanc–Lac Noir (France): The accident took place in 1934, while the hydroelectric power plant with a pumping station from Lac Blank to Lac Noir was tested before commissioning.

The power plant was fitted with four 30 MW turbines and with three 15 MW pumps. Hydraulic resonance appeared under the form of strong vibrations when the power plant operated as a pumping station, especially in the proximity of the pumps. The accident occurred when the power plant was operating with one pump, for a geodetic height of 100 m.

Several engineers who were testing the installation tried to eliminate the pump vibrations. The accident produced by the break of a pipe near a visiting window, followed by the crush of the pumping station ceiling, involved many victims. It was assumed that the disaster was generated by the vibrations produced by a guiding valve at the pumps.

The hydroelectric power plant of Kandergrund (Switzerland). The main characteristics of this power plant built between 1908 and 1911, are: geodetic height $H = 312$ m, steady flow rate $Q = 6\,m^3/sec$. After a long period of operation, in 1945 three zones of horizontal cracks about 20 m long, and several centimetres wide, appeared in the admission gallery (length $L = 4{,}214$ m). Infiltration from this gallery produced land slides in the region, involving casualties and material damages.

The situation was totally unaccountable as the power plant surge tank of relatively large dimensions and low head losses at the basis should stop the propagation of pressure waves from the penstock to the supply gallery. The accident was assumed to be caused by the hydraulic resonance generated by the vibration of a spring air valve.

The hydraulic project Bersimis II (Canada). The water supply of the plant consists of an admission gallery ($L_1 = 720$ m; $d_1 = 11.6$ m), a penstock ($L_2 = 123$ m; $d_2 = 11.6$ m) and five turbine distribution pipes ($L_3 = 153$ m; $d_3 = 5.20$ m). Resonance appeared at the plant in 1961 and consisted in large pressure variations in the penstock for several combinations of turbine operation.

This is one of the cases of resonance produced at hydroelectric power plants deeply studied for its theoretical and practical aspects. Many accurate measurements of hydraulic parameters involving up-to-date technical devices were performed for different operation situations, as well as numerical simulation of the phenomenon. Finally, it was established that the hydraulic resonance was generated by the five valves of the turbines. From the conclusions of the work carried out to understand this case it followed that high harmonics in the penstock can induce dangerous situations in the entire system and that pressure waves can enter the admission gallery "by-passing" the surge tank, even if well located and well designed. The phenomenon of resonance disappeared once the cause has been removed.

In all the above-mentioned cases, it should be noted that the vibrations generating hydraulic resonance appeared as a periodic perturbation in the flow rate in a section immediately upstream the turbine.

General theory usually defines the phenomenon of hydraulic resonance in close concordance with the eigenfrequencies of the hydraulic circuit. However, from a practical point of view, when studying hydraulic resonance it is essential to determine the excitation frequencies that induce periodic oscillations of pressure and flow involving high peak amplitudes in one of the characteristic hydraulic parameters of the system (for example, the water pressure at a point) [2].

Papers address the problem of hydraulic resonance either by different methods of computation [1, 8, 15, 20], by analysing and describing the cases of hydraulic resonance phenomena produced at hydroelectric power plants in operation [3, 7, 10, 11] or by laboratory experiments [2]. In rock-bored tunnels, hydraulic resonance requires a special approach, as pointed out in reference [21].

In this chapter, a theory on hydraulic resonance in hydroelectric power plants and pumping stations is developed by treating the general equations of flow with the method of characteristics and assuming various hypotheses on the perturbative factors that may generate hydraulic resonance phenomena in the system.

The possibilities of appearance of hydraulic resonance at several Romanian hydroelectric power plants and pumping stations are analysed as practical applications. The ranges of the dangerous frequencies for some perturbative factors are established and in the hypothesis of resonance phenomena appearance and development, the values of the maximum pressures occurring in the system are computed by numerical methods [4, 13–19].

4.1. The computation of resonance in hydroelectric power plants for the basic hydraulic scheme

The mathematical model

The water flow general equations in the system consisting of feed pipe–surge tank–penstock, are the following (Fig. 4.1)

$$\frac{\partial H_1}{\partial t} + a_1 c_1^2 \frac{\partial Q_1}{\partial x} = 0, \quad \frac{\partial H_1}{\partial x} + a_1 \frac{\partial Q_1}{\partial t} = 0$$

$$F\frac{\mathrm{d}Z}{\mathrm{d}t} + Q_2(0,t) = Q_1(\ell_1, t) \tag{4.1}$$

$$\frac{\partial H_2}{\partial t} + a_2 c_2^2 \frac{\partial Q_2}{\partial x} = 0, \quad \frac{\partial H_2}{\partial x} + a_2 \frac{\partial Q_2}{\partial t} = 0$$

with the boundary conditions

(a) $H_1(0,t) = H$

$$H_1(\ell_1, t) = H + Z(t) + R' \left|\frac{\mathrm{d}Z(t)}{\mathrm{d}t}\right| \frac{\mathrm{d}Z(t)}{\mathrm{d}t} = H_2(0,t) \tag{4.2}$$

$$\eta g Q_2\,(\ell_2, t)\,[H + Z\,(t)] = N$$

(assuming the power plant is operating at constant power), or

(b) $Q_2(\ell_2, t) = \alpha\sqrt{H_2(\ell_2, t)}$

(admitting a given law for the flow rate at the power plant), where the following notation is involved: H, head in steady flow; H_i, instantaneous piezometric head in conduit i

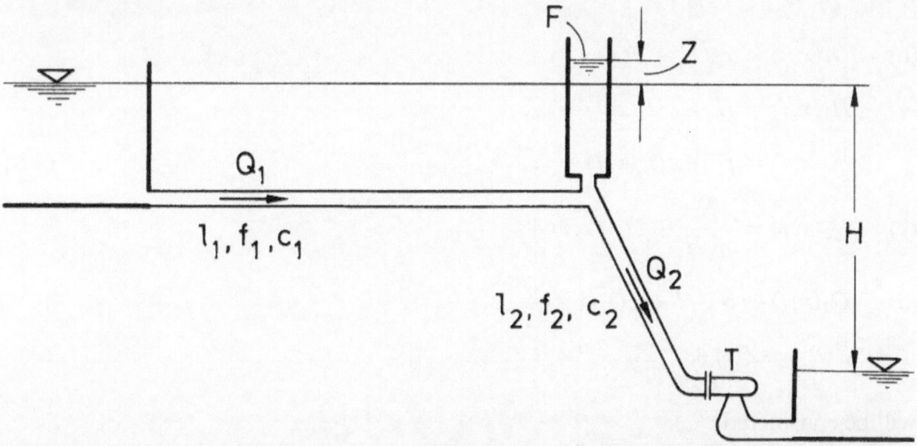

Figure 4.1. Scheme of a hydraulic power plant equipped with a surge tank.

against a horizontal reference plane at the turbine level; Z, water level in the surge tank; Q, discharge of the plant in steady flow; Q_i, instantaneous discharge in conduit i; Q_c, the inflow/outflow rate in the surge tank; R, local coefficient of head loss in the restricted area of the surge tank inlet; R', local coefficient of head loss; ℓ_i, length of conduit i; f_i, cross-section area of conduit i; N, turbine power; t, time; x, the distance measured in the flow direction; η, turbine output; g, gravitational acceleration; α, coefficient of discharge; c, celerity of pressure waves; ω, angular frequency; τ, theoretical half-period of the conduit ($2\,\ell/c$); $a_i = \frac{1}{gf_i}$; A, constant; ε, dimensionless constant.

The corresponding steady flow is given by the relations

$$\begin{aligned} &\text{(a)}\quad H_1 = H_2 = H, \quad Z = 0, \quad Q_1 = Q_2 = \frac{N}{\eta g H} \\ &\text{(b)}\quad H_1 = H_2 = H, \quad Z = 0, \quad Q_1 = Q_2 = \alpha\sqrt{H} \end{aligned} \tag{4.3}$$

The problem of resonance

It is supposed that for the boundary conditions (a) and (b) respectively, perturbations of the form

$$\text{(a)}\quad \eta g Q_2(\ell_2, t)\left[H + Z(t)\right] = N + \varepsilon A \sin \omega t$$

and

$$\text{(b)}\quad Q_2(\ell_2, t) = (\alpha + \varepsilon A \sin \omega t)\sqrt{H_2(\ell_2, t)} \tag{4.4}$$

respectively appear.

As a result of such perturbations, the perturbations corresponding to the steady flow, such as

$$\begin{aligned} &\text{(a)}\quad Q_1(x,t) = \frac{N}{\eta g H} + \varepsilon \tilde{Q}_1(x,t) + \ldots \\ &\text{(b)}\quad Q_1(x,t) = x\sqrt{H} + \varepsilon \tilde{Q}_1(x,t) + \ldots \\ &\qquad H_1(x,t) = H + \varepsilon \tilde{H}_1(x,t) + \ldots \\ &\qquad H_2(x,t) = H + \varepsilon \tilde{H}_2(x,t) + \ldots \\ &\text{(c)}\quad Q_2(x,t) = \frac{N}{\eta g H} + \varepsilon \tilde{Q}_2(x,t) + \ldots \\ &\text{(d)}\quad Q_2(x,t) = \alpha\sqrt{H} + \varepsilon \tilde{Q}_2(x,t) + \ldots \\ &\qquad Z(t) = \varepsilon \tilde{Z}(t) + \ldots \end{aligned} \tag{4.5}$$

will be considered.

For the new unknowns $\tilde{Q}_1, \tilde{H}_1, \tilde{H}_2, \tilde{Q}_2, \tilde{Z}$, one obtains the linear equations in the first approximation (where, to facilitate computations, the harmonic perturbations are written in the complex form $\Sigma A e^{i\omega t}$)

$$\frac{\partial \tilde{H}_i}{\partial t} + a_i c_i^2 \frac{\partial \tilde{Q}_i}{\partial x} = 0, \quad \frac{\partial \tilde{H}_i}{\partial x} + a_i \frac{\partial \tilde{Q}_i}{\partial t} = 0, \quad i = 1, 2$$

$$H_1(0, t) = 0$$

$$F \frac{d\tilde{Z}(t)}{dt} + \tilde{Q}_2(0, t) = \tilde{Q}_1(\ell_1, t) \tag{4.6}$$

$$\tilde{H}_1(\ell_1, t) = \tilde{Z}(t) = \tilde{H}_2(0, t)$$

$$\text{(a)} \quad \eta H \tilde{Q}_2(\ell_2, t) + \frac{N}{H} \tilde{Z}(t) = A e^{i\omega t}$$

$$\text{(b)} \quad \tilde{Q}_2(\ell_2, t) = \frac{\alpha}{2\sqrt{H}} \tilde{H}_2(\ell_2, t) + A\sqrt{H} e^{i\omega t}$$

We look for the ω frequencies for which $|H_2(\ell_2, t)|$ is maximum, as these are the ones that may generate dangerous situations in the system.

The method of characteristics

By performing a change of variables

$$u_i = \frac{1}{\sqrt{a_i}} \left(\tilde{H}_i + a_i c_i \tilde{Q}_i \right), \quad v_i = \frac{1}{\sqrt{a_i}} \left(a_i c_i \tilde{Q}_i - \tilde{H}_i \right) \tag{4.7}$$

hence

$$\tilde{H}_i = \frac{\sqrt{a_i}}{2} (u_i - v_i), \quad \tilde{Q}_i = \frac{1}{2c_i\sqrt{a_i}} (u_i + v_i) \tag{4.8}$$

the propagation equations become

$$\frac{\partial u_i}{\partial t} + e_i \frac{\partial u_i}{\partial x} = 0, \quad \frac{\partial v_i}{\partial t} - e_i \frac{\partial v_i}{\partial x} = 0, \quad i = 1, 2 \tag{4.9}$$

By introducing the functions $\eta_i(t) = u_i(0, t), \zeta_i(t) = v_i(\ell_i, t)$ it follows that

$$\tilde{H}_i(x, t) = \frac{\sqrt{a_i}}{2} \left[\eta_i \left(t - \frac{x}{c_i} \right) - \zeta_i \left(t - \frac{\ell_i - x}{c_i} \right) \right]$$

$$\tilde{Q}_i(x, t) = \frac{1}{2c_i\sqrt{a_i}} \left[\eta_i \left(t - \frac{x}{c_i} \right) + \zeta_i \left(t - \frac{\ell_i - x}{c_i} \right) \right] \tag{4.10}$$

$$\tilde{Z}(t) + \frac{\sqrt{a_1}}{2} \left[\eta_1 \left(t - \frac{\ell_1}{c_1} \right) - \zeta_1(t) \right] = \frac{\sqrt{a_2}}{2} \left[\eta_2(t) - \zeta_2 \left(t - \frac{\ell_2}{c_2} \right) \right]$$

In order to write the boundary conditions in the new variables we shall use the relations

$$u_i(\ell_i, t) = \eta_i \left(t - \frac{\ell_i}{c_i} \right)$$
$$v_i(0, t) = \zeta_i \left(t - \frac{\ell_i}{c_i} \right) \tag{4.11}$$

obtained from eqn. (4.9). In the mathematical model we should have the initial conditions included as well. Since in what follows we are interested in the periodic response of the system to periodic excitations, we are not interested in the initial conditions actually influencing only the transient regime.

In the new variables u_i and v_i, the boundary conditions are

$$\frac{\sqrt{a_1}}{2}\left[u_1(0,t)-v_1(0,t)\right]=0$$

$$F\frac{\sqrt{a_2}}{2}\frac{\mathrm{d}}{\mathrm{d}t}\left[u_2(0,t)-v_2(0,t)\right]+\frac{1}{2c_2\sqrt{a_2}}\left[u_2(0,t)+v_2(0,t)\right]$$

$$=\frac{1}{2c_1\sqrt{a_1}}[u_1(\ell_1,t)+v_1(\ell_1,t)]$$

$$\frac{\sqrt{a_1}}{2}\left[u_1(\ell_1,t)-v_1(\ell_1,t)\right]=\frac{\sqrt{a_2}}{2}\left[u_2(0,t)-v_2(0,t)\right] \tag{4.12}$$

$$\text{(a)}\quad \frac{1}{2c_2\sqrt{a_2}}\left[u_2(\ell_2,t)+v_2(l_2,t)\right]+\frac{Q}{H}\frac{\sqrt{a_2}}{2}\left[u_2(0,t)-v_2(0,t)\right]=A\mathrm{e}^{\mathrm{i}\omega t}$$

$$\text{(b)}\quad \frac{1}{2c_2\sqrt{a_2}}\left[u_2(\ell_2,t)+v_2(l_2,t)\right]=\frac{\alpha}{2\sqrt{H}}\frac{\sqrt{a_2}}{2}\left[u_2(\ell_2,t)-v_2(\ell_2,t)\right]$$

$$+A\sqrt{H}\mathrm{e}^{\mathrm{i}\omega t}$$

For the functions η_i, ζ_i the following relations are obtained:

$$\eta_1\,(t)=\zeta_1\left(t-\frac{\ell_1}{c_1}\right)$$

$$F\frac{\mathrm{d}}{\mathrm{d}t}\left[\eta_2(t)-\zeta_2\left(t-\frac{\ell_2}{c_2}\right)\right]+\frac{1}{c_2a_2}\left[\eta_2(t)+\zeta_2\left(t-\frac{\ell_2}{c_2}\right)\right]$$

$$=\frac{1}{c_1\sqrt{a_1a_2}}\left[\eta_1\left(t-\frac{\ell_1}{c_1}\right)+\zeta_1(t)\right]$$

$$\sqrt{a_1}\left[\eta_1\left(t-\frac{\ell_1}{c_1}\right)-\zeta_1(t)\right]=\sqrt{a_2}\left[\eta_2\,(t)-\zeta_2\left(t-\frac{\ell_2}{c_2}\right)\right] \tag{4.13}$$

$$\text{(a)}\quad \frac{1}{c_2a_2}\left[\eta_2\left(t-\frac{\ell_2}{c_2}\right)+\zeta_2(t)\right]+\frac{Q}{H}\left[\eta_2(t)-\zeta_2\left(t-\frac{\ell_2}{c_2}\right)\right]=\frac{2A}{\sqrt{a_2}}\mathrm{e}^{\mathrm{i}\omega t}$$

$$\text{(b)}\quad \frac{1}{c_2a_2}\left[\eta_2\left(t-\frac{\ell_2}{c_2}\right)+\zeta_2(t)\right]=\frac{\alpha}{2\sqrt{H}}\left[\eta_2(t-\frac{\ell_2}{c_2})-\zeta_2\,(t)\right]+\frac{2A\sqrt{H}}{\sqrt{a_2}}\mathrm{e}^{\mathrm{i}\omega t}$$

The periodic response

We look for the functions η_i, ζ_i satisfying eqns. (4.13) in the form $\eta_i(t)=\mathrm{e}^{\mathrm{i}\omega t}\tilde{\eta}_i(\omega)$, $\zeta_i(t)=\mathrm{e}^{\mathrm{i}\omega t}\tilde{\zeta}_i(\omega)$.

To compute these unknown functions $\tilde{\eta}_i(\omega), \tilde{\zeta}_i(\omega)$ the following system of algebraic equations is obtained from the boundary conditions

$$
\begin{aligned}
&\tilde{\eta}_1(\omega) = \mathrm{e}^{-\mathrm{i}\omega\frac{\ell_1}{c_1}}\tilde{\zeta}_1(\omega) \\
&i\omega F\left[\tilde{\eta}_1(\omega) - \mathrm{e}^{-\mathrm{i}\omega\frac{\ell_2}{c_2}}\tilde{\zeta}_1(\omega)\right] + \frac{1}{c_2 a_2}\left[\eta_2(\omega) + \mathrm{e}^{-\mathrm{i}\omega\frac{\ell_2}{c_2}}\zeta_2(\omega)\right] \\
&\quad = \frac{1}{c_1\sqrt{a_1 a_2}}\left[\mathrm{e}^{-\mathrm{i}\omega\frac{\ell_1}{c_1}}\tilde{\eta}_1(\omega) + \tilde{\zeta}_1(\omega)\right] \\
&\sqrt{a_1}\left[\mathrm{e}^{-\mathrm{i}\omega\frac{\ell_1}{c_1}}\tilde{\eta}_1(\omega) - \tilde{\zeta}_1(\omega)\right] = \sqrt{a_2}\left[\tilde{\eta}_2(\omega) - \mathrm{e}^{-\mathrm{i}\omega\frac{\ell_2}{c_2}}\tilde{\zeta}_2(\omega)\right] \qquad (4.14) \\
&\text{(a)}\quad \frac{1}{c_2 a_2}\left[\mathrm{e}^{-\mathrm{i}\omega\frac{\ell_2}{c_2}}\tilde{\eta}_2(\omega) + \tilde{\zeta}_2(\omega)\right] + \frac{Q}{H}\left[\tilde{\eta}_2(\omega) - \mathrm{e}^{-\mathrm{i}\omega\frac{\ell_2}{c_2}}\tilde{\zeta}_2(\omega)\right] = \frac{2A}{\sqrt{a_2}} \\
&\text{(b)}\quad \frac{1}{c_2 a_2}\left[\mathrm{e}^{-\mathrm{i}\omega\frac{\ell_2}{c_2}}\tilde{\eta}_2(\omega) + \tilde{\zeta}_2(\omega)\right] = \frac{\alpha}{2\sqrt{H}}\left[\mathrm{e}^{-\mathrm{i}\omega\frac{\ell_2}{c_2}}\tilde{\eta}_2(\omega) - \tilde{\zeta}_2(\omega)\right] + \frac{2A\sqrt{H}}{\sqrt{a_2}}
\end{aligned}
$$

Excitation frequencies that may generate dangerous phenomena in the hydraulic system can be determined by focussing our attention on function $\tilde{H}_2(\ell_2, t)$, which describes the pressure variations in the penstock, in the area of the turbines.

The computations show that

$$
\text{(a)}\quad \left|\tilde{H}_2(\ell_2, t)\right| = Ac_2 a_2\left|\frac{f_1(\omega)}{\mathrm{i}f_2(\omega) - \dfrac{Q}{H}\sin\omega\dfrac{\ell_1}{c_1}}\right|
$$

$$
\begin{aligned}
\text{(b)}\quad \left|\tilde{H}_2(\ell_2, t)\right| &= Ac_2 a_2\sqrt{H}\left|\frac{f_1(\omega)}{\mathrm{i}f_2(\omega) - \dfrac{\alpha c_2 a_2}{2\sqrt{H}}f_1\omega}\right| \qquad (4.15) \\
&= Ac_2 a_2\sqrt{H}\left|\frac{f_1(\omega)}{\mathrm{i}f_2(\omega) - \dfrac{c_2 a_2}{2}\dfrac{Q}{H}f_1(\omega)}\right|
\end{aligned}
$$

where

$$
\begin{aligned}
&\text{(a)}\quad f_1(\omega) = \left(\omega F\sin\omega\frac{\ell_1}{c_1} - \frac{1}{c_1 a_1}\cos\omega\frac{\ell_1}{c_1}\right)\sin\omega\frac{\ell_2}{c_2} - \frac{1}{c_2 a_2}\sin\omega\frac{\ell_1}{c_1}\cos\omega\frac{\ell_2}{c_2} \\
&\qquad\qquad\qquad (4.16) \\
&\text{(b)}\quad f_2(\omega) = \left(\omega F\sin\omega\frac{\ell_1}{c_1} - \frac{1}{c_1 a_1}\cos\omega\frac{\ell_1}{c_1}\right)\cos\omega\frac{\ell_2}{c_2} + \frac{1}{c_2 a_2}\sin\omega\frac{\ell_1}{c_1}\sin\omega\frac{\ell_2}{c_2}
\end{aligned}
$$

The resonance effect developed at different points of the admission gallery or penstock may be analysed—for instance for case (b)—with the following relationship:

$$
\left|\tilde{H}_1(x, \omega)\right| = A\sqrt{H}\left|\frac{\sin\omega\dfrac{x}{c_1}}{\mathrm{i}f_2(\omega) - \dfrac{\alpha c_2 a_2}{2\sqrt{H}}f_1(\omega)}\right|
$$

$$\left|\tilde{H}_2(x,\omega)\right| = Ac_2a_2\sqrt{H}\left|\frac{f_x(\omega)}{if_2(\omega) - \dfrac{\alpha c_2 a_2}{2\sqrt{H}} f_1(\omega)}\right| \tag{4.17}$$

where

$$f_x(\omega) = \left(\omega F \sin\frac{\ell_1}{c_1} - \frac{1}{c_1a_1}\cos\omega\frac{\ell_1}{c_1}\right)\sin\omega\frac{x}{c_2} - \frac{1}{c_a a_2}\sin\omega\frac{\ell_1}{c_1}\cos\omega\frac{x}{c_2}$$

Remark
In case (a), the denominator in the expression of $\left|\tilde{H}_2(\ell_2,t)\right|$ can be cancelled if:

$$\omega\frac{\ell_1}{c_1} = k_1\pi, \quad \omega\frac{\ell_2}{c_2} = (2k_2+1)\frac{\pi}{2}$$

where k_1, k_2 are integers, or, by denoting

$$\frac{\ell_1}{c_1} = \frac{\tau_1}{2}, \quad \frac{\ell_2}{c_2} = \frac{\tau_2}{2}$$

this situation is described by

$$\frac{\tau_1}{\tau_2} = \frac{2k_1}{2k_2+1} \tag{4.18}$$

In this case, the formula for $\left|\tilde{H}_2(\ell_2,t)\right|$ is senseless and the computations have to be repeated in a form specific to the presence of secular terms. This is the case where

$$\omega = \frac{k_1\pi c_1}{\ell_1} = \frac{2k_1\pi}{\tau_1}$$

are the eigenfrequencies of the hydraulic system.

The case (b) contains also the assumption of a law giving the flow rate at the power plant as a linear function of pressure.

Numerical simulation and applications
To illustrate the practical applicability of the method to the simulation of hydraulic resonance, computations have been performed for two Romanian hydroelectrical power plants.

The HPP Somes–Mariselu. The main characteristics of the power plant are

$\ell_1 = 8130\,\mathrm{m}$;	$\ell_2 = 740\,\mathrm{m}$;	$F = 28.3\,\mathrm{m}$;
$H = 450\,\mathrm{m}$;	$N = 215\,\mathrm{MW}$;	$d_1 = 4.40\,\mathrm{m}$;
$d_2 = 3.40\,\mathrm{m}$;	$R = 0.00416Q_c^2$;	$Q = 50\,\mathrm{m^3/s}$.

The power plant operates steadily if set on constant flow.

In order to detect the dangerous frequencies in the assumption of excitations with boundary conditions of the type (a) or (b), functions $|H_2(\ell_2,t)|$ have been computed and plotted in Fig. 4.2 as a function of ω. Thus the ranges of dangerous frequencies of excitation have been exhibited. Since these frequencies have been obtained from the linearised model, it is important to know what is their effect in the real system.

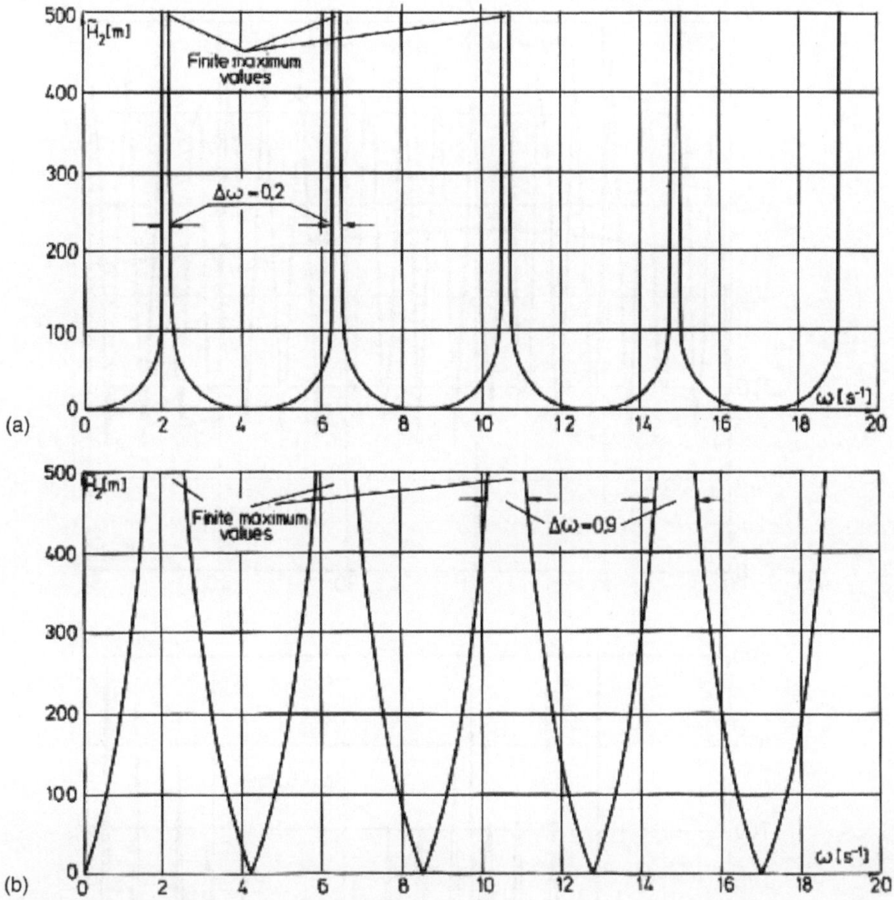

Figure 4.2. Somes–Mariselu hydroelectric power plant. The graph of the quantity $|\tilde{H}_2(\ell_2, t)|$ as a function of ω under boundary conditions (4.2).

In the hypothesis that hydraulic resonance phenomena appear and develop in the system, it is of practical importance to know the maximum pressures that can develop under these conditions.

By using a computation procedure based on the characteristic method, the non-linear system of eqns. (4.1) has been integrated with boundary conditions of the type (4.2) and the behaviour of the hydraulic system under the excitation frequencies detected as dangerous by the resonance analysis, has been numerically simulated. Some of the results are given in Fig. 4.3, and they show that the resonance frequencies detected by linear analysis are really dangerous.

The HPP Bicaz. Among the characteristics of the power plant, the following are mentioned:

$\ell_1 = 4.650\,\text{m}$; $\quad f_1 = 38\,\text{m}^2$;
$\ell_2 = 157\,\text{m}$; $\quad f_2 = 28\,\text{m}^2$;
$F = 530\,\text{m}$; $\quad R = 0.001Q_c^2$;
$H = 140\,\text{m}$; $\quad Q = 175\,\text{m}^3/\text{s}$;
$N = 225\,\text{MW}$.

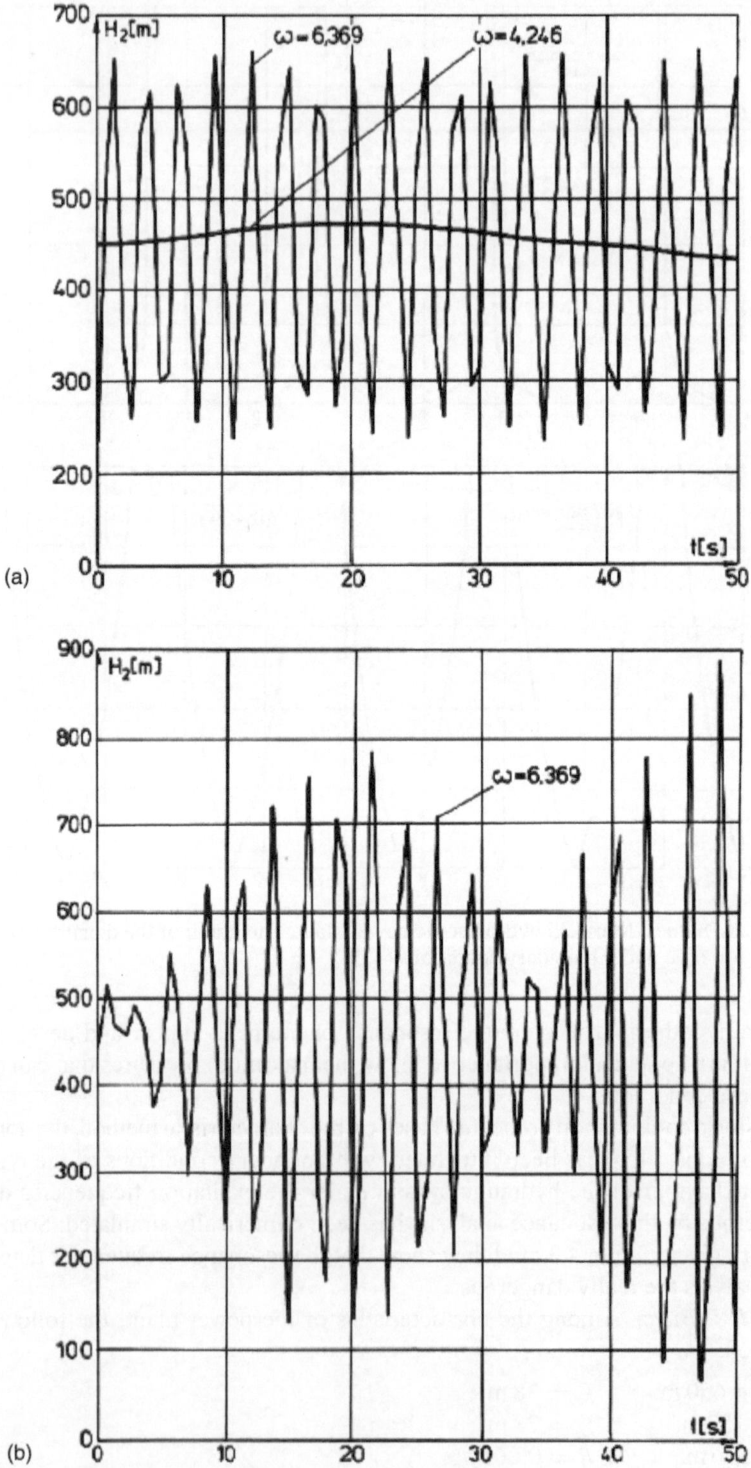

Figure 4.3. Somes–Mariselu hydroelectric power plant. Curves of variation of the pressure $H_2(\ell_2,t)$ as a function of t under boundary conditions (4.2).

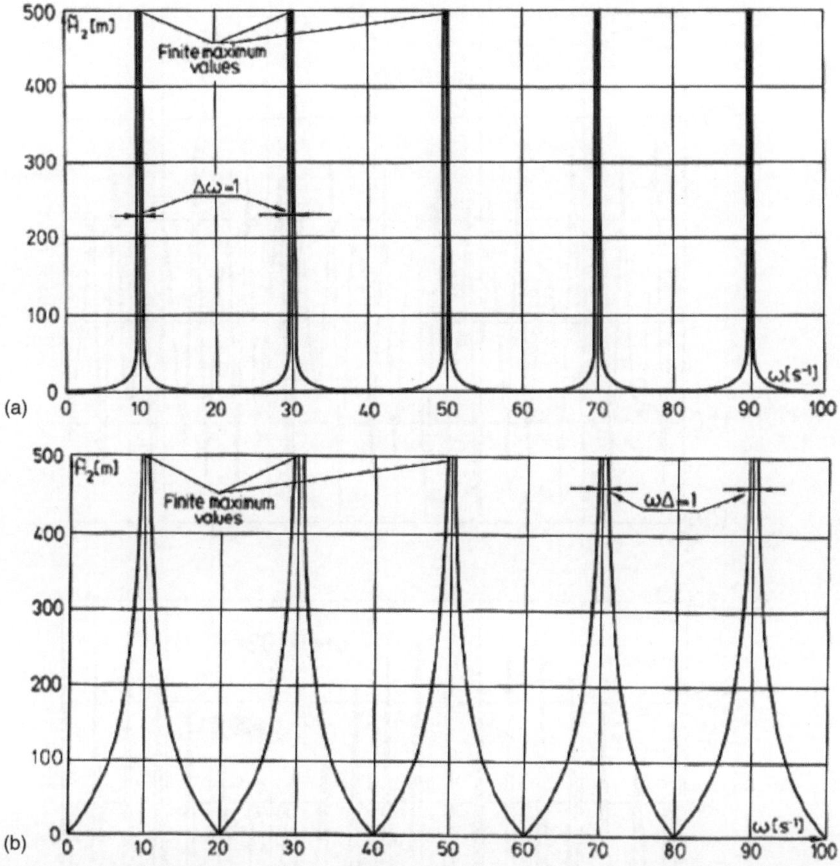

Figure 4.4. Bicaz hydroelectric power plant. The graph of the quantity $|\tilde{H}_2(\ell_2, t)|$ as a function of ω under boundary conditions (4.2).

The power plant is hydraulically stable for a perfect setting on constant power. The computations have been performed similarly.

The ranges of dangerous frequencies for the power plant have been obtained (Fig. 4.4) and the maximum pressures have been computed occurring along the penstock, when some resonance phenomena are set in and develop in the system (Fig. 4.5).

4.2. The resonance computation for hydroelectric power plants in complex schemes

In the design of modern hydroelectric power plants, more and more, complex schemes are now being adopted which contain several reservoirs, admission galleries, surge tanks and turbines. The hydraulic computation of the unsteady flow in such schemes, concerning the maximum amplitudes of water level oscillations inside the surge tank or the hydraulic stability in operation may be performed by numerical methods. In this section we are concerned with the computation of hydraulic resonance in hydroelectric

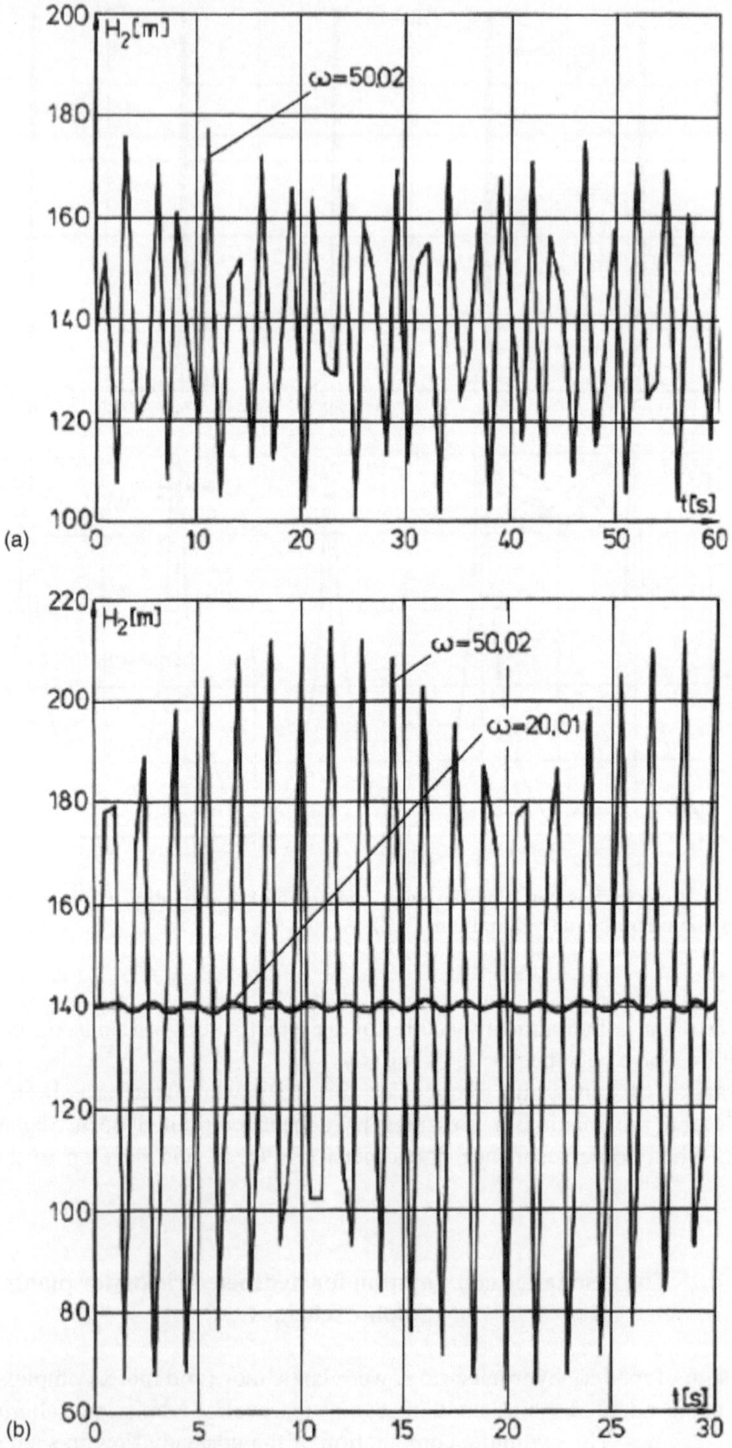

Figure 4.5. Bicaz hydroelectric power plant. Curves of variation of the pressure $H_2(\ell_2,t)$ as a function of t under boundary conditions (4.2).

power plants in the case of some complex hydraulic schemes, more frequently used in engineering practice. We shall adopt a simplified reference method by denoting them by the letters A, B, C, D (see also Chapter 8). The governing water flow equations will be considered, by assuming small periodic perturbations at the power plant.

For these equations, linearised around the steady flow, the response to harmonic perturbations is analysed and the resonance frequencies corresponding to the maximum amplitudes of water pressure variations in the penstock at the turbines section are obtained. The frequencies thus obtained, are then tested by numerical simulation of the full non-linear model. The general computation method is demonstrated in a practical application to one of the basic schemes considered. Performed computations are basically equivalent to the "impedance method".

4.2.1. Hydraulic scheme with two surge tanks in the system

Before considering complex hydraulic schemes, we shall analyse the basic hydraulic scheme with two surge tanks in the system (Fig. 4.6).

The mathematical model

The general water flow equations in the system are

$$\frac{\partial H_i}{\partial t} + a_i c_i^2 \frac{\partial Q_i}{\partial x} = 0, \qquad \frac{\partial H_i}{\partial x} + a_i \frac{\partial Q_i}{\partial x} = 0; \quad i = 1;2;3;4$$

$$F_1 \frac{dZ_1(t)}{dt} + Q_2(0,t) = Q_1(\ell_1, t) \tag{4.19}$$

$$F_2 \frac{dZ_2(t)}{dt} + Q_4(0,t) = Q_3(\ell_3, t)$$

with the boundary conditions

$$H_1(0,t) = H, \quad H_4(\ell_4, t) = 0, \quad Q_2(\ell_2, t) = Q_3(0,t)$$

$$H_1(\ell_1, t) = H + Z_1(t) + R_1' \left| \frac{dZ_1(t)}{dt} \right| \frac{dZ_1(t)}{dt} = H_2(0,t)$$

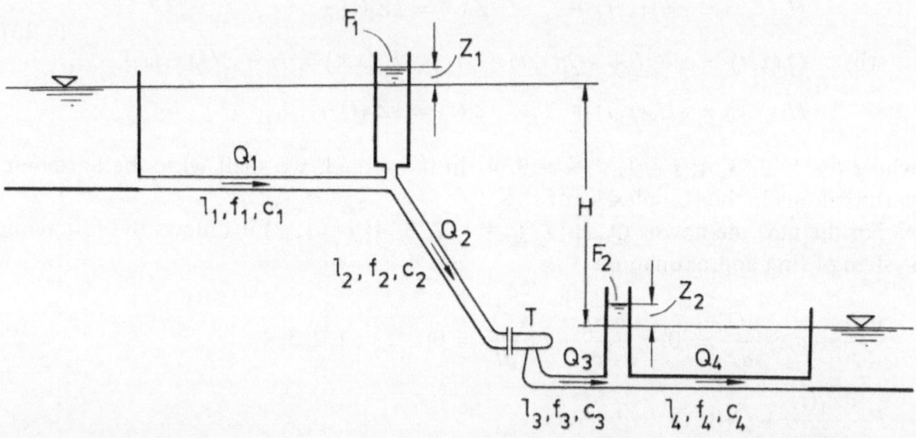

Figure 4.6. The scheme of a hydroelectric power plant equipped with two surge tanks arranged in a system.

$$H_3(\ell_3, t) = Z_2(t) + R_2' \left|\frac{\mathrm{d}Z_2(t)}{\mathrm{d}t}\right| \frac{\mathrm{d}Z_2(t)}{\mathrm{d}t} = H_4(0, t)$$

(a) $\eta g Q_2(\ell_2, t)\left[H + Z_1(t) - Z_2(t)\right] = N$

(assuming the power plant is set at constant power) or (4.20)

(b) $Q_2(\ell_2, t) = \alpha\sqrt{H_2(\ell_2, t) - H_3(0, t)}$

(assuming a flow law at the power plant).

The corresponding steady flow is given by

$$\begin{aligned} &\text{(a)} \quad H_1 = H_2 = H, \quad Z_1 = Z_2 = 0, \quad H_3 = H_4 = 0 \\ &\qquad Q_1 = Q_2 = Q_3 = Q_4 = q = \frac{N}{\eta g H} \\ &\text{(b)} \quad H_1 = H_2 = H, \quad Z_1 = Z_2, \quad H_3 = H_4 = 0 \\ &\qquad Q_1 = Q_2 = Q_3 = Q_4 = q = \alpha\sqrt{H} \end{aligned} \tag{4.21}$$

The resonance problem

We assume that in the boundary conditions of the types (a) or (b), respectively, the perturbations appear in the form

$$\begin{aligned} &\text{(a)} \quad \eta g Q_2(\ell_2, t)\left[H + Z_1(t) - Z_2(t)\right] = N + \varepsilon A \sin\omega t \\ &\text{(b)} \quad Q_2(\ell_2, t) = (\alpha + \varepsilon A \sin\omega t)\sqrt{H_2(\ell_2, t) - H_3(0, t)} \end{aligned} \tag{4.22}$$

As a result, we shall consider the corresponding perturbations of the steady flow in the form

$$\begin{aligned} &\text{(a)} \quad Q_i(x,t) = \frac{N}{\eta g H} + \varepsilon\tilde{Q}_i(x,t) + \ldots, \quad H_j(x,t) = H + \varepsilon\tilde{H}_j(x,t) + \ldots, \\ &\qquad H_k(x,t) = \varepsilon\tilde{H}_k(x,t) + \ldots, \quad Z_j(t) = \varepsilon\tilde{Z}_j(t) + \ldots, \\ &\text{(b)} \quad Q_i(x,t) = \alpha\sqrt{H} + \varepsilon\tilde{Q}_i(x,t) + \ldots, \quad H_j(x,t) = H + \varepsilon\tilde{H}_j(x,t) + \ldots, \\ &\qquad H_k(x,t) = \varepsilon\tilde{H}_k(x,t) + \ldots, \quad Z_j(t) = \varepsilon\tilde{Z}_j(t) + \ldots, \end{aligned} \tag{4.23}$$

where i = 1, 2, 3, 4; j = 1, 2; k = 3, 4. In the sequel, we shall write the harmonic perturbations in the complex form $\varepsilon A e^{i\omega t}$.

For the new unknowns $\tilde{Q}_i, \tilde{H}_i, \tilde{Z}_j$ (i = 1, 2, 3, 4; j = 1, 2) it follows the following system of first approximation:

$$\frac{\partial\tilde{H}_i}{\partial t} + a_i c_i^2 \frac{\partial\tilde{Q}_i}{\partial x} = 0, \quad \frac{\partial\tilde{H}_i}{\partial x} + a_i \frac{\partial\tilde{Q}_i}{\partial t} = 0, \quad i = 1,2,3,4$$

$$F_1 \frac{\mathrm{d}\tilde{Z}_1(t)}{\mathrm{d}t} + \tilde{Q}_2(0,t) = \tilde{Q}_1(\ell_1, t)$$

$$F_2 \frac{\mathrm{d}\tilde{Z}_2(t)}{\mathrm{d}t} + \tilde{Q}_4(0,t) = \tilde{Q}_3(\ell_3, t)$$

$$\tilde{H}_1(0,t) = 0, \quad \tilde{H}_4(\ell_4,t) = 0, \quad \tilde{Q}_3(0,t) = \tilde{Q}_2(\ell_2,t) \tag{4.24}$$

$$\tilde{H}_1(\ell_1,t) = \tilde{Z}_1(t) = \tilde{H}_2(0,t), \quad \tilde{H}_3(\ell_3,t) = \tilde{Z}_2(t) = \tilde{H}_4(0,t)$$

$$\text{(a)} \quad \eta g H \tilde{Q}_2(\ell_2,t) + \frac{N}{H}\left[\tilde{Z}_1(t) - \tilde{Z}_2(t)\right] = A\mathrm{e}^{\mathrm{i}\omega t}$$

$$\text{(b)} \quad \tilde{Q}_2(\ell_2,t) - \frac{\alpha}{2H}\left[\tilde{H}_2(\ell_2,t) - \tilde{H}_3(0,t)\right] = A\sqrt{H}\mathrm{e}^{\mathrm{i}\omega t}$$

To select the dangerous frequencies, we look for those values of ω for which $|\mathrm{H}_2(\ell_2,\mathrm{t})|$ is maximum.

The method of characteristics
Performing the change of variables

$$u_i = \frac{1}{\sqrt{a_i}}\left(\tilde{H}_i + a_i c_i \tilde{Q}_i\right), \quad v_i = \frac{1}{\sqrt{a_i}}\left(a_i c_i \tilde{Q}_i - \tilde{H}_i\right) \tag{4.25}$$

hence

$$\tilde{H}_i = \frac{\sqrt{a_i}}{2}\left(u_i - v_i\right), \quad \tilde{Q}_i = \frac{1}{2c_i\sqrt{a_i}}\left(u_i + v_i\right) \tag{4.26}$$

the equations of propagation become

$$\frac{\partial u_i}{\partial t} + c_i\frac{\partial u_i}{\partial x} = 0, \quad \frac{\partial v_i}{\partial t} - c_i\frac{\partial u_i}{\partial x} = 0, \quad i = 1,2,3,4 \tag{4.27}$$

By introducing the functions $\eta_i(t) = u_i(0,t)$ and $\zeta_i(t) = v_i(\ell_i,t)$ it follows that

$$\begin{aligned}
\tilde{H}_i(x,t) &= \frac{\sqrt{a_i}}{2}\left[\eta_i\left(t - \frac{x}{c_i}\right) - \zeta_i\left(t - \frac{\ell_i - x}{c_i}\right)\right] \\
\tilde{Q}_i(x,t) &= \frac{1}{2c_i\sqrt{a_i}}\left[\eta_i\left(t - \frac{x}{c_i}\right) + \zeta_1\left(t - \frac{\ell_i - x}{c_i}\right)\right] \\
\tilde{Z}_1(t) &= \frac{\sqrt{a_1}}{2}\left[\eta_1\left(t - \frac{\ell_1}{c_1}\right) - \zeta_1(t)\right] = \frac{\sqrt{a_2}}{2}\left[\eta_2(t) - \zeta_2\left(t - \frac{\ell_2}{c_2}\right)\right] \\
\tilde{Z}_2(t) &= \frac{\sqrt{a_3}}{2}\left[\eta_3\left(t - \frac{\ell_3}{c_3}\right) - \zeta_3(t)\right] = \frac{\sqrt{a_4}}{2}\left[\eta_4(t) - \zeta_4\left(t - \frac{\ell_4}{c_4}\right)\right]
\end{aligned} \tag{4.28}$$

In this case, the boundary conditions are

$$\eta_1(t) - \zeta_1\left(t - \frac{\ell_1}{c_1}\right) = 0; \quad \eta_4\left(t - \frac{\ell_4}{c_4}\right) - \zeta_4(t) = 0$$

$$\frac{1}{2c_3\sqrt{a_3}}\left[\eta_3(t) + \zeta_3\left(t - \frac{\ell_3}{c_3}\right)\right] = \frac{1}{2c_2\sqrt{a_2}}\left[\eta_2\left(t - \frac{\ell_2}{c_2}\right) + \zeta_2(t)\right]$$

$$\frac{\sqrt{a_1}}{2}\left[\eta_1\left(t-\frac{\ell_1}{c_1}\right)-\zeta_1(t)\right]=\frac{\sqrt{a_2}}{2}\left[\eta_2(t)-\zeta_2\left(t-\frac{\ell_2}{c_2}\right)\right]=\tilde{Z}_1(t)$$
$$\frac{\sqrt{a_3}}{2}\left[\eta_3\left(t-\frac{\ell_3}{c_3}\right)-\zeta_3(t)\right]=\frac{\sqrt{a_4}}{2}\left[\eta_4(t)-\zeta_4\left(t-\frac{\ell_4}{c_4}\right)\right]=\tilde{Z}_2(t) \quad (4.29)$$
$$F_1\frac{\mathrm{d}}{\mathrm{d}t}Z(t)+\frac{1}{2c_2\sqrt{a_2}}\left[\eta_2(t)+\zeta_2\left(t-\frac{\ell_2}{c_2}\right)\right]=\frac{1}{2c_1\sqrt{a_1}}\left[\eta_1\left(t-\frac{\ell_1}{c_1}\right)+\zeta_1(t)\right]$$
$$F_2\frac{\mathrm{d}}{\mathrm{d}t}Z_2(t)+\frac{1}{2c_4a_4}\left[\eta_4(t)+\zeta_4\left(t-\frac{\ell_4}{c_4}\right)\right]=\frac{1}{2c_3\sqrt{a_3}}\left[\eta_3\left(t-\frac{\ell_3}{c_3}\right)+\zeta_3(t)\right]$$

The mathematical model thus obtained and the following conditions

$$\text{(a)}\quad \eta g H\frac{1}{2c_2\sqrt{a_2}}\left[\eta_2\left(t-\frac{\ell_2}{c_2}\right)+\zeta_2(t)\right]+\eta g q\left[Z_1(t)-Z_2(t)\right]=A\mathrm{e}^{\mathrm{i}\omega t}$$
$$\text{(b)}\quad \frac{1}{2c_2\sqrt{a_2}}\left[\eta_2\left(t-\frac{\ell_2}{c_2}\right)+\zeta_2(t)\right]-\frac{q}{2H}\left\{\frac{\sqrt{a_2}}{2}\left[\eta_2\left(t-\frac{\ell_2}{c_2}\right)-\zeta_2\right]\right. \quad (4.30)$$
$$\left.-\frac{\sqrt{a_3}}{2}\left[\eta_3(t)-\zeta_3\left(t-\frac{\ell_3}{c_3}\right)\right]\right\}=A\sqrt{H}\mathrm{e}^{\mathrm{i}\omega t}$$

can be reduced to a system of two differential equations with retarded argument in ζ_1 and η_3.

The periodic response

We shall look for the function

$$\eta_k(t)=\mathrm{e}^{\mathrm{i}\omega t}\tilde{\eta}_k(\omega),\quad \zeta_k(t)=\mathrm{e}^{\mathrm{i}\omega t}\tilde{\zeta}_k(\omega)$$

For $\tilde{\eta}_k(\omega)$, $\tilde{\zeta}_k(\omega)$ from the boundary conditions, the following system of algebraic equations follows

$$\tilde{\eta}_1(\omega)=\mathrm{e}^{-\mathrm{i}\omega\frac{\ell_1}{c_1}}\tilde{\zeta}_1(\omega),\quad \tilde{\zeta}_4(\omega)=\mathrm{e}^{-\mathrm{i}\omega\frac{\ell_4}{c_4}}\tilde{\eta}_4(\omega)$$
$$\frac{1}{c_3\sqrt{a_3}}\left[\tilde{\eta}_3(\omega)+\mathrm{e}^{-\mathrm{i}\omega\frac{\ell_3}{c_3}}\tilde{\zeta}_3(\omega)\right]=\frac{1}{c_2\sqrt{a_2}}\left[\mathrm{e}^{-\mathrm{i}\omega\frac{\ell_2}{c_2}}\tilde{\eta}_2(\omega)+\tilde{\zeta}_2(\omega)\right]$$
$$\sqrt{a_1}\left[\tilde{\eta}_1(\omega)\mathrm{e}^{-\mathrm{i}\omega\frac{\ell_1}{c_1}}-\tilde{\zeta}_1(\omega)\right]=\sqrt{a_2}\left[\tilde{\eta}_2(\omega)-\mathrm{e}^{-\mathrm{i}\omega\frac{\ell_2}{c_2}}\tilde{\zeta}_2(\omega)\right]$$
$$\sqrt{a_3}\left[\tilde{\eta}_3(\omega)\mathrm{e}^{-\mathrm{i}\omega\frac{\ell_3}{c_3}}-\tilde{\zeta}_3(\omega)\right]=\sqrt{a_4}\left[\tilde{\eta}_4(\omega)-\mathrm{e}^{-\mathrm{i}\omega\frac{\ell_4}{c_4}}\tilde{\zeta}_4(\omega)\right] \quad (4.31)$$
$$\mathrm{i}\omega F_1\left[\tilde{\eta}_1(\omega)\mathrm{e}^{-\mathrm{i}\omega\frac{\ell_1}{c_1}}-\tilde{\zeta}_1(\omega)\right]+\frac{1}{c_2a_2}\left[\tilde{\eta}_2(\omega)+\mathrm{e}^{-\mathrm{i}\omega\frac{\ell_2}{c_2}}\tilde{\zeta}_2(\omega)\right]$$
$$=\frac{1}{c_1\sqrt{a_1c_1}}\left[\mathrm{e}^{-\mathrm{i}\omega\frac{\ell_1}{c_1}}\tilde{\eta}_1(\omega)+\tilde{\zeta}_1(\omega)\right]$$

$$\mathrm{i}\omega F_2\left[\tilde{\eta}_3(\omega)\mathrm{e}^{-\mathrm{i}\omega\frac{\ell_3}{c_3}}-\tilde{\zeta}_3(\omega)\right]+\frac{1}{c_4a_4}\left[\tilde{\eta}_4(\omega)+\mathrm{e}^{-\mathrm{i}\omega\frac{\ell_4}{c_4}}\tilde{\zeta}_4(\omega)\right]$$

$$=\frac{1}{c_3\sqrt{a_3c_3}}\left[\mathrm{e}^{-\mathrm{i}\omega\frac{\ell_3}{c_3}}\tilde{\eta}_3(\omega)+\tilde{\zeta}_3(\omega)\right]$$

to which the one deduced from eqn. (4.30) for the cases (a) or (b) has to be added.

From the basic formulae (4.28) we have

$$\tilde{H}_k(x,t)=\frac{\sqrt{a_k}}{2}\mathrm{e}^{\mathrm{i}\omega\left(t-\frac{x}{c_k}\right)}\tilde{\eta}_k(\omega)-\mathrm{e}^{\mathrm{i}\omega\left(t-\frac{\ell_k-x}{c_k}\right)}\tilde{\zeta}_k(\omega)=\mathrm{e}^{\mathrm{i}\omega t}\hat{H}_k(x,\omega)$$

$$\tilde{Q}_k(x,t)=\frac{1}{2c_k\sqrt{a_k}}\mathrm{e}^{\mathrm{i}\omega\left(t-\frac{x}{c_k}\right)}\tilde{\eta}_k(\omega)+\mathrm{e}^{\mathrm{i}\omega\left(t-\frac{\ell_k-x}{c_k}\right)}\tilde{\zeta}_k(\omega)=\mathrm{e}^{\mathrm{i}\omega t}\hat{Q}_k(x,\omega) \tag{4.31'}$$

and correspondingly $\tilde{Z}_k(t)=\mathrm{e}^{\mathrm{i}\omega t}\hat{Z}_k(\omega)$.

The transfer functions

We look for the formulae expressing $\hat{Q}_2(\ell_2,\omega), \hat{Z}_1(\omega), \hat{Z}_2(\omega), \hat{H}_3(0,\omega)$ as functions of $\hat{H}_2(\ell_2,\omega)$; thereafter, by making the relevant substitutions in the conditions (a), (b) of system (4.24) we get $\left|\tilde{H}_2(\ell_2,t)\right|=\left|\hat{H}_2(\ell_2,\omega)\right|$.

Performing the computations, from the equations of system (4.31) we successively obtain

$$\hat{Q}_2(\ell_2,\omega)=\frac{\mathrm{i}f_2(\omega)}{c_2a_2f_1(\omega)}\hat{H}_2(\ell_2,\omega)$$

$$\hat{Z}_2(\omega)=\frac{f_2(\omega)\sin\omega\frac{\ell_4}{c_4}}{c_2a_2g_2(\omega)f_1(\omega)}\hat{H}_2(\ell_2,\omega)$$

$$\hat{Z}_2(\omega)=\frac{\sin\omega\frac{\ell_1}{c_1}}{c_2a_2f_1(\omega)}\hat{H}_2(\ell_2,\omega) \tag{4.32}$$

$$\hat{H}_3(0,\omega)=\frac{c_3a_3}{c_2a_2}\frac{g_1(\omega)f_2(\omega)}{g_2(\omega)f_1(\omega)}\hat{H}_2(\ell_2,\omega)$$

where

$$f_1(\omega)=\left(\omega F_1\sin\omega\frac{\ell_1}{c_1}-\frac{1}{c_1a_1}\cos\omega\frac{\ell_1}{c_1}\right)\sin\omega\frac{\ell_2}{c_2}-\frac{1}{c_2a_2}\sin\omega\frac{\ell_1}{c_1}\cos\omega\frac{\ell_2}{c_2}$$

$$f_2(\omega)=\left(\omega F_1\sin\omega\frac{\ell_1}{c_1}-\frac{1}{c_1a_1}\cos\omega\frac{\ell_1}{c_1}\right)\cos\omega\frac{\ell_2}{c_2}+\frac{1}{c_2a_2}\sin\omega\frac{\ell_1}{c_1}\sin\omega\frac{\ell_2}{c_2}$$

$$g_1(\omega)=\left(\omega F_2\sin\omega\frac{\ell_4}{c_4}-\frac{1}{c_4a_4}\cos\omega\frac{\ell_4}{c_4}\right)\sin\omega\frac{\ell_3}{c_3}-\frac{1}{c_3a_3}\sin\omega\frac{\ell_4}{c_4}\cos\omega\frac{\ell_3}{c_3} \tag{4.33}$$

$$g_2(\omega)=\left(\omega F_2\sin\omega\frac{\ell_4}{c_4}-\frac{1}{c_4a_4}\cos\omega\frac{\ell_4}{c_4}\right)\cos\omega\frac{\ell_4}{c_4}+\frac{1}{c_3a_3}\sin\omega\frac{\ell_4}{c_4}\sin\omega\frac{\ell_3}{c_3}$$

The resonance computations

The relations in system (4.32) allow us to deduce the following expressions for $|\tilde{H}_2(\ell_2,t)|$ in the two considered cases

$$
\begin{aligned}
&\text{(a)}\quad |\tilde{H}_2(\ell_2,t)| = \left| \frac{Ac_2a_2f_1(\omega)}{if_2(\omega) - \dfrac{Q}{H}\left[\sin\omega\dfrac{\ell_1}{c_1} - \dfrac{f_2(\omega)}{g_2(\omega)}\sin\omega\dfrac{\ell_4}{c_4}\right]} \right| \\
&\text{(b)}\quad |\tilde{H}_2(\ell_2,t)| = \left| \frac{Ac_2a_2\sqrt{H}f_1(\omega)}{if_2(\omega) - \dfrac{Q}{2H}\left[c_2a_2f_1(\omega) - c_3a_3\dfrac{g_1(\omega)}{g_2(\omega)}f_2(\omega)\right]} \right|
\end{aligned}
\tag{4.34}
$$

where functions $f_1(\omega), f_2(\omega), g_1(\omega), g_2(\omega)$ are defined in system (4.33).

Remark

In case (a) we see that we may get a classical resonance by cancelling the denominator whence

$$
\sin\omega\frac{\ell_1}{c_1} = 0, \quad \sin\omega\frac{\ell_4}{c_4} = 0, \quad \cos\omega\frac{\ell_2}{c_2} = 0 \tag{4.35}
$$

On the other hand, for case (b) it follows that situations that make $|H_2(\ell_2,t)|$ infinite can never appear.

Simplifying these relations, the results obtained for the basic hydraulic scheme are once again encountered:

$$
\begin{aligned}
&\text{(a)}\quad \tilde{H}_2(\ell_2,t) = Ac_2a_2\left| \frac{f_1(\omega)}{if_2(\omega) - \dfrac{Q}{H}\sin\omega\dfrac{\ell_1}{c_1}} \right| \\
&\text{(b)}\quad \tilde{H}_2(\ell_2,t) = Ac_2a_2\sqrt{H}\left| \frac{f_1(\omega)}{if_2(\omega) - \dfrac{c_2a_2}{2}\dfrac{Q}{H}f_1(\omega)} \right|
\end{aligned}
\tag{4.36}
$$

Application

We shall consider the case of a hydroelectric power plant discussed with the following main characteristics:

$\ell_1 = 5000\,\text{m};$	$f_1 = 38.5\,\text{m}^2;$	$F_1 = 400\,\text{m}^2;$
$\ell_2 = 300\,\text{m};$	$f_2 = 24.4\,\text{m}^2;$	$F_2 = 100\,\text{m}^2;$
$\ell_3 = 60\,\text{m};$	$f_3 = 38.5\,\text{m}^2;$	$R_1 = R_2 = 0.001Q_c^2$
$\ell_4 = 500\,\text{m};$	$f_4 = 38.5\,\text{m}^2;$	
$Q = 110\,\text{m}^3/\text{s};$	$H = 130\,\text{m}.$	

The power plant has a stable hydraulic operation when set at constant power. Similar to the basic hydraulic scheme, in order to make evident the dangerous frequencies $|\tilde{H}_2(\ell_2,t)|$, assuming perturbing factors of excitation with boundary conditions of type (a) or (b), the plots in Fig. 4.7 are given. Thus, the dangerous frequency ranges have been more clearly emphasized.

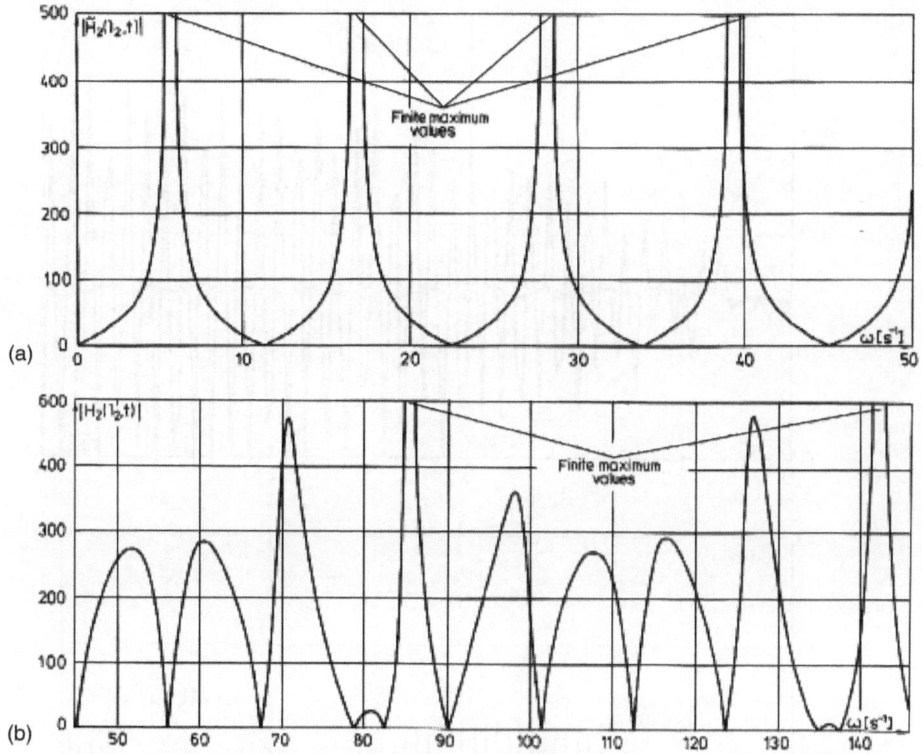

Figure 4.7. $|\tilde{H}_2(\ell_2, t)|$ versus ω under boundary conditions (4.20).

Following the general methodology for the non-linear system of eqns. (4.19), the behaviour of the hydraulic system working under the previously computed dangerous excitation frequencies, was considered. Some of the results are plotted in Fig. 4.8, which show that the frequencies exhibited by linear analysis are really dangerous. In the same way, we have checked the frequencies that actually do not affect the system.

4.2.2. Hydraulic scheme A

In this case, the partial differential equations of flow are the following

$$\frac{\partial H_i}{\partial t} + a_i c_i^2 \frac{\partial Q_i}{\partial x} = 0, \qquad \frac{\partial H_i}{\partial x} + a_i \frac{\partial Q_i}{\partial t} = 0, \qquad i = 1, 2, 3, 4, 5 \tag{4.37}$$

to which the boundary conditions are added.

After a linearisation around the steady flow and considering harmonic perturbations, we get (Fig. 4.9a)

$$\begin{aligned}
&H_1(0,t) = 0; \quad H_1(\ell_1,t) = H_2(0,t) = H_4(0,t) \\
&Q_1(\ell_1,t) = Q_2(0,t) + Q_4(0,t) \\
&F_1 \frac{\mathrm{d}Z_1(t)}{\mathrm{d}t} + Q_3(0,t) = Q_2(\ell_2,t); \quad F_2 \frac{\mathrm{d}Z_2(t)}{\mathrm{d}t} + Q_5(0,t) = Q_4(\ell_4,t) \\
&H_2(\ell_2,t) = Z_1(t) = H_3(0,t); \quad H_4(\ell_4,t) = Z_2(t) = H_5(0,t)
\end{aligned} \tag{4.38}$$

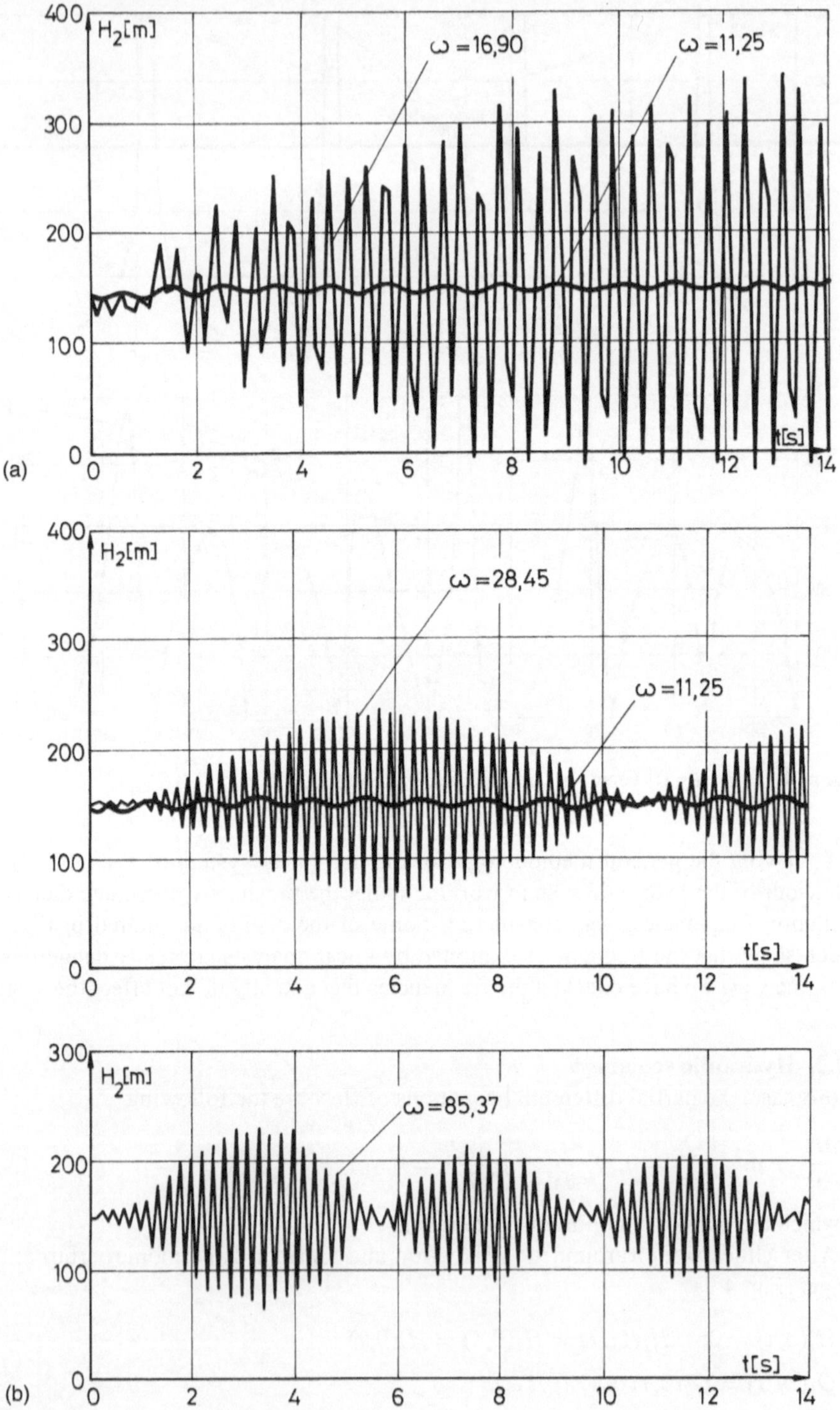

Figure 4.8. Curves of pressure variation $H_2(\ell_2,t)$ with respect to t under boundary conditions (4.20).

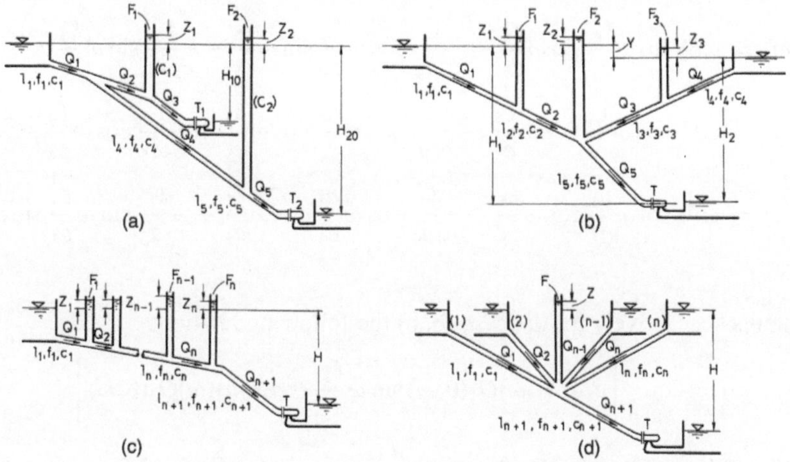

Figure 4.9. Hydroelectric power plant conceived on complex hydraulic schemes.

$$\begin{aligned}
&(\text{a, a}) \quad Q_3(\ell_3, t) + \beta H_3(0, t) = A_1 e^{i\omega t}\\
&\qquad\quad\;\; Q_5(\ell_5, t) + \beta' H_5(0, t) = A_2 e^{i\omega t}\\
&(\text{b, b}) \quad Q_3(\ell_3, t) + \beta H_3(\ell_3, t) = A_1 e^{i\omega t}\\
&\qquad\quad\;\; Q_5(\ell_5, t) + \beta' H_5(\ell_5, t) = A_2 e^{i\omega t}
\end{aligned}$$

where we have denoted

$$\begin{aligned}
\beta &= \frac{N_1}{\eta g H_{20}^2}; \quad \beta' = \frac{N_2}{\eta g H_{20}^2} \text{ for case (a, a) and}\\
\beta &= \frac{\alpha_1}{2\sqrt{H_{20}}}; \quad \beta' = \frac{\alpha_2}{2\sqrt{H_{20}}} \text{ for case (b, b).}
\end{aligned} \tag{4.39}$$

It should be mentioned that the notation (a, a) and (b, b) means that the same type of perturbing factors has been considered for both power plants, set for constant power or for a certain flow law respectively.

By using the same computation method as in the preceding case and denoting

$$h_1(\omega) = \omega F_1 \cos\omega\frac{\ell_2}{c_2}\sin\omega\frac{\ell_3}{c_3} + \frac{1}{c_2 a_2}\sin\omega\frac{\ell_2}{c_2}\sin\omega\frac{\ell_3}{c_3} - \frac{1}{c_3 a_3}\cos\omega\frac{\ell_2}{c_2}\cos\omega\frac{\ell_3}{c_3}$$

$$h_2(\omega) = \omega F_1 \sin\omega\frac{\ell_2}{c_2}\sin\omega\frac{\ell_3}{c_3} - \frac{1}{c_2 a_2}\cos\omega\frac{\ell_2}{c_2}\sin\omega\frac{\ell_3}{c_3} - \frac{1}{c_3 a_3}\sin\omega\frac{\ell_2}{c_2}\cos\omega\frac{\ell_3}{c_3}$$

$$g_1(\omega) = \omega F_1 \cos\omega\frac{\ell_2}{c_2}\cos\omega\frac{\ell_3}{c_3} + \frac{1}{c_2 a_2}\sin\omega\frac{\ell_2}{c_2}\cos\omega\frac{\ell_3}{c_3} + \frac{1}{c_3 a_3}\cos\omega\frac{\ell_2}{c_2}\sin\omega\frac{\ell_3}{c_3}$$

$$g_2(\omega) = \omega F_1 \sin\omega\frac{\ell_2}{c_2}\cos\omega\frac{\ell_3}{c_3} - \frac{1}{c_2 a_2}\cos\omega\frac{\ell_2}{c_2}\cos\omega\frac{\ell_3}{c_3} + \frac{1}{c_3 a_3}\sin\omega\frac{\ell_2}{c_2}\sin\omega\frac{\ell_3}{c_3}$$

$$\hat{h}_1(\omega) = \omega F_2 \cos\omega\frac{\ell_4}{c_4}\sin\omega\frac{\ell_5}{c_5} + \frac{1}{c_4 a_4}\sin\omega\frac{\ell_4}{c_4}\sin\omega\frac{\ell_5}{c_5} - \frac{1}{c_5 a_5}\cos\omega\frac{\ell_4}{c_4}\cos\omega\frac{\ell_5}{c_5}$$

$$h_4(\omega) = \omega F_2 \sin\omega\frac{\ell_4}{c_4}\sin\omega\frac{\ell_5}{c_5} - \frac{1}{c_4a_4}\cos\omega\frac{\ell_4}{c_4}\sin\omega\frac{\ell_5}{c_5} - \frac{1}{c_5a_5}\sin\omega\frac{\ell_4}{c_4}\cos\omega\frac{\ell_5}{c_5}$$

$$\hat{g}_1(\omega) = \omega F_2 \cos\omega\frac{\ell_4}{c_4}\cos\omega\frac{\ell_5}{c_5} + \frac{1}{c_4a_4}\sin\omega\frac{\ell_4}{c_4}\cos\omega\frac{\ell_5}{c_5} + \frac{1}{c_5a_5}\cos\omega\frac{\ell_4}{c_4}\sin\omega\frac{\ell_5}{c_5}$$

$$g_4(\omega) = \omega F_2 \sin\omega\frac{\ell_4}{c_4}\cos\omega\frac{\ell_5}{c_5} - \frac{1}{c_4a_4}\cos\omega\frac{\ell_4}{c_4}\cos\omega\frac{\ell_5}{c_5} + \frac{1}{c_5a_5}\sin\omega\frac{\ell_4}{c_4}\sin\omega\frac{\ell_5}{c_5} \tag{4.40}$$

one obtains successively for the case (b, b) the following relations

$$\begin{aligned}
\hat{H}_3(\ell_3,\omega) &= \mathrm{i}c_3a_3\left[c_1a_1h_1(\omega)\hat{Q}_1(0,\omega)\sin\omega\frac{\ell_1}{c_1} + a_2c_2h_2(\omega)\hat{Q}_2(0,\omega)\right] \\
\hat{Q}_3(\ell_3,\omega) &= -c_1a_1g_1(\omega)\hat{Q}_1(0,\omega)\sin\omega\frac{\ell_1}{c_1} - c_2a_2g_2(\omega)\hat{Q}_2(0,\omega) \\
\hat{H}_5(\ell_5,\omega) &= \mathrm{i}c_5a_5\left[c_1a_1\hat{h}_1(\omega)\hat{Q}_1(0,\omega)\sin\omega\frac{\ell_1}{c_1} + c_4a_4h_4(\omega)\hat{Q}_4(0,\omega)\right] \\
\hat{Q}_5(\ell_5,\omega) &= -c_1a_1\hat{g}_1(\omega)\hat{Q}_1(0,\omega)\sin\omega\frac{\ell_1}{c_1} - c_2a_2g_4(\omega)\hat{Q}_4(0,\omega)
\end{aligned} \tag{4.41}$$

then

$$\begin{aligned}
\hat{Q}_2(0,\omega) &= -\frac{A_1}{c_2a_2} - \frac{c_1a_1}{c_2a_2}\frac{g_1(\omega) - \mathrm{i}\beta c_3a_3h_1(\omega)}{g_2(\omega) - \mathrm{i}\beta c_3a_3h_2(\omega)}\hat{Q}_1(0,\omega)\sin\omega\frac{\ell_1}{c_1} \\
\hat{Q}_4(0,\omega) &= -\frac{A_2}{c_4a_4} - \frac{c_1a_1}{c_4a_4}\frac{\hat{g}_1(\omega) - \mathrm{i}\beta' c_5a_5\hat{h}_1(\omega)}{g_4(\omega) - \mathrm{i}\beta' c_5a_5h_4(\omega)}\hat{Q}_1(0,\omega)\sin\omega\frac{\ell_1}{c_1}
\end{aligned} \tag{4.42}$$

and finally

$$\left\{\frac{1}{c_1a_1}\cos\omega\frac{\ell_1}{c_1} + \left[\frac{1}{c_2a_2}\frac{g_1(\omega) - \mathrm{i}\beta c_3a_3h_1(\omega)}{g_2(\omega) - \mathrm{i}\beta c_3a_3h_2(\omega)} + \frac{1}{c_4a_4}\frac{\hat{g}_1(\omega) - \mathrm{i}\beta' c_5a_5h_1(\omega)}{g_4(\omega) - \mathrm{i}\beta' c_5a_5h_4(\omega)}\right]\right.$$
$$\left.\times\sin\omega\frac{\ell_1}{c_1}\right\}\hat{Q}_1(0,\omega) = \frac{1}{c_1a_1}\left(\frac{A_1}{c_2a_2} + \frac{A_2}{c_4a_4}\right) \tag{4.43}$$

The resonance computation for the two power plants is performed step by step, as follows: computations start by getting $\hat{Q}_1(0,\omega)$ from eqn. (4.43), then $\hat{Q}_2(0,\omega)$ and $\hat{Q}_4(0,\omega)$ from eqns. (4.42) and then eqns. (4.41) allow computing $\hat{H}_3(\ell_3,\omega)$ and $\hat{H}_5(\ell_5,\omega)$. Taking into account that $\left|\tilde{H}_3(\ell_3,t)\right| = \left|\hat{H}_3(\ell_3,\omega)\right|$ and $\left|\tilde{H}_5(\ell_5,t)\right| = \left|\hat{H}_5(\ell_5,\omega)\right|$, the obtained formulae give the resonance frequencies.

4.2.3. Hydraulic scheme B

The differential equations and boundary conditions are written as in the preceding cases and after linearisation they become (Fig. 4.9b)

$$H_1(0,t) = 0, \quad H_4(\ell_4,t) = 0$$

$$F_1\frac{\mathrm{d}Z_1(t)}{\mathrm{d}t} + Q_2(0,t) = Q_1(\ell_1,t)$$

$$
\begin{aligned}
&F_2\frac{\mathrm{d}Z_2(t)}{\mathrm{d}t}+Q_5(0,t)=Q_2(\ell_2,t)+Q_3(\ell_3,t)\\
&F_3\frac{\mathrm{d}Z_3(t)}{\mathrm{d}t}+Q_3(0,t)=Q_4(\ell_4,t)\\
&H_1(\ell_1,t)=Z_1(t)=H_2(0,t)\\
&H_2(\ell_2,t)=Z_2(t)=H_3(\ell_3,t)=H_5(0,t)\\
&H_3(0,t)=Z_3(t)=H_4(\ell_4,t)\\
\text{(a)}\quad &Q_5(\ell_5,t)+\beta H_5(0,t)=A\mathrm{e}^{\mathrm{i}\omega t}\\
\text{(b)}\quad &Q_5(\ell_5,t)+\beta H_5(\ell_5,t)=A\mathrm{e}^{\mathrm{i}\omega t}
\end{aligned}
\tag{4.44}
$$

With $f_1(\omega)$ and $f_2(\omega)$ defined in eqns. (4.33) and further introducing the notation below

$$
\begin{aligned}
f_3(\omega)&=\omega F_3\sin\omega\frac{\ell_3}{c_3}\sin\omega\frac{\ell_4}{c_4}+\frac{1}{c_3a_3}\cos\omega\frac{\ell_3}{c_3}\sin\omega\frac{\ell_4}{c_4}+\frac{1}{c_4a_4}\cos\omega\frac{\ell_4}{c_4}\sin\omega\frac{\ell_3}{c_3}\\
g_3(\omega)&=\omega F_3\cos\omega\frac{\ell_3}{c_3}\sin\omega\frac{\ell_4}{c_4}-\frac{1}{c_3a_3}\sin\omega\frac{\ell_3}{c_3}\sin\omega\frac{\ell_4}{c_4}+\frac{1}{c_4a_4}\cos\omega\frac{\ell_4}{c_4}\cos\omega\frac{\ell_3}{c_3}
\end{aligned}
\tag{4.45}
$$

it follows, for example, in case (b) that $\hat{H}_5(\ell_5,\omega)$ can be computed from

$$
\hat{H}_5(\ell_5,\omega)\left[\beta+\frac{\mathrm{i}}{c_5a_5}\frac{\Phi(\omega)\cos\omega\dfrac{\ell_5}{c_5}-\Psi(\omega)\sin\omega\dfrac{\ell_5}{c_5}}{\Psi(\omega)\cos\omega\dfrac{\ell_5}{c_5}+\Phi(\omega)\sin\omega\dfrac{\ell_5}{c_5}}\right]=A
\tag{4.46}
$$

where

$$
\begin{aligned}
\Phi(\omega)&=f_1(\omega)\left[\omega F_2-\frac{1}{c_3a_3}g_3(\omega)+\frac{1}{c_2a_2}f_2(\omega)f_3(\omega)\right]\\
\Psi(\omega)&=\frac{1}{c_5a_5}f_1(\omega)f_3(\omega)
\end{aligned}
$$

4.2.4. Hydraulic scheme C

The propagation equations are the usual ones and the boundary conditions become after liniarisation (Fig. 4.9c)

$$
\begin{aligned}
&H_1(0,t)=0\\
&F_k\frac{\mathrm{d}Z_k(t)}{\mathrm{d}t}+Q_{k+1}(0,t)=Q_k(\ell_k,t)\\
&H_k(\ell_k,t)=Z_k(t)=H_{k+1}(0,t),\quad k=1,2,\ldots,n\\
\text{(a)}\quad &Q_{n+1}(\ell_{n+1},t)+\beta H_{n+1}(0,t)=A\mathrm{e}^{\mathrm{i}\omega t}\\
\text{(b)}\quad &Q_{n+1}(\ell_{n+1},t)+\beta H_{n+1}(\ell_{n+1},t)=A\mathrm{e}^{\mathrm{i}\omega t}
\end{aligned}
\tag{4.47}
$$

Since we are interested in the penstock section upstream the turbines we shall study the behaviour of $\hat{H}_{n+1}(\ell_{n+1},t)$.

The basic formulae given by the method of characteristics are

$$\begin{aligned} H_x(x,t) &= \frac{\sqrt{a_k}}{2}\left[\eta_k\left(t-\frac{x}{c_k}\right)-\zeta_k\left(t-\frac{\ell_k-x}{c_k}\right)\right] \\ Q_x(x,t) &= \frac{1}{2c_k\sqrt{a_k}}\left[\eta_k\left(t-\frac{x}{c_k}\right)+\zeta_k\left(t-\frac{\ell_k-x}{c_k}\right)\right] \end{aligned} \tag{4.48}$$

The general strategy of computation is based on the relations between $\hat{H}_k(0,\omega)$, $\hat{Q}_k(0,\omega)$, and $\hat{H}_k(\ell_k,\omega)$, $\hat{Q}_k(\ell_k,\omega)$.

By using the boundary conditions we get

$$\begin{aligned} \hat{H}_k(\ell_k,\omega) &= \hat{H}_k(0,\omega)\cos\omega\frac{\ell_k}{c_k} - \mathrm{i}c_k a_k \hat{Q}_k(0,\omega)\sin\omega\frac{\ell_k}{c_k} \\ \hat{Q}_k(\ell_k,\omega) &= \hat{Q}_k(0,\omega)\cos\omega\frac{\ell_k}{c_k} - \frac{\mathrm{i}}{c_k a_k}\hat{H}_k(0,\omega)\sin\omega\frac{\ell_k}{c_k} \end{aligned} \tag{4.49}$$

In the nodes with surge tanks we have

$$\begin{aligned} \hat{H}_{k+1}(0,\omega) &= \hat{H}_k(\ell_k,\omega) \\ \hat{Q}_{k+1}(0,\omega) &= \hat{Q}_k(\ell_k,\omega) - \mathrm{i}\omega F_k \hat{H}_k(\ell_k,\omega) \end{aligned} \tag{4.50}$$

From the relations (4.49) and (4.50) we obtain the formula:

$$\frac{\hat{Q}_{k+1}(\ell_{k+1},\omega)}{\hat{H}_{k+1}(\ell_{k+1},\omega)} = \frac{\mathrm{i}}{c_{k+1}a_{k+1}} \frac{\omega F_k\cos\omega\dfrac{\ell_{k+1}}{c_{k+1}} + \dfrac{1}{c_{k+1}a_{k+1}}\sin\omega\dfrac{\ell_{k+1}}{c_{k+1}} + \mathrm{i}\dfrac{\hat{Q}_k(\ell_k,\omega)}{\hat{H}_k(\ell_k,\omega)}\cos\omega\dfrac{\ell_{k+1}}{c_{k+1}}}{\omega F_k\sin\omega\dfrac{\ell_{k+1}}{c_{k+1}} - \dfrac{1}{c_{k+1}a_{k+1}}\cos\omega\dfrac{\ell_{k+1}}{c_{k+1}} + \mathrm{i}\dfrac{\hat{Q}_k(\ell_k,\omega)}{\hat{H}_k(\ell_k,\omega)}\sin\omega\dfrac{\ell_{k+1}}{c_{k+1}}} \tag{4.51}$$

to which we have to add the computation for $k = 1$.

Since $\hat{H}_1(0,\omega) = 0$, we have

$$\frac{\hat{Q}_1(\ell_1,\omega)}{\hat{H}_1(\ell_1,\omega)} = \frac{\mathrm{i}}{c_1 a_1}\frac{\cos\omega\dfrac{\ell_1}{c_1}}{\sin\omega\dfrac{\ell_1}{c_1}} \tag{4.52}$$

hence

$$\frac{\hat{Q}_2(\ell_2,\omega)}{\hat{H}_2(\ell_2,\omega)} = \frac{\mathrm{i}}{c_2 a_2} \times \frac{\omega F_1\cos\omega\dfrac{\ell_2}{c_2} + \dfrac{1}{c_2 a_2}\sin\omega\dfrac{\ell_2}{c_2} - \dfrac{1}{c_1 a_1}\dfrac{\cos\omega\dfrac{\ell_1}{c_1}}{\sin\omega\dfrac{\ell_1}{c_1}}\cos\omega\dfrac{\ell_2}{c_2}}{\omega F_1\sin\omega\dfrac{\ell_2}{c_2} - \dfrac{1}{c_2 a_2}\cos\omega\dfrac{\ell_2}{c_2} - \dfrac{1}{c_1 a_1}\dfrac{\cos\omega\dfrac{\ell_1}{c_1}}{\sin\omega\dfrac{\ell_1}{c_1}}\sin\omega\dfrac{\ell_2}{c_2}} \tag{4.53}$$

If we denote

$$\frac{\hat{Q}_k(\ell_k,\omega)}{\hat{H}_k(\ell_k,\omega)} = \frac{1}{c_k a_k}\varphi_k(\omega) \tag{4.54}$$

we obtain the recurrence formula

$$\varphi_{k+1}(\omega) = \frac{\omega F_k \cos\omega\dfrac{\ell_{k+1}}{c_{k+1}} + \dfrac{1}{c_{k+1}a_{k+1}}\sin\omega\dfrac{\ell_{k+1}}{c_{k+1}} - \dfrac{1}{c_k a_k}\varphi_k(\omega)\cos\omega\dfrac{\ell_{k+1}}{c_{k+1}}}{\omega F_k \sin\omega\dfrac{\ell_{k+1}}{c_{k+1}} - \dfrac{1}{c_{k+1}a_{k+1}}\cos\omega\dfrac{\ell_{k+1}}{c_{k+1}} - \dfrac{1}{c_k a_k}\varphi_k(\omega)\sin\omega\dfrac{\ell_{k+1}}{c_{k+1}}} \tag{4.55}$$

In order to compute the hydraulic resonance—for example, in case (b)—we have

$$\hat{H}_{n+1}(\ell_{n+1},\omega) = \frac{A}{\beta + \dfrac{\mathrm{i}}{c_{n+1}a_{n+1}}\varphi_{n+1}(\omega)} \tag{4.56}$$

and studying the minima of function

$$\left|\beta + \frac{\mathrm{i}}{c_{n+1}a_{n+1}}\varphi_{n+1}(\omega)\right|$$

we obtain the resonance frequencies.

For case (a) the relation

$$\hat{H}_{n+1}(\ell_{n+1},\omega) = \frac{A c_{n+1}a_{n+1}}{\beta + \mathrm{i}\varphi_{n+1}(\omega)\left[\omega F_n \sin\omega\dfrac{\ell_{n+1}}{c_{n+1}} - \dfrac{1}{c_{n+1}a_{n+1}}\cos\omega\dfrac{\ell_{n+1}}{c_{n+1}} - \dfrac{1}{c_n a_n}\varphi_n(\omega)\sin\omega\dfrac{\ell_{n+1}}{c_{n+1}}\right]} \tag{4.57}$$

is used.

In particular, for $n = 2$—corresponding to a basic hydraulic scheme with two surge tanks—along the admission gallery, we have

$$\varphi_2(\omega) = \frac{f_2(\omega)}{f_1(\omega)}$$

$$\varphi_3(\omega) = \frac{\left[\omega F_2 \cos\omega\dfrac{\ell_3}{c_3} + \dfrac{1}{c_3 a_3}\sin\omega\dfrac{\ell_3}{c_3}\right] f_1(\omega) - \dfrac{1}{c_2 a_2} f_2(\omega)\cos\omega\dfrac{\ell_3}{c_3}}{\left[\omega F_2 \sin\omega\dfrac{\ell_3}{c_3} - \dfrac{1}{c_3 a_3}\cos\omega\dfrac{\ell_3}{c_3}\right] f_1(\omega) - \dfrac{1}{c_3 a_3} f_2(\omega)\sin\omega\dfrac{\ell_3}{c_3}} \tag{4.58}$$

and the resonance frequencies can be computed with formula (4.57).

4.2.5. Hydraulic scheme D

In this case, the boundary conditions are

$$\begin{aligned}
&H_k(0,t) = 0, \quad k = 1,2,\ldots,n \\
&F\frac{\mathrm{d}Z(t)}{\mathrm{d}t} + Q_{n+1}(0,t) = Q_1(\ell_1,t) + Q_2(\ell_2,t) + \cdots + Q_n(\ell_n,t) \\
&H_k(\ell_k,t) = Z(t) = H_{n+1}(0,t)
\end{aligned} \tag{4.59}$$

(a) $Q_{n+1}(\ell_{n+1}, t) + \beta H_{n+1}(0, t) = A e^{i\omega t}$

(b) $Q_{n+1}(\ell_{n+1}, t) + \beta H_{n+1}(\ell_{n+1}, t) = A e^{i\omega t}$

Performing the computations in the same manner, it follows that

$$\frac{\hat{Q}_{n+1}(\ell_{n+1},\omega)}{\hat{H}_{n+1}(\ell_{n+1},\omega)} = \frac{i}{c_{n+1}a_{n+1}} \times \frac{\left(\omega F - \sum_{k=1}^{n} \frac{1}{c_k a_k} \frac{\cos \omega \frac{\ell_k}{c_k}}{\sin \omega \frac{\ell_k}{c_k}}\right) \cos \omega \frac{\ell_{n+1}}{c_{n+1}} + \frac{1}{c_{n+1}a_{n+1}} \sin \omega \frac{\ell_{n+1}}{c_{n+1}}}{\left(\omega F - \sum_{k=1}^{n} \frac{1}{c_k a_k} \frac{\cos \omega \frac{\ell_k}{c_k}}{\sin \omega \frac{\ell_k}{c_k}}\right) \sin \omega \frac{\ell_{n+1}}{c_{n+1}} - \frac{1}{c_{n+1}a_{n+1}} \cos \omega \frac{\ell_{n+1}}{c_{n+1}}} \tag{4.60}$$

and the effective computation of the hydraulic resonance may be performed for case (a) by means of the formula

$$\left[\frac{i}{c_{n+1}a_{n+1}} \frac{\left(\omega F - \sum_{k=1}^{n} \frac{1}{c_k a_k} \frac{\cos \omega \frac{\ell_k}{c_k}}{\sin \omega \frac{\ell_k}{c_k}}\right) \cos \omega \frac{\ell_{n+1}}{c_{n+1}} + \frac{1}{c_{n+1}a_{n+1}} \sin \omega \frac{\ell_{n+1}}{c_{n+1}}}{\left(\omega F - \sum_{k=1}^{n} \frac{1}{a_k c_k} \frac{\cos \omega \frac{\ell_k}{c_k}}{\sin \omega \frac{\ell_k}{c_k}}\right) \sin \omega \frac{\ell_{n+1}}{c_{n+1}} - \frac{1}{c_{n+1}a_{n+1}} \cos \omega \frac{\ell_{n+1}}{c_{n+1}}} \right.$$
$$\left. - \frac{\beta}{c_{n+1}a_{n+1}} \frac{1}{\left(\omega F - \sum_{k=1}^{n} \frac{1}{c_k a_k} \frac{\cos \omega \frac{\ell_k}{c_k}}{\sin \omega \frac{\ell_k}{c_k}}\right) \sin \omega \frac{\ell_{n+1}}{c_{n+1}} - \frac{1}{c_{n+1}a_{n+1}} \cos \omega \frac{\ell_{n+1}}{c_{n+1}}}\right]$$
$$\times \hat{H}_{n+1}(\ell_{n+1}, \omega) = A \tag{4.61}$$

and for case (b) by means of the formula

$$\left[\beta + \frac{1}{c_{n+1}a_{n+1}} \times \frac{\left(\omega F - \sum_{k=1}^{n} \frac{1}{c_k a_k} \frac{\cos \omega \frac{\ell_k}{c_k}}{\sin \omega \frac{\ell_k}{c_k}}\right) \cos \omega \frac{\ell_{n+1}}{c_{n+1}} + \frac{1}{c_{n+1}a_{n+1}} \sin \omega \frac{\ell_{n+1}}{c_{n+1}}}{\left(\omega F - \sum_{k=1}^{n} \frac{1}{c_k a_k} \frac{\cos \omega \frac{\ell_k}{c_k}}{\sin \omega \frac{\ell_k}{c_k}}\right) \sin \omega \frac{\ell_{n+1}}{c_{n+1}} - \frac{1}{c_{n+1}a_{n+1}} \cos \omega \frac{\ell_{n+1}}{c_{n+1}}}\right]$$
$$\times \hat{H}_{n+1}(\ell_{n+1}, \omega) = A \tag{4.61'}$$

For $n = 1$, the formulae for the computation of hydraulic resonance in the case of the basic hydraulic scheme with one surge tank are regained.

4.3. Computation of hydraulic resonance at pumping stations for the basic hydraulic scheme

We reconsider the problem of the hydraulic resonance, this time for pumping stations, starting with the results previously obtained for hydroelectric power plants and taking into account the constructive and operation features specific to pumping stations. In this way first we shall analyse the hydraulic resonance at pumping stations for the case of a basic hydraulic scheme endowed with air chamber or surge tank (Fig. 4.10) [17].

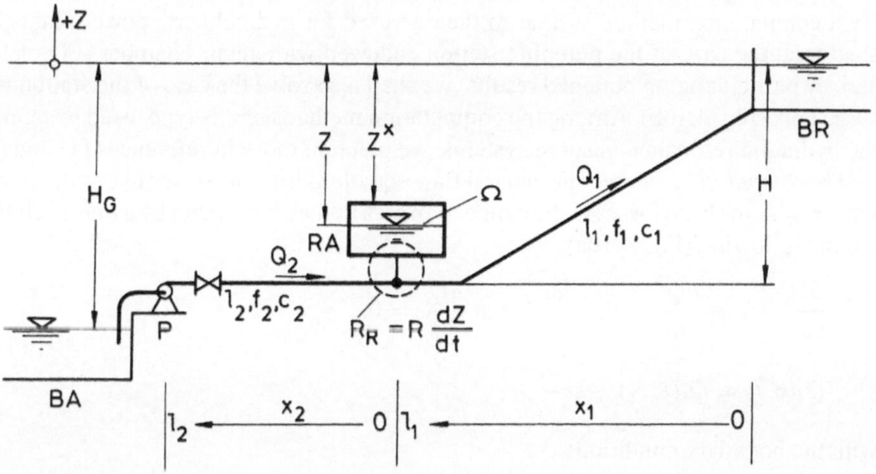

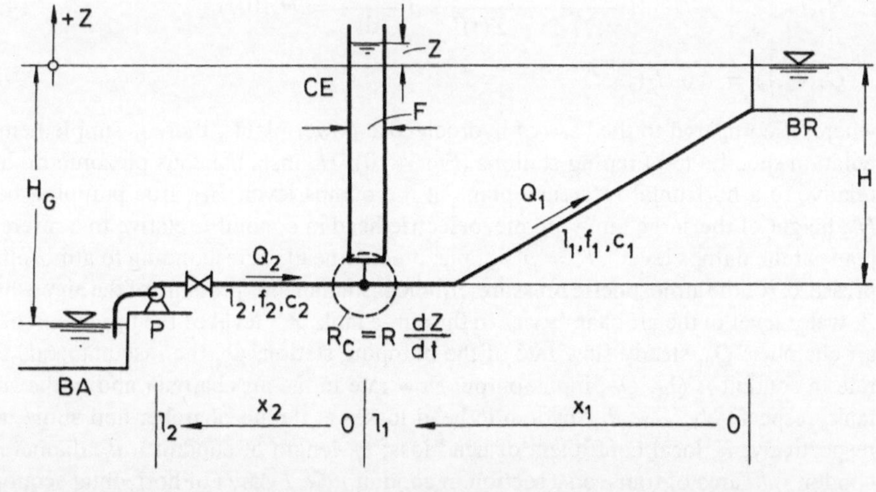

Figure 4.10. Scheme of a pumping station equipped with: (a) air chamber, or (b) surge tank.

The general equations governing the water flow in this hydraulic system are analysed under the hypothesis of some types of small periodic flow-rate perturbations at the station (pumps, valves, check valves, etc.).

Using the model linearised around the steady state, the response to harmonic perturbations is studied and resonance frequencies corresponding to maximum amplitudes of pressure variations of the water in the discharge pipe in the pumps section are obtained. The obtained frequencies are tested by numerical simulation on the non-linear mathematical model of the hydraulic system.

On the basis of the obtained theoretical results specific to pumping stations, we have analysed the possibility of appearance of hydraulic resonance phenomena at two pumping stations in Romania and we have used numerical methods to compute the maximum pressures which may occur in the system under these resonance conditions.

Presentation of the computation method

By a computation method similar to the one used for hydroelectric power plants, we shall treat the case of the pumping station endowed with an air chamber (Fig. 4.10a) and, by particularizing obtained results, we shall also solve the case of the station with surge tank (Fig. 4.10b). Among the computation methods of this type, used to estimate the hydraulic resonance in surge systems, we mention those in references [1] and [7].

The computation model. The general flow equations in the analysed hydraulic system that express the head losses in the connection area of the air chamber by a linear relation, are the following (Fig. 4.10a)

$$\begin{aligned} &\frac{\partial H_j}{\partial t} + a_j c_j^2 \frac{\partial Q_j}{\partial x} = 0, \quad \frac{\partial H_j}{\partial x} + a_j \frac{\partial Q_j}{\partial t} = 0; \quad j = 1,2 \\ &Q_2(0,t) = Q_1(\ell_1,t) + \Omega \frac{\partial Z(t)}{dt} \end{aligned} \tag{4.62}$$

with the boundary conditions

$$\begin{aligned} H_1(0,t) &= H \\ H_1(\ell_1,t) &= -H^* + \frac{C}{\gamma\Omega\left[Z^* - Z(t)\right]} + R\frac{\mathrm{d}Z(t)}{\mathrm{d}t} = H_2(0,t) \\ Q_2(\ell_2,t) &= \alpha\sqrt{H_2(\ell_2,t)} \end{aligned} \tag{4.63}$$

where, if compared to the case of hydroelectric power plants, there is supplementary notation specific to pumping stations (Fig. 4.10): H, instantaneous piezometric head relative to a horizontal reference plane at the pumps level; H_G, true pumping head; H_c, height of the surge tank; H_j, piezoelectric head in conduit j relative to a reference plane at the pumps level; $H^* = p^*/\zeta$, piezometric head corresponding to atmospheric pressure; p^*, the atmospheric pressure; p_a, the instantaneous pressure of the air cushion; Z, water level in the air chamber or in the surge tank; Z^*, level of the upper end of the air chamber; Q_o, steady flow rate of the pumping station; Q_j, the instantaneous flow rate in conduit j; Q_R, Q_C, input–output flow rate in the air chamber and in the surge tank, respectively; R_R, R_C, hydraulic head losses at the air chamber and surge tank, respectively; R, local coefficient of head loss; ℓ_j, length of conduit j; d_j, diameter of conduit j; f_j, area of transverse section in conduit j; Ω, F, area of horizontal section in the air chamber and surge tank, respectively; V_t, total volume of the air chamber; V_a, instantaneous volume of the air cushion in the air chamber; V_o, steady-state volume of the air cushion; $C = (p_a + p^*)V_a$, elastic constant of the air cushion in the air chamber.

The admitted flow law of type (4.63) can also be applied to pumping stations to study the unsteady flow in the system, assuming small perturbations in the flow rate at the station. This allows a direct comparison of the final results obtained for pumping stations with those previously obtained for hydroelectric power plants.

The steady flow. This corresponds to constant values of the flow rates (Q), pressures (H) and levels (Z), which are denoted by the subscript 0

$$\begin{aligned} &Q_1 = Q_2 = \alpha\sqrt{H} = Q_o \\ &H_1 = H_2 = H = H_o \\ &H^* + H_o = \frac{C}{\gamma\Omega(Z^* - Z_o)} \end{aligned} \tag{4.64}$$

The harmonic perturbation in the first approximation. A harmonic perturbation of the flow rate is admitted in the section of the pumps and α is replaced by $(\alpha + \varepsilon A e^{i\omega t})$.

For the equation system (4.62), with the boundary conditions (4.63), a solution in the form

$$\begin{aligned} H_j(x,j) &= H + \varepsilon\hat{H}_j(x,t) + \cdots \\ Q_j(x,t) &= \alpha\sqrt{H} + \varepsilon\hat{Q}_j(x,t) + \cdots \\ Z(t) &= Z_o + \varepsilon\hat{Z}(t) + \cdots; \quad j = 1,2 \end{aligned} \tag{4.65}$$

is sought for, from which the equations, in the first approximation, follow (only the leading terms are retained)

$$\begin{aligned} &\frac{\partial\hat{H}_j(x,t)}{\partial t} + a_j c_j^2 \frac{\partial\hat{Q}_j(x,t)}{\partial x} = 0 \\ &\frac{\partial\hat{H}_j(x,t)}{\partial x} + a_j \frac{\partial\hat{Q}_j(x,t)}{\partial t} = 0 \\ &\hat{H}_1(0,t) = 0 \\ &\hat{Q}_2(0,t) = \hat{Q}_1(\ell_1,t) + \Omega\frac{\mathrm{d}\hat{Z}(t)}{\mathrm{d}t} \\ &\hat{H}_1(\ell_1,t) = \hat{H}_2(0,t) = R\frac{\mathrm{d}\hat{Z}(t)}{\mathrm{d}t} + \frac{H^* + H_o}{Z^* - Z_o}Z(t) \\ &\hat{Q}_2(\ell_2,t) = \frac{\alpha}{2\sqrt{H}}\hat{H}_2(\ell_2,t) + A\sqrt{H}e^{i\omega t} \end{aligned} \tag{4.66}$$

The harmonic response

A response of the form

$$\begin{aligned} \hat{H}_j(x,t) &= e^{i\omega t}\tilde{H}_j(x) \\ \hat{Q}_j(x,t) &= e^{i\omega t}\hat{Q}_j(x) \\ \hat{Z}(t) &= e^{i\omega t}\tilde{Z}(t) \end{aligned} \tag{4.67}$$

is sought.

The propagation equation becomes

$$\begin{aligned} &\mathrm{i}\omega\tilde{H}_j(x) + a_j c_j^2 \tilde{Q}_j'(x) = 0, \quad \tilde{H}_j'(x) + \mathrm{i}\omega a_j \tilde{Q}_j(x) = 0 \\ &\tilde{H}''(x) - \mathrm{i}\omega\frac{\mathrm{i}\omega}{c_j^2}\tilde{H}_j(x) = 0 \end{aligned} \tag{4.68}$$

from which it follows

$$\begin{aligned} &\tilde{H}_j(x) = A_j \mathrm{e}^{(\mathrm{i}\omega x/c_j)} + B_j \mathrm{e}^{-(\mathrm{i}\omega x/c_j)} \\ &\tilde{Q}_j(x) = \frac{1}{a_j c_j}\left(B_j \mathrm{e}^{-(\mathrm{i}\omega x/c_j)} - A_j \mathrm{e}^{(\mathrm{i}\omega x/c_j)}\right); \quad j = 1,2 \end{aligned} \tag{4.69}$$

and from the boundary conditions we get

$$\begin{aligned} \tilde{H}_1(0) &= 0, \quad \tilde{Q}_2(0) = \tilde{Q}_1(\ell_1) + \mathrm{i}\omega\Omega\tilde{Z}(t) \\ \tilde{H}_1(\ell_1) &= \tilde{H}_2(0) = \left(\mathrm{i}\omega R + \frac{H^* + H_o}{Z^* - Z_o}\right)\tilde{Z}(t) \\ \tilde{Q}_2(\ell_2) &= \frac{\alpha}{2\sqrt{H}}\tilde{H}_2(\ell_2) + A\sqrt{H} \end{aligned} \tag{4.70}$$

The resonance computation
In order to point out the excitation frequencies which can have a dangerous effect in the analysed hydraulic system, we shall compute and analyse the water pressure variation in the discharge pipe, at the pump section, following the behaviour of function $\tilde{H}_2(\ell_2)$.

From the formula of $\tilde{H}_j(x)$ and $\tilde{Q}_j(x)$ it follows

$$\begin{aligned} A_j &= \frac{1}{2}\left[\tilde{H}_j(\ell_j) - a_j c_j \tilde{Q}_j(\ell_j)\right] \mathrm{e}^{-(\mathrm{i}\omega\ell_j/c_j)} \\ B_j &= \frac{1}{2}\left[\tilde{H}_j(\ell_j) + a_j c_j \tilde{Q}_j(\ell_j)\right] \mathrm{e}^{(\mathrm{i}\omega\ell_j/c_j)}; \quad j = 1,2 \end{aligned} \tag{4.71}$$

The condition $\tilde{H}_1(0) = 0$ stands for $A_1 + B_1 = 0$, hence

$$\tilde{H}_1(\ell_1)\cos\omega\frac{\ell_1}{c_1} + \mathrm{i}a_1 c_1 \tilde{Q}_1(\ell_1)\sin\omega\frac{\ell_1}{c_1} = 0 \tag{4.72}$$

Further, we have

$$\begin{aligned} \tilde{Z}(t) &= \frac{1}{\mathrm{i}\omega R + \dfrac{H^* + H_o}{Z^* - Z_o}}\tilde{H}_2(0) \\ \tilde{Q}_2(0) &- \frac{\mathrm{i}\omega R}{\mathrm{i}\omega R + \dfrac{H^* + H_o}{Z^* - Z_o}}\tilde{H}_2(0) = Q_1(\ell_1) \\ \tilde{H}_2(0) &= H_1(\ell_1) \end{aligned} \tag{4.73}$$

$$\left[\frac{1}{a_1c_1}\cos\frac{\omega\ell_1}{c_1}+\frac{\omega\Omega}{i\omega R+\dfrac{H^*+H_o}{Z^*-Z_o}}\sin\frac{\omega\ell_1}{c_1}\right]\tilde{H}_2(0)+i\tilde{Q}_2(0)\sin\frac{\omega\ell_1}{c_1}=0$$

But

$$\begin{aligned}\tilde{H}_2(0)&=A_2+B_2=\tilde{H}_2(\ell_2)\cos\frac{\omega\ell_2}{c_2}+ia_2c_2\tilde{Q}_2(\ell_2)\sin\frac{\omega\ell_2}{c_2}\\ \tilde{Q}_2(0)&=\frac{1}{a_2c_2}(B_2-A_2)=\frac{i}{a_2c_2}\tilde{H}_2(\ell_2)\sin\frac{\omega\ell_2}{c_2}+\tilde{Q}_2(\ell_2)\cos\frac{\omega\ell_2}{c_2}\end{aligned}\tag{4.74}$$

It follows that

$$\begin{aligned}&\left[\frac{1}{a_1c_1}\cos\frac{\omega\ell_1}{c_1}+\frac{\omega\Omega}{i\omega R+\dfrac{H^*+H_o}{Z^*-Z_o}}\sin\frac{\omega\ell_1}{c_1}\right]\\ &\times\left[\tilde{H}_2(\ell_2)\cos\frac{\omega\ell_2}{c_2}+ia_2c_2\tilde{Q}_2(\ell_2)\sin\frac{\omega\ell_2}{c_2}\right]\\ &+i\left[\frac{i}{a_2c_2}\tilde{H}_2(\ell_2)\sin\frac{\omega\ell_2}{c_2}-\tilde{Q}_2(\ell_2)\cos\frac{\omega\ell_2}{c_2}\right]\sin\frac{\omega\ell_1}{c_1}=0\end{aligned}\tag{4.75}$$

hence,

$$\begin{aligned}&\left[\frac{1}{a_1c_1}\cos\frac{\omega\ell_1}{c_1}\cos\frac{\omega\ell_2}{c_2}-\frac{1}{a_2c_2}\sin\frac{\omega\ell_2}{c_2}\sin\frac{\omega\ell_1}{c_1}+\frac{\omega\Omega\sin\omega\dfrac{\ell_1}{c_1}\cos\omega\dfrac{\ell_2}{c_2}}{i\omega R+\dfrac{H^*+H_o}{Z^*-Z_o}}\right]\tilde{H}_2(\ell_2)\\ &+ia_2c_2\left[\frac{1}{a_1c_1}\cos\frac{\omega\ell_1}{c_1}\sin\frac{\omega\ell_2}{c_2}+\frac{1}{a_2c_2}\sin\frac{\omega\ell_1}{c_1}\cos\frac{\omega\ell_2}{c_2}\right.\\ &\left.+\frac{\omega\Omega\sin\omega\dfrac{\ell_1}{c_1}\sin\omega\dfrac{\ell_2}{c_2}}{i\omega R+\dfrac{H^*+H_o}{Z^*-Z_o}}\right]\tilde{Q}_2(\ell_2)=0\end{aligned}\tag{4.76}$$

In relation (4.76), $\tilde{Q}_2(\ell_2)$ is substituted by using the last relation from the boundary conditions brought to the form (4.70), and we obtain

$$\begin{aligned}&\left[\frac{1}{a_1c_1}\cos\frac{\omega\ell_1}{c_1}\cos\frac{\omega\ell_2}{c_2}-\frac{1}{a_2c_2}\sin\frac{\omega\ell_1}{c_1}\sin\frac{\omega\ell_2}{c_2}+\frac{\omega\Omega\sin\omega\frac{\ell_1}{c_1}\sin\omega\frac{\ell_2}{c_2}}{i\omega R+\dfrac{H^*+H_o}{Z^*-Z_o}}\right]\tilde{H}_2(\ell_2)\\ &+\frac{ia_2c_2\alpha}{2\sqrt{H}}\left[\frac{1}{a_1c_1}\cos\frac{\omega\ell_1}{c_1}\sin\frac{\omega\ell_2}{c_2}+\frac{1}{a_2c_2}\sin\frac{\omega\ell_1}{c_1}\cos\frac{\omega\ell_2}{c_2}\right.\end{aligned}$$

$$+\left.\frac{\omega\Omega\sin\frac{\omega\ell_1}{c_1}\sin\frac{\omega\ell_2}{c_2}}{\mathrm{i}\omega R+\frac{H^*+H_o}{Z^*-Z_o}}\right]\tilde{H}_2(\ell_2)+\mathrm{i}a_2c_2A\sqrt{H}\left[\frac{1}{a_1c_1}\cos\frac{\omega\ell_1}{c_1}\sin\frac{\omega\ell_2}{c_2}\right.$$

$$\left.+\frac{1}{a_2c_2}\sin\frac{\omega\ell_1}{c_1}\cos\frac{\omega\ell_2}{c_2}+\frac{\omega\Omega\sin\frac{\omega\ell_1}{c_1}\sin\frac{\omega\ell_2}{c_2}}{\mathrm{i}\omega R+\frac{H^*+H_o}{Z^*-Z_o}}\right]=0 \tag{4.77}$$

Performing the computations, we have successively

$$\left\{\left(\mathrm{i}\omega R+\frac{H^*+H_o}{Z^*-Z_o}\right)\left(\frac{1}{a_1c_1}\cos\frac{\omega\ell_1}{c_1}\cos\frac{\omega\ell_2}{c_2}-\frac{1}{a_2c_2}\sin\frac{\omega\ell_1}{c_1}\sin\frac{\omega\ell_2}{c_2}\right)\right.$$

$$+\omega\Omega\sin\frac{\omega\ell_1}{c_1}\cos\frac{\omega\ell_2}{c_2}+\frac{\mathrm{i}a_2c_2\alpha}{2\sqrt{H}}\left[\left(\mathrm{i}\omega R+\frac{H^*+H_o}{Z^*-Z_o}\right)\right.$$

$$\left.\left.\times\left(\frac{1}{a_1c_1}\cos\frac{\omega\ell_1}{c_1}\sin\frac{\omega\ell_2}{c_2}+\frac{1}{a_2c_2}\sin\frac{\omega\ell_1}{c_1}\cos\frac{\omega\ell_2}{c_2}\right)-\omega\Omega\sin\frac{\omega\ell_1}{c_1}\sin\frac{\omega\ell_2}{c_2}\right]\right\}$$

$$\times\tilde{H}_2(\ell_2)=-\mathrm{i}a_2c_2A\sqrt{H}\left[\left(\mathrm{i}\omega R+\frac{H^*+H_o}{Z^*-Z_o}\right)\left(\frac{1}{a_1c_1}\cos\frac{\omega\ell_1}{c_1}\cos\frac{\omega\ell_2}{c_2}\right.\right.$$

$$\left.\left.+\frac{1}{a_2c_2}\sin\frac{\omega\ell_1}{c_1}\sin\frac{\omega\ell_2}{c_2}\right)+\omega\Omega\sin\frac{\omega\ell_1}{c_1}\sin\frac{\omega\ell_2}{c_2}\right] \tag{4.78}$$

$$\left\{\left[\frac{H^*+H_o}{Z^*-Z_o}\left(\frac{1}{a_1c_1}\cos\frac{\omega\ell_1}{c_1}\cos\frac{\omega\ell_2}{c_2}-\frac{1}{a_2c_2}\sin\frac{\omega\ell_1}{c_1}\sin\frac{\omega\ell_2}{c_2}\right)\right.\right.$$

$$+\omega\Omega\sin\frac{\omega\ell_1}{c_1}\cos\frac{\omega\ell_2}{c_2}-\frac{a_2c_2\alpha\omega R}{2\sqrt{H}}\left(\frac{1}{a_1c_1}\cos\frac{\omega\ell_1}{c_1}\sin\frac{\omega\ell_2}{c_2}+\frac{1}{a_2c_2}\sin\frac{\omega\ell_1}{c_1}\right.$$

$$\left.\left.\times\cos\frac{\omega\ell_2}{c_2}\right)\right]+\mathrm{i}\left[\frac{a_2c_2\alpha\omega\Omega}{2\sqrt{H}}\sin\frac{\omega\ell_1}{c_1}\sin\frac{\omega\ell_2}{c_2}+\frac{a_2c_2\alpha}{2\sqrt{H}}\cdot\frac{H^*+H_o}{Z^*-Z_o}\right.$$

$$\times\left(\frac{1}{a_1c_1}\cos\frac{\omega\ell_1}{c_1}\sin\frac{\omega\ell_2}{c_2}+\frac{1}{a_2c_2}\sin\frac{\omega\ell_1}{c_1}\cos\frac{\omega\ell_2}{c_2}\right)+\omega R$$

$$\left.\left.\times\left(\frac{1}{a_1c_1}\cos\frac{\omega\ell_1}{c_1}\cos\frac{\omega\ell_2}{c_2}-\frac{1}{a_2c_2}\sin\frac{\omega\ell_1}{c_1}\sin\frac{\omega\ell_2}{c_2}\right)\right]\right\}\tilde{H}_2(\ell_2)=-\mathrm{i}a_2c_2A\sqrt{H}$$

$$\times\left[\frac{H^*+H_o}{Z^*-Z_o}\left(\frac{1}{a_1c_1}\cos\frac{\omega\ell_1}{c_1}\sin\frac{\omega\ell_2}{c_2}+\frac{1}{a_2c_2}\sin\frac{\omega\ell_2}{c_2}\cos\frac{\omega\ell_2}{c_2}\right)+\omega\Omega\sin\right.$$

$$\left.\times\frac{\omega\ell_1}{c_1}\sin\frac{\omega\ell_2}{c_2}+\mathrm{i}\omega\Omega\left(\frac{1}{a_1c_1}\cos\frac{\omega\ell_1}{c_1}\sin\frac{\omega\ell_2}{c_2}+\frac{1}{a_2c_2}\sin\frac{\omega\ell_1}{c_1}\cos\frac{\omega\ell_2}{c_2}\right)\right] \tag{4.79}$$

Denoting

$$\begin{aligned}\varphi(\omega) &= \frac{1}{a_1c_1}\cos\frac{\omega\ell_1}{c_1}\cos\frac{\omega\ell_2}{c_2} - \frac{1}{a_2c_2}\sin\frac{\omega\ell_1}{c_1}\sin\frac{\omega\ell_2}{c_2}\\ \psi(\omega) &= \frac{1}{a_1c_1}\cos\frac{\omega\ell_1}{c_1}\sin\frac{\omega\ell_2}{c_2} + \frac{1}{a_2c_2}\sin\frac{\omega\ell_1}{c_1}\cos\frac{\omega\ell_2}{c_2}\end{aligned} \tag{4.80}$$

it follows that

$$\begin{aligned}&\left\{\frac{H^*+H_o}{Z^*-Z_o}\varphi(\omega) + \omega\Omega\sin\frac{\omega\ell_1}{c_1}\cos\frac{\omega\ell_2}{c_2} - \frac{a_2c_2\alpha\omega R}{2\sqrt{H}}\psi(\omega)\right.\\ &\left.+\mathrm{i}\left[\frac{a_2c_2\alpha\omega R}{2\sqrt{H}}\sin\frac{\omega\ell_1}{c_1}\cos\frac{\omega\ell_2}{c_2} + \frac{a_2c_2\alpha R}{2\sqrt{H}}\frac{H^*+H_o}{Z^*-Z_o}\psi(\omega) + \omega R\varphi(\omega)\right]\right\}\tilde{H}_2(\ell_2)\\ &= -\mathrm{i}a_2c_2A\sqrt{H}\left[\frac{H^*+H_o}{Z^*-Z_o}\psi(\omega) + \omega\Omega\sin\frac{\omega\ell_1}{c_1}\sin\frac{\omega\ell_2}{c_2} + \mathrm{i}\omega R\psi(\omega)\right]\end{aligned} \tag{4.81}$$

New notation is introduced

$$\begin{aligned}f(\omega) &= \frac{H^*+H_o}{Z^*-Z_o}\varphi(\omega) + \omega\Omega\sin\frac{\omega\ell_1}{c_1}\cos\frac{\omega\ell_2}{c_2}\\ g(\omega) &= \frac{H^*+H_o}{Z^*-Z_o}\psi(\omega) + \omega\Omega\sin\frac{\omega\ell_1}{c_1}\sin\frac{\omega\ell_2}{c_2}\end{aligned} \tag{4.82}$$

One obtains

$$\begin{aligned}&\left[f(\omega) - \frac{a_2c_2\alpha\omega R}{2\sqrt{H}}\psi(\omega) + \mathrm{i}\left(\frac{a_2c_2\alpha}{2\sqrt{H}}g(\omega) + \omega R\varphi(\omega)\right)\right]\tilde{H}_2(\ell_2)\\ &= -\mathrm{i}a_2c_2A\sqrt{H}\left[g(\omega) + \mathrm{i}\omega R\psi(\omega)\right]\end{aligned} \tag{4.83}$$

hence

$$\tilde{H}_2(\ell_2) = -\mathrm{i}a_2c_2A\sqrt{H}\frac{g(\omega) + \mathrm{i}\omega R\psi(\omega)}{f(\omega) - \dfrac{a_2c_2\alpha\omega R}{2\sqrt{H}}\psi(\omega) + \mathrm{i}\left[\dfrac{a_2c_2\alpha}{2\sqrt{H}}g(\omega) + \omega R\varphi(\omega)\right]} \tag{4.84}$$

and

$$\left|\tilde{H}_2(\ell_2)\right| = a_2c_2A\sqrt{H}\sqrt{\frac{g^2(\omega) + \omega^2R^2\psi^2(\omega)}{\left[f(\omega) - \dfrac{a_2c_2\alpha\omega R}{2\sqrt{H}}\psi(\omega)\right]^2 + \left[\dfrac{a_2c_2\alpha}{2\sqrt{H}}g(\omega) + \omega R\varphi(\omega)\right]^2}} \tag{4.85}$$

respectively.

Remarks

(1) If the pumping station is equipped with surge tank instead of air chamber, the formula of $\left|\tilde{H}_2(\ell_2)\right|$ is the same, functions $\varphi(\omega)$ and $\psi(\omega)$ remain unchanged and

functions $f(\omega)$ and $g(\omega)$ become

$$
\begin{aligned}
f(\omega) &= \varphi(\omega) + \omega F \sin\frac{\omega \ell_1}{c_1}\cos\frac{\omega \ell_2}{c_2} \\
g(\omega) &= \psi(\omega) + \omega F \sin\frac{\omega \ell_1}{c_1}\sin\frac{\omega \ell_2}{c_2}
\end{aligned}
\tag{4.86}
$$

(2) If instead of a pumping station, a hydroelectric power plant is considered, the formulae are written in the same way, replacing the values Ω and F by $-\Omega$ and $-F$, respectively.

(3) If the case of a hydroelectric power plant is considered and R is assumed to be 0 the previously obtained results on hydraulic resonance at hydroelectric power plant, are once again obtained for the basic hydraulic scheme.

Applications

To illustrate the practical applicability of the method in the study of hydraulic resonance that can appear in the operation of pumping stations computations have been performed for two Romanian pumping stations.

Pumping station in the Sinoe system. The station is fitted with a protection system with air chamber and has the following main characteristics

$$
\begin{aligned}
\ell_1 &= 1600\,\text{m}, & d_1 &= 1.80\,\text{m}, \\
\ell_2 &= 400\,\text{m}, & d_2 &= 1.80\,\text{m}, \\
V_o &= 30\,\text{m}^3, & V_t &= 60\,\text{m}^3, \\
H_G &= 70.5\,\text{m}, & R_R &= 98 Q_R^2. \\
Q_o &= 5\,\text{m}^3/\text{s}, & &
\end{aligned}
$$

In order to point out the dangerous frequencies, in the case of small perturbations appearing in the flow rate at the station, the function $\left|\tilde{H}_2(\ell_2)\right|$ has been computed and the results were plotted in Fig. 4.11. Thus, the ranges of dangerous excitation frequencies for this station are exhibited. As these frequencies have been obtained from the linearised model, it is important to know their effect in the real hydraulic system.

From a practical point of view, in the hypothesis of appearance and development of some hydraulic resonance phenomena it is important to know the maximum pressures that can appear in such circumstances.

Using a numerical computation technique, based on the method of characteristics, by numerical integration of the non-linear equation system of the type (4.62)—with quadratic head—losses and for realistic boundary conditions of the type (4.63) the behaviour of the hydraulic system for the previously computed dangerous excitation frequencies was analysed. Some of the results of the numerical computations are shown in a graphical form in Fig. 4.12. They show that the frequencies detected by linearised analysis are really dangerous.

The pumping station of the Razelm–Tulcea system. The station is fitted with a protection system of surge tank type and has the following main characteristics:

$$
\begin{aligned}
\ell_1 &= 300\,\text{m}, & \text{d}_1 &= 1.25\,\text{m}, \\
\ell_2 &= 120\,\text{m}, & \text{d}_2 &= 1.25\,\text{m}, \\
H_C &= 17\,\text{m}, & F &= 2.85\,\text{m}^2,
\end{aligned}
$$

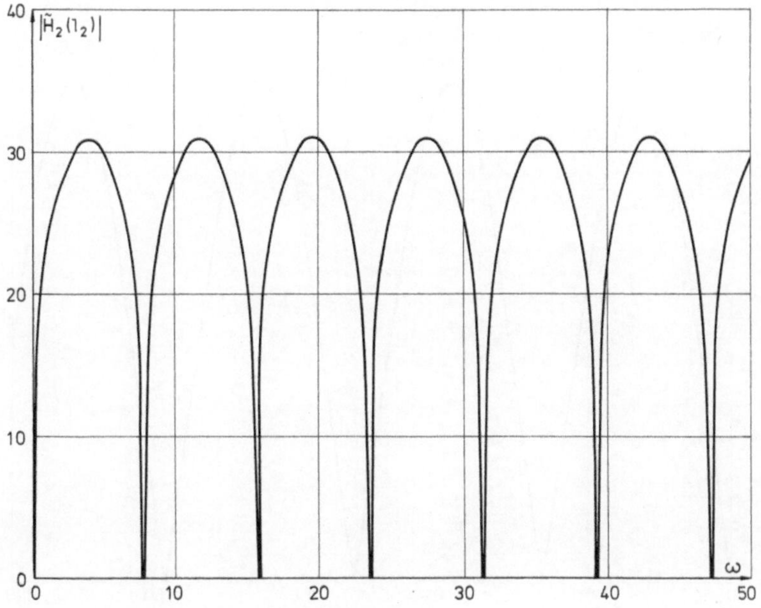

Figure 4.11. Sinoe pumping station. $|\tilde{H}_2(\ell_2)|$ versus ω.

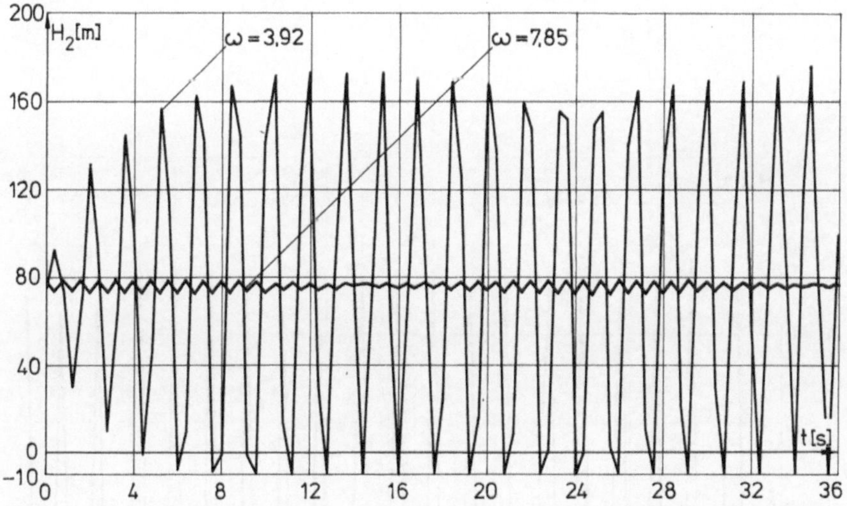

Figure 4.12. Sinoe pumping station. Curves of time variation of the pressure $H_2(\ell_2/t)$.

$$H_G = 21\,\text{m}, \qquad Q_o = 3\,\text{m}^3/\text{s}.$$
$$R_e = 0.25Q_c^2,$$

The computations have been performed in a similar manner.

The ranges of dangerous excitation frequencies are pointed out (Fig. 4.13) and the maximum pressures that can appear in the discharge pipes in the event of resonance phenomena occurring in the system (Fig. 4.14) are calculated.

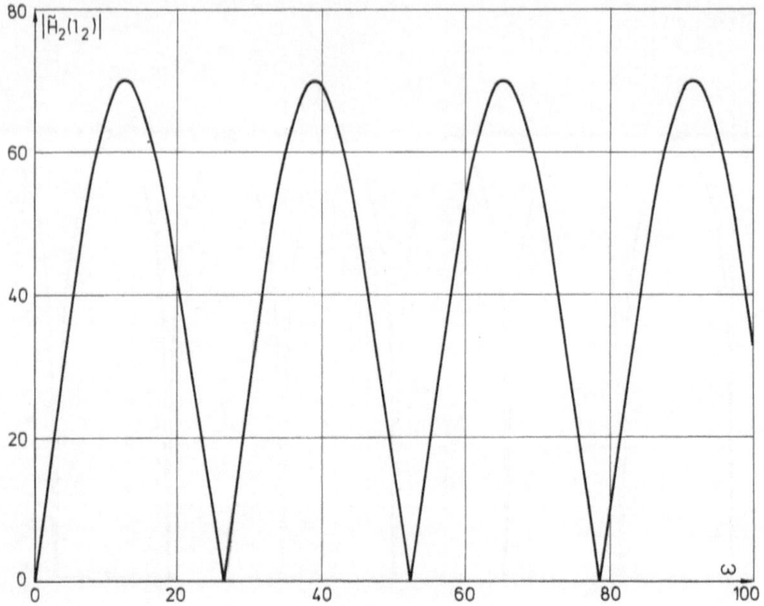

Figure 4.13. Razelm–Tulcea pumping station. $|\tilde{H}_2(\ell_2)|$ versus ω.

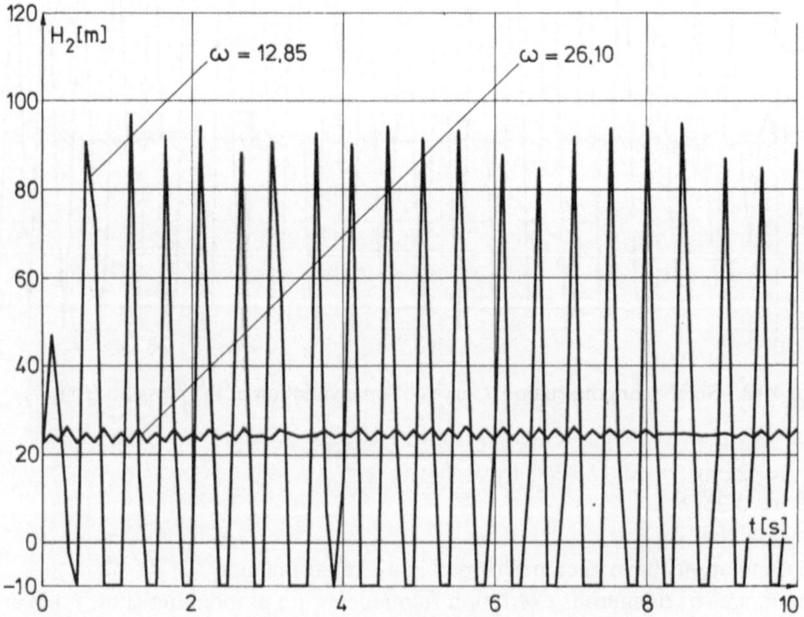

Figure 4.14. Razelm–Tulcea pumping station. Curves of time variation of the pressure $H_2(\ell_2/t)$.

4.4. Computation of hydraulic resonance for pumping stations provided with double protection system against waterhammer

The schemes of pumping stations with double protection system against waterhammer proved to be very efficient in engineering practice, especially when the discharge pipe exhibits changes in the slope in longitudinal section at certain high points either in the variant with two air chambers (Fig. 4.15I), or in the variant with air chamber and surge tank (Fig 4.15II) when, the latter can be designed with a surge tank height of 20–25 m.

In the following, the problem of hydraulic resonance for this kind of pumping station will be analysed by an extension of the computation method previously presented for the basic hydraulic scheme.

The computation model. The general flow equations in the hydraulic system I (Fig. 4.15), expressing the head losses in the area of the air chambers connections chambers, by linearised relations, are the following [19]:

$$\frac{\partial H_j}{\partial t} + a_j c_j^2 \frac{\partial Q_j}{\partial x} = 0, \quad \frac{\partial H_j}{\partial x} + a_j \frac{\partial H_j}{\partial t} = 0, \quad j = 1,2,3$$
$$\Omega_j \frac{\mathrm{d}Z_j}{\mathrm{d}t} + Q_j(\ell_j, t) = Q_{j+1}(0, t), \quad j = 1,2 \tag{4.87}$$

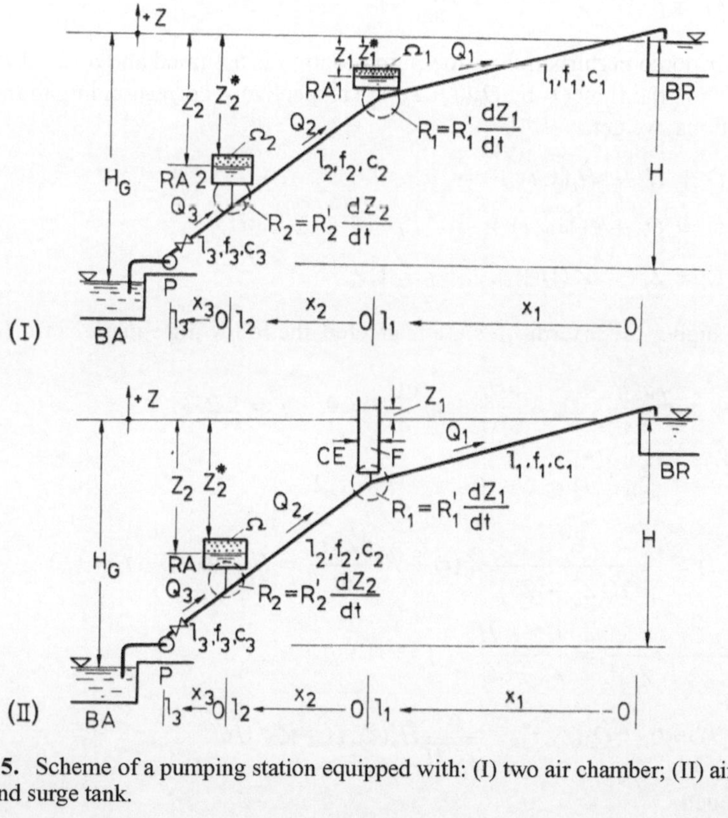

Figure 4.15. Scheme of a pumping station equipped with: (I) two air chamber; (II) air chamber and surge tank.

with the boundary conditions

$$
\begin{aligned}
H_1(0,t) &= H \\
H_j(\ell_j,t) &= -H^* + \frac{C_j}{\gamma * \Omega_j\left[Z_j^* - Z_j(t)\right]} + R_j\frac{\mathrm{d}Z_j(t)}{\mathrm{d}t} \\
&= H_{j+1}(0,t), \quad j=1,2 \\
Q_3(\ell_3,t) &= \alpha\sqrt{H_3(\ell_3,t)}
\end{aligned}
\tag{4.88}
$$

A more general flow law was admitted at the valve in the form (4.88). It should be mentioned that in current computations, a small perturbation in the flow rate at the pump can be simulated by using the simple relation of steady flow:

$$
Q_3(\ell_3,t) = Q_o \tag{4.89}
$$

similar to a step in the flow rate.

The corresponding steady flow is given by the relations

$$
\begin{aligned}
&H_1 = H_2 = H_3 = H, \quad Q_1 = Q_2 = Q_3 = Q_o\left(=\alpha\sqrt{H}\right) \\
&H + H^* = \frac{C_1}{\gamma\Omega_2\left(Z_1^* - Z_1^o\right)} = \frac{C_2}{\gamma\Omega_2\left(Z_2^* - Z_2^o\right)} \\
&Z_j^* - Z_j^o = \frac{\gamma\Omega_j\left(H^* + H\right)}{C_j}, \quad j=1,2
\end{aligned}
\tag{4.90}
$$

If a harmonic perturbation in the pump section is assumed and α is substituted by $(\alpha + \varepsilon A\mathrm{e}^{\mathrm{i}\varpi t})$, the flow Q_o by $Q_o(1 + \varepsilon\mathrm{e}^{\mathrm{i}\varpi t})$ respectively, corresponding to this kind of perturbations, we get

$$
\begin{aligned}
H_j(x,t) &= H + \varepsilon\hat{H}_j(x,t) + \cdots \\
Q_j(x,t) &= Q_o + \varepsilon\hat{Q}_j(x,t) + \ldots, \quad j=1,2,3 \\
Z_j(t) &= Z_j^o + \varepsilon\hat{Z}_j(t) + \ldots, \quad j=1,2
\end{aligned}
\tag{4.91}
$$

If the higher-order terms in ε are neglected, the following equations for $\hat{H}_j$, $\hat{Q}_j$, $\hat{Z}_j$

$$
\begin{aligned}
&\frac{\partial\hat{H}_j}{\partial t} + a_j c_j^2\frac{\partial\hat{Q}_j}{\partial x} = 0, \quad \frac{\partial\hat{H}_j}{\partial x} + a_j\frac{\partial\hat{Q}_j}{\partial t} = 0, \quad j=1,2,3 \\
&\Omega_j\frac{\mathrm{d}\hat{Z}(t)}{\mathrm{d}t} + \hat{Q}_j(\ell_j,t) = \hat{Q}_{j+1}(0,t), \quad j=1,2 \\
&\hat{H}_j(\ell_j,t) = \frac{C_j}{\gamma\Omega_j\left(Z_j^* - Z_j^o\right)^2}\hat{Z}_j(t) + R_j'\frac{\mathrm{d}\hat{Z}_j(t)}{\mathrm{d}t} = \hat{H}_{j+1}(0,t), \quad j=1,2 \\
&\frac{C_j}{\gamma\Omega_j\left(Z_j^* - Z_j^o\right)^2} = \frac{H^* + H}{Z_j^* - Z_j^o}, \quad j=1,2 \\
&\hat{H}_1(0,t) = 0, \quad \hat{Q}_3(\ell_3,t) = \frac{\alpha}{2\sqrt{H}}\hat{H}_3(\ell_3,t) + A\sqrt{H}\mathrm{e}^{\mathrm{i}\omega t}
\end{aligned}
\tag{4.92}
$$

are obtained.

For the case of a step in the flow-rate, the last above condition becomes

$$Q_3(\ell_3, t) = Q_o e^{i\omega t} \tag{4.93}$$

In the following we shall treat the general case, with boundary conditions of the type (4.88).

The harmonic response will be analysed in the form

$$\begin{aligned} \hat{H}_j(x,t) &= e^{i\omega t}\tilde{H}_j(x), \quad \hat{Q}_j(x,t) = e^{i\omega t}\tilde{Q}_j(x), \quad j = 1,2,3 \\ \hat{Z}_j(t) &= e^{i\omega t}\tilde{Z}_j, \quad j = 1,2 \end{aligned} \tag{4.94}$$

For this response the equations

$$\begin{aligned} & i\omega\tilde{H}_j(x) + a_j c_j^2 \tilde{Q}_j'(x) = 0 \\ & \tilde{H}_j'(x) + i\omega a_j \tilde{Q}_j(x) = 0, \quad j = 1,2,3 \end{aligned} \tag{4.95}$$

follow, yielding

$$\begin{aligned} \tilde{H}_j(x) &= \alpha_j e^{\frac{i\omega x}{c_j}} + \beta_j e^{-\frac{i\omega x}{c_j}} \\ \tilde{Q}_j(x) &= \frac{1}{a_j c_j}\left[\beta_j e^{-\frac{i\omega x}{c_j}} - \alpha_j e^{\frac{i\omega x}{c_j}}\right], \quad j = 1,2,3 \end{aligned} \tag{4.96}$$

and from the boundary conditions one obtains

$$\begin{aligned} \tilde{H}_1(0) &= 0, \quad i\omega\Omega_j\tilde{Z}_j + \tilde{Q}_j(\ell_j) = \tilde{Q}_{j+1}(0), \quad j = 1,2 \\ \tilde{H}_j(\ell_j) &= \frac{H^* + H}{Z_j^* \quad Z_j^o}\tilde{Z}_j + i\omega R_j'\tilde{Z}_j = \tilde{H}_{j+1}(0), \quad j = 1,2 \\ \tilde{Q}_3(\ell_3) &= \frac{\alpha}{2\sqrt{H}}\tilde{H}_3(\ell_3) + A\sqrt{H} \end{aligned} \tag{4.97}$$

The dangerous frequencies are those for which function $|\tilde{H}_3(\ell_3)|$ takes the maximum value. This means that in the analysis of resonance phenomena, the pressure amplitude in the valve section was considered the most significant.

Computation of resonance frequencies. Denoting

$$\rho_j(\omega) = \frac{\Omega_j\omega}{\dfrac{H^* + H}{Z_j^* - Z_j^o} + i\omega R_j'}$$

$$S_1(\omega) = \frac{1}{a_3c_3}\sin\frac{\omega\ell_2}{c_2}\cos\frac{\omega\ell_3}{c_3} + \frac{1}{a_2c_2}\cos\frac{\omega\ell_2}{c_2}\sin\frac{\omega\ell_3}{c_3}$$

$$S_3(\omega) = \frac{1}{a_1c_1}\cos\frac{\omega\ell_1}{c_1}\sin\frac{\omega\ell_2}{c_2} + \frac{1}{a_2c_2}\sin\frac{\omega\ell_1}{c_1}\cos\frac{\omega\ell_2}{c_2}$$

$$C_1(\omega) = \frac{1}{a_2c_2}\cos\frac{\omega\ell_2}{c_2}\cos\frac{\omega\ell_3}{c_3} - \frac{1}{a_3c_3}\sin\frac{\omega\ell_2}{c_2}\sin\frac{\omega\ell_3}{c_3}$$

$$C_3(\omega) = \frac{1}{a_1c_1}\cos\frac{\omega\ell_1}{c_1}\cos\frac{\omega\ell_2}{c_2} - \frac{1}{a_2c_2}\sin\frac{\omega\ell_2}{c_2}\cos\frac{\omega\ell_1}{c_1} \tag{4.98}$$

$$S(\omega) = \frac{1}{a_3c_3}S_3(\omega)\cos\frac{\omega\ell_3}{c_3} + \frac{1}{a_2c_2}C_3(\omega)\sin\frac{\omega\ell_3}{c_3}$$

$$C(\omega) = \frac{1}{a_2c_2}C_3(\omega)\cos\frac{\omega\ell_3}{c_3} - \frac{1}{a_3c_3}S_3(\omega)\sin\frac{\omega\ell_3}{c_3}$$

$$R(\omega) = \rho_1(\omega)\rho_2(\omega)\sin\frac{\omega\ell_1}{c_1}\sin\frac{\omega\ell_2}{c_2}\sin\frac{\omega\ell_3}{c_3}$$

and performing the computations in the same manner as for the basic hydraulic scheme, one obtains

$$\frac{1}{\left|\tilde{H}_3(\ell_3)\right|} = \left|\frac{\alpha}{AH} + \frac{1}{\mathrm{i}A\sqrt{H}a_3c_3}\right.$$
$$\left.\times\frac{C(\omega) + \rho_1(\omega)C_1(\omega)\sin\dfrac{\omega\ell_1}{c_1} + \rho_2(\omega)S_3(\omega)\cos\dfrac{\omega\ell_3}{c_3} + R(\omega)}{S(\omega) + \rho_1(\omega)S_1(\omega)\sin\dfrac{\omega\ell_1}{c_1} + \rho_2(\omega)S_3(\omega)\sin\dfrac{\omega\ell_3}{c_3} + R(\omega)}\right| \tag{4.99}$$

while in the case of step perturbation in the flow rate one assumes $\alpha = 0$, $A\sqrt{H} = 0$ and the previous relation becomes

$$\frac{1}{a_3c_3Q_o}\left|\tilde{H}_3(\ell_3)\right|$$
$$= \left|\frac{S(\omega) + \rho_1(\omega)S_1(\omega)\sin\dfrac{\omega\ell_1}{c_1} + \rho_2(\omega)S_3(\omega)\sin\dfrac{\omega\ell_3}{c_3} + R(\omega)}{C(\omega) + \rho_1(\omega)C_1(\omega)\sin\dfrac{\omega\ell_1}{c_1} + \rho_2(\omega)S_3(\omega)\cos\dfrac{\omega\ell_3}{c_3} + R(\omega)}\right| \tag{4.100}$$

For a pumping station provided with air chamber and surge tank (Fig. 4.15II), the computations show that relations (4.99) and (4.100) remain unchanged, but the expressions for $\rho_1(\omega)$ and $\rho_2(\omega)$ become

$$\rho_1(\omega) = \frac{F\omega}{1 + \mathrm{i}\omega R_i'}, \quad \rho_2(\omega) = \frac{\Omega\omega}{\dfrac{H^* + H}{Z_2^* - Z_2^o} + \mathrm{i}\omega R_2'} \tag{4.101}$$

Applications. To illustrate the practical applicability of the method to the study of hydraulic resonance that may occur in the operation of pumping stations demonstration computations have been performed for a pumping station equipped with two air chambers, in several protection alternatives.

In the following table some of the main technical characteristics are given of the pumping station SRPA 1-6 belonging to the Sinoe system that makes the object of our analysis:

$$\begin{aligned}
\ell &= 2{,}313\,\text{m}, & d &= 1.6\,\text{m}, & \ell_3 &= 20\,\text{m},\\
\ell_1 &= 1{,}313\,\text{m}, & \ell_2 &= 980\,\text{m}, & R_1 &= 2Q_R^2,\\
V_{t1} &= 30\,\text{m}^3, & V_{01} &= 5\,\text{m}^3, & R_2 &= 3.9Q_R^2.\\
V_{t2} &= 60\,\text{m}^3, & V_{02} &= 30\,\text{m}^3, & &\\
H_G &= 64.5\,\text{m}, & Q_o &= 4.5\,\text{m}^3/\text{s}, & &
\end{aligned}$$

In order to gain a general view on the behaviour of such a hydraulic system to resonance, several location alternatives for the two air chambers along the discharge pipe were investigated

(1) $\ell_1 = 1{,}313\,\text{m}$, $\ell_2 = 200\,\text{m}$, $\ell_3 = 800\,\text{m}$,
(2) $\ell_1 = 1{,}313\,\text{m}$, $\ell_2 = 400\,\text{m}$, $\ell_3 = 600\,\text{m}$,
(3) $\ell_1 = 1{,}313\,\text{m}$, $\ell_2 = 800\,\text{m}$, $\ell_3 = 200\,\text{m}$,

as well as several variants of head-losses values in the connection areas of the air chambers

(a) $R_1 = 60Q_R$, $R_2 = 78Q_R$,
(b) $R_1 = 60Q_R$, $R_2 = 780Q_R$,
(c) $R_1 = 60Q_R$, $R_2 = 1{,}560Q_R$.

In the following, the procedure and the computation steps are identical to those previously used. Thus, to point out the dangerous frequencies, when a small perturbation in the flow rate appears at the station, the function $|H_3(\ell_3)|$ was computed from relation (4.100) and plotted for the adopted alternatives (Figs. 4.16–4.18). The dangerous ranges of the excitation frequencies were indicated for all these alternatives.

Assuming the appearance and development of hydraulic resonance phenomena, from a practical view point, it is important to know the maximum pressures that may occur in the system under such conditions.

To simulate the behaviour of the real hydraulic system under resonance conditions, the non-linear system of eqns. (4.87) was numerically integrated with quadratic head-losses, with realistic boundary conditions of the type (4.88)–(4.89). Part of the numerical computation results have been graphically processed and presented in Figs. 4.16–4.21.

4.5. The sensitivity to resonance of hydroelectric power plants and pumping stations

We agree with Jaeger's remark [9] that the resonance that may appear in hydraulic pressure systems is one of the problems to be taken into account by the modern designers involved in hydraulic engineering. But, while in the present literature devoted to hydraulics much is said about the causes and the prevention of dangerous frequencies, we start from a computation procedure and pay special attention to the effects produced by resonance. The main purpose of our approach of the problem is to plead for a combined strategy of resonance computations coupled with practical surveillance actions of the operation of resonance sensitive hydraulic systems [18]. Within the presented

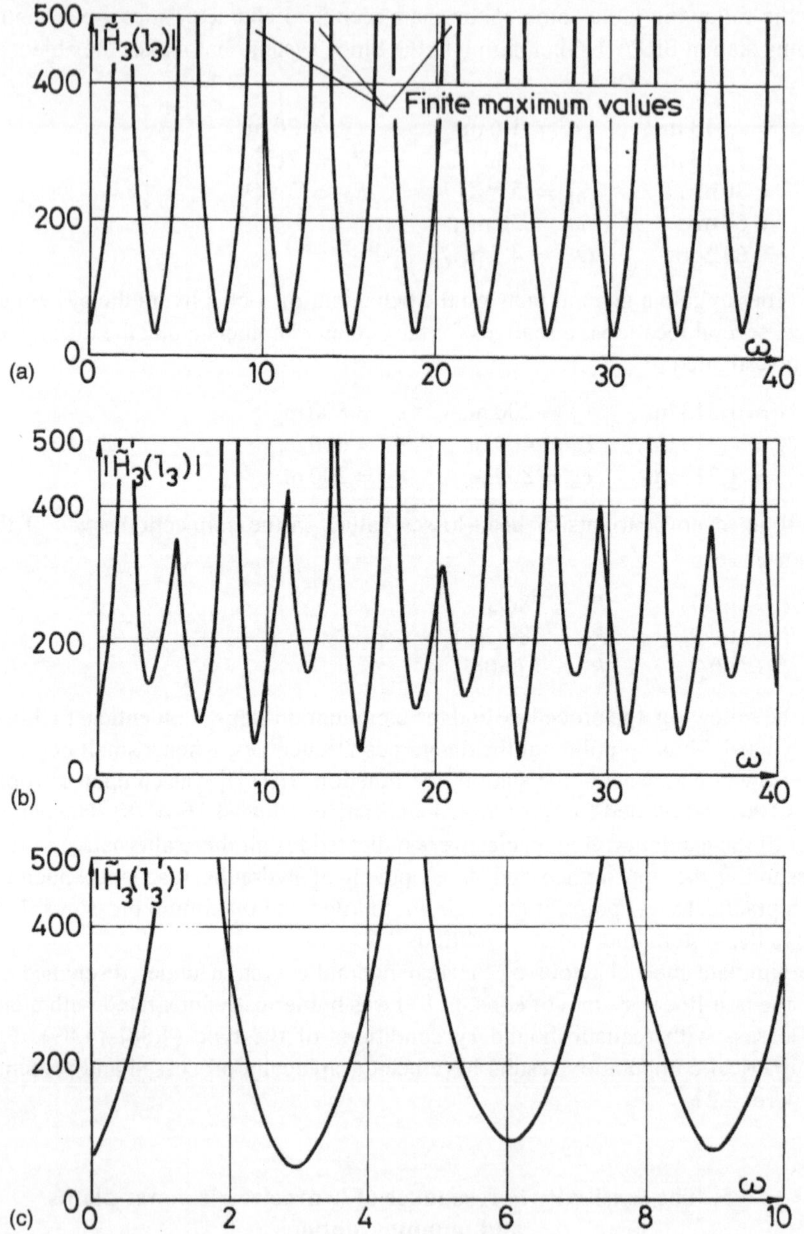

Figure 4.16. $\tilde{H}_3(\ell_3)$ versus ω (variant 1).

methods of computation, the linearised analysis revealed the dangerous frequencies from the point of view of hydraulic resonance. For these frequencies (and also taking into account some harmless frequencies for comparison) a numerical computation was performed for the entire non-linear model of the flow equations.

When analysing the aspect of the plots of hydraulic resonance for the HPP Somes–Mariselu and Bicaz (Figs. 4.3, 4.5) and for the pumping stations Sinoe and

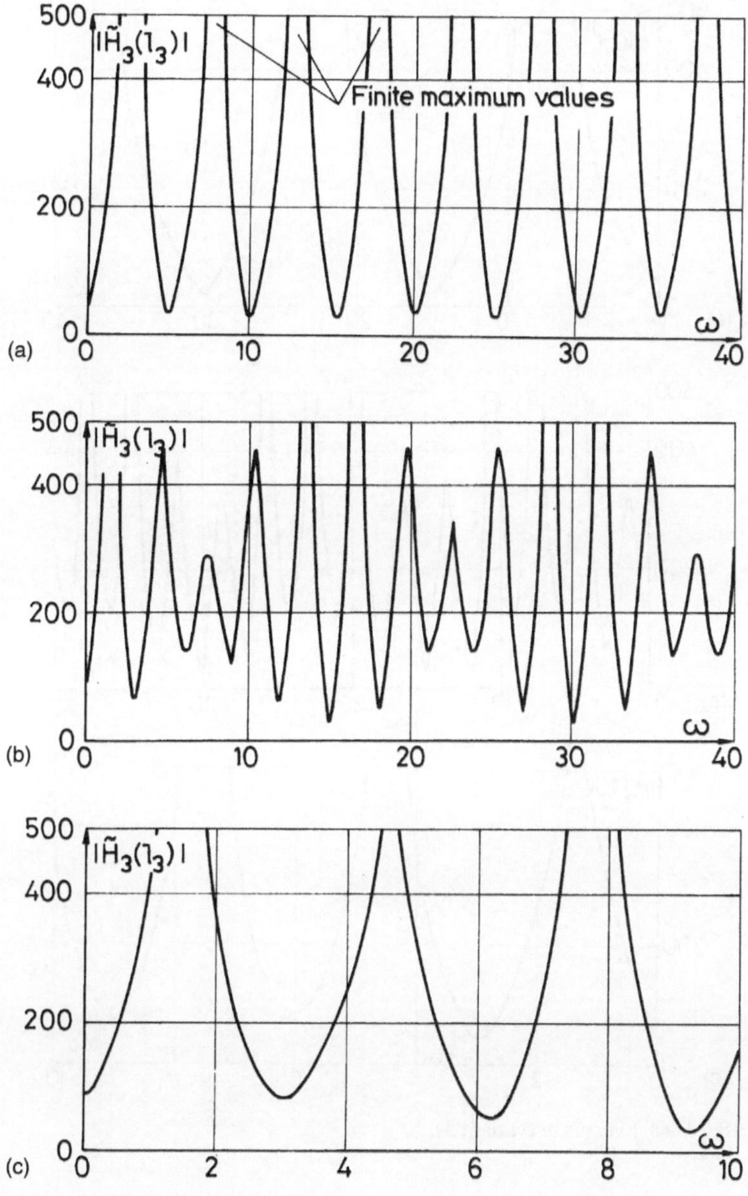

Figure 4.17. $\tilde{H}_3(\ell_3)$ versus ω (variant 2).

Razelm–Tulcea (Figs. 4.12, 4.14), a significant difference in the maximum pressures appearing under resonance conditions can be found. That is why it is natural to measure the *resonance sensitivity* of various hydraulic pressure systems by the ratio between the maximum overpressure produced by resonance and the steady flow pressure

$$S = \frac{\Delta H}{H_o} \tag{4.102}$$

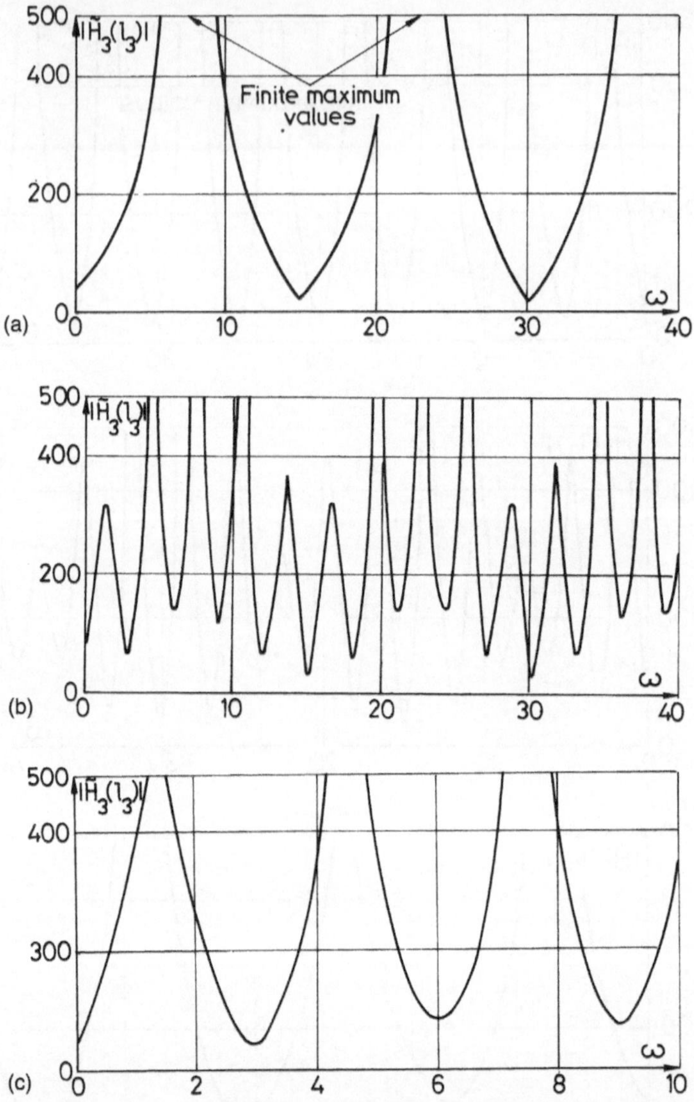

Figure 4.18. $\tilde{H}_3(\ell_3)$ versus ω (variant 3).

In the hydroelectric power plants and pumping stations analysed by us a different range of resonance sensitivity is noticed, starting from $S = 0.5$, $S = 0.8$ and going up to $S = 3.5$.

Practically, the hydraulic resonance in hydraulic pressure systems acts by vibrations of the pressure pipes and hydraulic devices, quite similar to common mechanical vibrations. Therefore, to avoid the development of hydraulic resonance, which can generate accidents in resonance sensitive hydraulic systems, a continuous surveillance of the vibrations in the pressure pipes and in the hydraulic devices of the hydraulic system in operation is necessary.

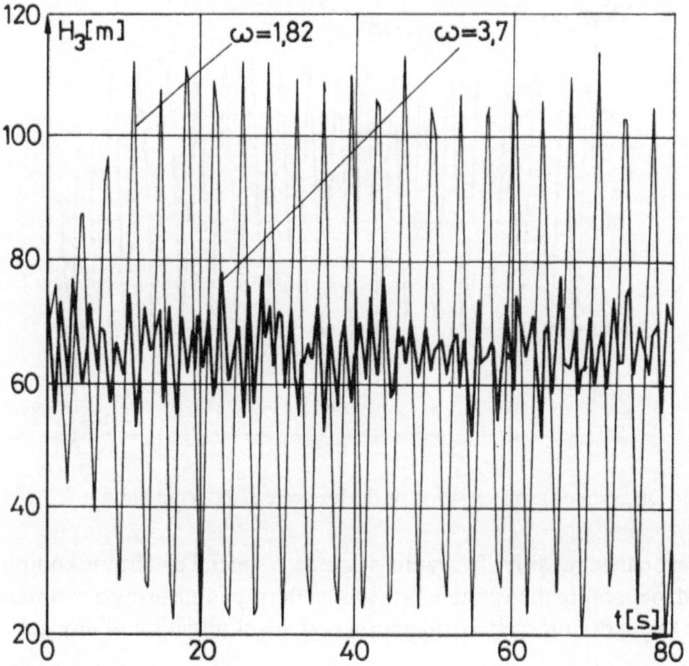

Figure 4.19. Diagram of time variation of the pressure $H_3(\ell_3,t)$ in variant 1a.

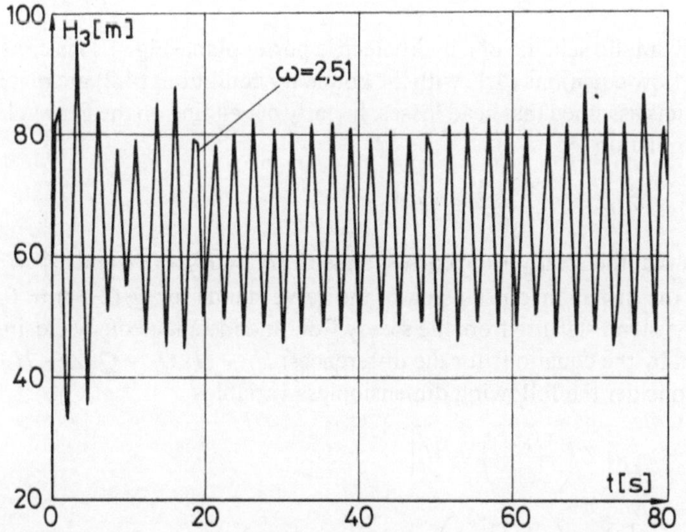

Figure 4.20. Diagram of time variation of the pressure $H_3(\ell_3,t)$ in variant 2a.

In conclusion, the resonance sensitivity exhibited by our analyses suggests the following practical measures:

- for each hydraulic pressure system, it is necessary to perform a special resonance analysis using the previously described procedure in order to determine the sensitivity to resonance;

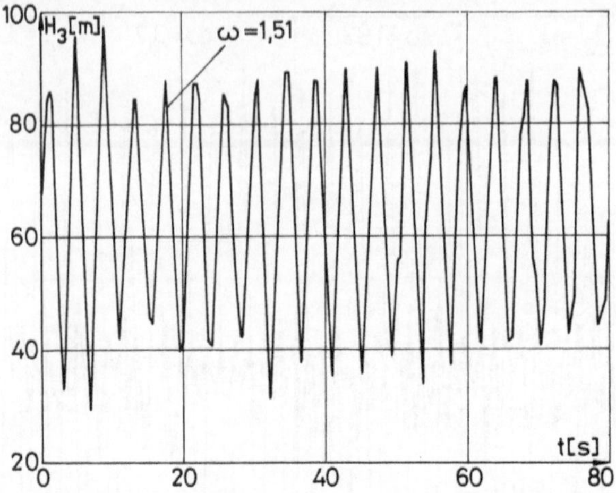

Figure 4.21. Diagram of time variation of the pressure $H_3(\ell_3,t)$ in variant 2c.

- for the resonance sensitive hydraulic systems, a special electronic equipment will be provided to measure the variable pressure in the pipes ensuring a continuous surveillance of the hydraulic system operation and the elimination of any accident risks.

4.6. An arithmetic property in the analysis of the behaviour to resonance of a hydraulic system equipped with a surge tank with strangulation

The basic hydraulic scheme of a hydroelectric power plant (Fig. 4.1) is considered, and the general flow equations (4.1) with the boundary conditions (4.2) are involved, where in addition it is assumed that head losses, linearly depending on the flow velocity, occur in the strangulation [5]

$$H_1(\ell_1, t) = Z(t) + R'_c \frac{\mathrm{d}Z(t)}{\mathrm{d}t} = H_2(0, t) \tag{4.103}$$

It is known that the steady flow corresponds to $H_1^\circ = H_2^\circ = H$ and $Q_1^\circ = Q_2^\circ = Q = \alpha\sqrt{H}$, and the instantaneous closing of the valve stands for $Q_2(\ell_2, t) = 0$. The water flow in the system, starting from the steady flow and instantaneously closing the valve, is described by the equations for the differences: $H_j - H, Q_j - Q, Z - H$.

Let us consider the following dimensionless variables

$$\begin{aligned}
y(\tau) &= \frac{1}{H}\left[Z\left(\frac{\tau f_1 \ell_1}{Q}\right) - H\right] \\
h_j(\xi, \tau) &= \frac{1}{H}\left[H_j\left(\ell_j \xi \frac{\tau f_1 \ell_1}{Q}\right) - H\right] \\
q_j(\xi, \tau) &= \frac{1}{Q}\left[Q_j\left(\ell_j \xi \frac{\tau f_1 \ell_1}{Q}\right) - Q\right], \quad j = 1,2
\end{aligned} \tag{4.104}$$

Thus, we obtain

$$\frac{\partial h_1}{\partial \tau}(\xi, \tau) + \frac{c_1^2}{gH}\frac{\partial q_1}{\partial \xi}(\xi, \tau) = 0$$

$$\frac{\partial h_1}{\partial \xi}(\xi,\tau) + \frac{Q^2}{gHf_1^2}\frac{\partial q_1}{\partial \xi}(\xi,\tau) = 0$$

$$\frac{\partial h_2}{\partial \tau}(\xi,\tau) + \frac{\ell_1 f_1}{\ell_2 f_2}\frac{c_2^2}{gH}\frac{\partial q_2}{\partial \xi}(\xi,\tau) = 0$$

$$\frac{\partial h_2}{\partial \xi}(\xi,\tau) + \frac{\ell_2}{\ell_1}\frac{Q^2}{gHf_1f_2}\frac{\partial q_2}{\partial \xi}(\xi,\tau) = 0$$

$$h_1(0,\tau) = 0 \tag{4.105}$$

$$y'(\tau) + \frac{f_1\ell_1}{FH}q_2(0,\tau) = \frac{f_1\ell_1}{FH}q_1(1,\tau)$$

$$h_1(1,\tau) = h_2(0,\tau) = y(\tau) + \frac{QR'_c}{f_1\ell_1}y'(\tau)$$

$$q_2(1,\tau) = -1$$

$$h_1(\xi,0) = h_2(\xi,0) = y(0) = q_1(\xi,0) = q_2(\xi,0) = 0$$

Making the change of variables

$$\begin{aligned} u_j &= \frac{1}{2}\left[q_j + \frac{gHf_j}{c_jQ}h_j\right] \\ v_j &= \frac{1}{2}\left[\frac{gHf_j}{c_jQ}h_j - q_j\right], \quad j = 1,2 \end{aligned} \tag{4.106}$$

we arrive at the system of equations

$$\begin{aligned} y'(\tau) &= \frac{\dfrac{gf_1\ell_1}{QF}\left(\dfrac{f_1}{c_1} + \dfrac{f_2}{f_1}\right)}{1 + \dfrac{gR'_c}{F}\left(\dfrac{f_1}{c_1} + \dfrac{f_2}{f_1}\right)}y(\tau) + \frac{\dfrac{2f_1\ell_1}{HF}}{1 + \dfrac{gR'_c}{F}\left(\dfrac{f_1}{c_1} + \dfrac{f_2}{c_2}\right)}[u(\tau - \varepsilon) + 1] \\ &\quad - \frac{\dfrac{2f_1\ell_1}{HF}}{1 + \dfrac{gR'_c}{F}\left(\dfrac{f_1}{c_1} + \dfrac{f_2}{c_2}\right)}v(\tau - \upsilon\varepsilon) \\ u(\tau) &= \frac{\dfrac{gHf_2}{c_2Q}}{1 + \dfrac{gR'_c}{F}\left(\dfrac{f_1}{c_1} + \dfrac{f_2}{c_2}\right)}y(\tau) - \frac{1 + \dfrac{gR'_c}{F}\left(\dfrac{f_1}{c_1} - \dfrac{f_2}{c_2}\right)}{1 + \dfrac{gR'_c}{F}\left(\dfrac{f_1}{c_1} + \dfrac{f_2}{c_2}\right)}[u(\tau - \varepsilon) + 1] \\ &\quad - \frac{\dfrac{2gR'_cf_2}{Fc_2}}{1 + \dfrac{gR'_c}{F}\left(\dfrac{f_1}{c_1} + \dfrac{f_2}{c_2}\right)}v(\tau - \upsilon\varepsilon) \end{aligned} \tag{4.107}$$

$$v(\tau) = \frac{\dfrac{gHf_2}{c_2 Q}}{1+\dfrac{gR'_c}{F}\left(\dfrac{f_1}{c_1}+\dfrac{f_2}{c_2}\right)} y(\tau) + \frac{\dfrac{2gR'_c f_1}{Fc_1}}{1+\dfrac{gR'_c}{F}\left(\dfrac{f_1}{c_1}+\dfrac{f_2}{c_2}\right)} [u(\tau-\varepsilon)+1]$$

$$+ \frac{1+\dfrac{gR'_c}{F}\left(\dfrac{f_2}{c_2}-\dfrac{f_1}{c_1}\right)}{1+\dfrac{gR'_c}{F}\left(\dfrac{f_1}{c_1}+\dfrac{f_2}{c_2}\right)} \nu(\tau-\upsilon\varepsilon)$$

where

$$\upsilon = \frac{\dfrac{\ell_1}{c_1}}{\dfrac{\ell_2}{c_2}}, \quad \varepsilon = \frac{2Q\ell_2}{c_2 f_1 \ell_1} \tag{4.108}$$

$$u(\tau) = u_2(0,\tau), \quad v(\tau) = v_1(1,\tau)$$

and the system is valid for $\tau \geq \varepsilon\upsilon$.

In the last equation, it is considered that

$$\lim_{\varepsilon\to 0} \frac{v(\tau)-\nu(\tau-\nu\varepsilon)}{\nu\varepsilon} \tag{4.109}$$

and when u (for $\varepsilon = 0$) is eliminated, the equation of slowly varied flow of the water in the surge tank is detached from the rest:

$$y''(\tau) + \frac{gR'_c f_1^2}{QF} y'(\tau) + \frac{g\ell_1 f_1^3}{Q^2 F} y(\tau) = 0 \tag{4.110}$$

The rapidly varying flow of the water in the pipe is described by the system of difference equations

$$\begin{aligned} u(\tau) &= \varepsilon A_{10} y(\tau) - (1-\varepsilon B_{11})[u(\tau-\varepsilon)+1] - \varepsilon B_{11} v(\tau-\nu\varepsilon) \\ v(\tau) &= \varepsilon A_{20} y(\tau) + \nu\varepsilon B_{22}[u(\tau-\varepsilon)+1] + (1-\nu\varepsilon B_{22})\, v(\tau-\nu\varepsilon) \end{aligned} \tag{4.111}$$

where the coefficients are positive and can be expressed by means of the basic constants of the hydraulic system.

We should like to underline an interesting property of this system of difference equations. For $w_1(\tau) = u(\varepsilon\tau)$, $w_o(\tau) = v(\varepsilon\tau)$ the linear difference system is

$$\begin{aligned} w_1(\tau) &= -(1-\varepsilon B_{11})\, w_1(\tau-1) - \varepsilon B_{11} w_2(\tau-\upsilon) \\ w_2(\tau) &= -\nu\varepsilon B_{22} w_1(\tau-1) + (1-\nu\varepsilon B_{22}) w_2(\tau-\upsilon) \end{aligned} \tag{4.112}$$

By the general theorem of J.K. Hale [6], this system admits a strong stability if and only if the spectral radius of the matrix

$$\begin{bmatrix} -(1-\varepsilon B_{11})\, e^{i\theta_1} & -\varepsilon B_{11} e^{i\theta_2} \\ \nu\varepsilon B_{22} e^{i\theta_1} & (1-\nu\varepsilon B_{22})\, e^{i\theta_2} \end{bmatrix} \tag{4.113}$$

is strictly smaller than 1 for every $\theta_1 \in [0, 2\pi]$ and $\theta_2 \in [0, 2\pi]$. But, for $\theta_1 = 0$, $\theta_2 = \pi$ the matrix assumes the value -1 and we have strict stability.

On the other hand, we shall exhibit that if $\nu = p/q$, the fraction being irreducible, then for p odd we have asymptotic stability with all its consequences, while for p even we have an oscillatory mode. Indeed, if $\nu = p/q$ the characteristic equation can be written as

$$\mu^{\frac{p+q}{q}} - a\mu^{\frac{p}{q}} + b\mu + 1 - b + a = 0$$

$$b = \frac{1 + \frac{gR'_c}{F}\left(\frac{f_2}{f_1} - \frac{f_1}{c_1}\right)}{1 + \frac{gR'_c}{F}\left(\frac{f_1}{c_1} + \frac{f_2}{c_2}\right)} \tag{4.114}$$

$$1 + a = \frac{\frac{2gR'_c}{F}\frac{f_2}{c_2}}{1 + \frac{gR'_c}{F}\left(\frac{f_1}{c_1} + \frac{f_2}{c_2}\right)} > 0$$

$$1 - b > 0, \quad 1 + b > 0, \quad b - a > 0$$

Then, an elementary computation shows that all the roots of this equation satisfy the stability criterion $|\mu| \leq 1$ and the only root with $|\mu| = 1$ is $\mu = -1$. If $\mu = -1$ is a root, then p is necessarily even. So, if p is odd, we have $|\mu| < 1$, that is asymptotic stability, while if p is even and υ irrational there is no asymptotic stability.

This system of difference equations is an example for the theory described in J.K.Hale's book [6] and, at the same time, it shows the role played by the arithmetic properties of the ratio $\nu = (\ell_1/c_1)/(\ell_2/c_2)$ (of the flow duration in the two pipes) in the behaviour of the hydraulic system. The computation shows

$$y(\tau) = \frac{f_1\ell_1}{\sigma II Fe^{-\rho\tau}}$$

$$\rho = \frac{gR'_c f_1^2}{2QF}, \quad \sigma = \frac{gf_1^2}{QF}\sqrt{\frac{\ell_1 F}{gf_1} - \frac{(R'_c)^2}{4}} \tag{4.115}$$

$$H_2(0,t) - H = \frac{c_2 Q}{gf_2}\left\{u_o\left(\frac{tQ}{f_1\ell_1}\right) - u_o\left(\frac{tQ}{f_1\ell_1} - \frac{2Q\ell_2}{c_2 f_1 \ell_1}\right)\right.$$
$$\left. + |G(r)|\left[y\left(\frac{tQ}{f_1\ell_1}\right) + y\left(\frac{tQ}{f_1\ell_1} - \frac{2Q\ell_2}{c_2 f_1\ell_1}\right)\right]\right\}$$

where $1 = -\rho + i\sigma$; G is a "transmition function" and u_o is defined by the system of difference equations whose stability was studied. In conclusion, we consider that the appearance of arithmetic conditions in the behaviour of this model is a phenomenon that deserves our attention.

References

1. Chaudhry, H.: Resonance in Pipe Systems. *Water Power and Dam Construction*, 7 (1970).
2. Fanelli, M.: Les Phénomènes de Résonance Hydraulique. *La Houille Blanche*, 4 (1975).

3. Guarga, R., Hiriart, G., Torres, J.: Oscillatory Problems at Mexico's La Angostura Plant. *Water Power and Dam Construction*, 10 (1983).
4. Halanay, A., Popescu, M.: Computation of Hydraulic Resonance for Hydroelectric Plants in the Case of Complex Schemes. *Revue Roumaine des Sciences Techniques-Méchanique Appliquée*, 3 (1982) 325–345.
5. Halanay, A., Popescu, M.: Une Proprietée Arithmétique dans l'Analyse du Comportement d'un Système Hydraulique Comprenant une Chambre d'Équilibre avec Étranglement. *C.R. Acad. Sci. Paris*, 305 (1987) 1227–1330.
6. Hale, J.K.: *Theory of Functional–Differential Equations*. Springer, Berlin, 1977, Ch. 12, sections 123–127.
7. Hudovernik, W., Lein, G.: Frequency Response Investigations in Highhead Plants. *Water Power and Dam Construction*, 8 (1970).
8. Jaeger, C.: Theory of Resonance in Hydropower Systems. *Water Power and Dam Construction*, 1–3 (1963).
9. Jaeger, C.: *Hydraulic Techniques*. Dunord, Paris, 1954.
10. McCain, I., Gibson, W.: *Some Measurements of Auto-Oscillation Initiated by Valve Characteristics*. IAHR Congress, London, 1963.
11. Plichon, J., Lescure, A.: Instabilités Hydrauliques dans les Centrales. *La Houille Blanche*. 2–3 (1979).
12. Popescu, M.: Modèle Mathématique pour l'Etude de la Stabilité d'un Système de Deux Chambres d'Equilibre. *C.R. Acad. Sci. Paris*, 273 (1971) 129–132.
13. Popescu, M., Alexander, F., Schneider, E.: A New Concept in Practical Tackling of the Problem of Hydraulic Resonance in Hydropower Plant. *Conference "Waterpower '99"*, Las Vegas, Nevada, U.S.A., 6–9 July 1999.
14. Popescu, M., Halanay, A.: A Computing Technique for Hydraulic Resonance in Hydropower Plants with Surge Tanks. *Revue Roumaine des Sciences Techniques-Méchanique Appliquée*. 3 (1981), 421–431.
15. Popescu, M.: Calculul Rezonantei Hidraulice la Uzine Hidroelectrice (The Computation of Hydraulic Resonance for Hydroelectric Power Plants). *Studii si Cercetari de Mecanica Aplicata*, 1 (1982) 23–41, 2 (1982) 233–256 (in Romanian).
16. Popescu, M., Halanay, A.: Le Calcul de la Résonance Hydraulique dans les Usines Hydroélectriques. *Proceedings of the XX IAHR Congress*, Vol. 6, Moscow, 1983, pp. 383–393.
17. Popescu, M., Halanay, A.: Le Calcul de la Résonance Hydraulique aux Stations de Pompage. *Revue Roumaine des Sciences Techniques-Méchanique Appliquée*, 6 (1984) 607–619.
18. Popescu, M., Halanay, A.: Resonance Sensitivity Hydropower and Pumping Stations. *Water Power and Dam Construction*, 9 (1984).
19. Popescu, M., Halanay, A.: Le Calcul de la Résonance Hydraulique aux Stations de Pompage. *XXème Journée de l'Hydraulique*, Question III, Rapport 5, Lyon, France, 4–6 Avril 1989.
20. Streeter, V., Wylie, E.: *Hydraulic Transients*. McGraw Hill, New York, 1967.
21. Suo, L., Wylie, E.: Hydraulic transients in Rock–Bored Tunnels. *Journal of Hydraulic Engineering*, 116 (2) (1990).

Chapter 5

Mass Oscillation in Hydraulic Surge Systems

Consider the scheme of a diversion hydroelectric power plant consisting of a reservoir pressure tunnel, surge tank, penstock, turbines and outflow canal (Fig. 5.1). Due to valve manoeuvres in the turbines (flow variations), the waterhammer appears in the penstock only, while in the pressure tunnel and in the surge tank the water flow is slowly varied.

The differential equations of the water motion in the gallery surge tank system are the following:

$$\frac{L}{g}\frac{\mathrm{d}W}{\mathrm{d}t} + Z + P'W|W| + R'V_{\mathrm{s}}|V_{\mathrm{s}}| = 0$$

$$Wf = F\frac{\mathrm{d}Z}{\mathrm{d}t} + Q_{\mathrm{T}} \tag{5.1}$$

where L is the pressure tunnel length; W, the water velocity in the tunnel; Z, the water level in the surge tank; P', R' are the longitudinal head loss coefficients in the tunnel and local head losses from the surge tank strangulation, respectively; V_{s} is the water velocity in the surge tank strangulation; f, the cross-sectional area of the gallery; F, the area of the surge tank horizontal section at the water line; Q_{T}, the turbined flow; t, time; g, the gravitational acceleration.

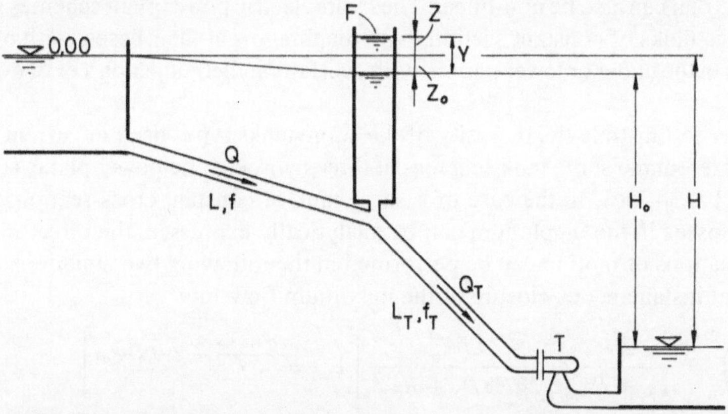

Figure 5.1. Scheme of a hydroelectric power plant with surge tank.

If the pressure tunnel consists of several segments of lengths L_i and cross-sections f_i, computations are performed by using the mean cross-section given by

$$f = \frac{L}{\sum \frac{L_i}{f_i}} \tag{5.2}$$

When deducing eqns. (5.1), the following hypotheses have been assumed:

- the pressure tunnel walls are rigid and the water is incompressible;
- the water inertia in the tank is small with respect to the one from the pressure tunnel and, consequently, can be neglected;
- the velocity distribution in the surge tank strangulation cross-section is uniform;
- in unsteady regime, the water kinetic energy in the tank is negligible;
- the longitudinal and local head losses are expressed in the unsteady regime by the same relations as in the steady one;
- the head losses due to the surge tank strangulation (diaphragm) are independent of the flow-rate distribution between the cross-sections at the tank basis.

To eqns. (5.1) the flow law of the turbine should be added, the form of which in the most general case is:

$$Q_T = f\left(Z, t, \frac{dZ}{dt}, \frac{dQ_T}{dt}\right) \tag{5.3}$$

Consequently, there are three available equations for the three unknown quantities Z, W, Q_T.

In a more general case when, for instance, n surge tanks are located on the pressure tunnel gallery, we have $2n$ equations of the (5.1) type and one equation of the (5.3) type, hence we have $(2n + 1)$ equations for the same number of unknown quantities, namely

- $n\, Z_i$ levels in the surge tanks;
- $n\, W_i$ velocities in the tunnel segments;
- the turbine discharge Q_T.

In the general case, the system of differential eqns. (5.1)–(5.3) cannot be analytically solved due to the non-linearity of some of the equation terms and to the boundary conditions that can also be non-linear. The hydroelectric power plant schemes generally adopt surge tanks of constant section, with diaphragms at their bases, with or without chambers in the upper or lower region, with overfalls at their upper part or of differential type.

In order to illustrate the diversity of the surge tanks types used in current practice, we shall present the surge tank schemes at three hydroelectric power plants (Figs. 5.2–5.4) [15, 17, 34, 36]. In the case of a surge tank of constant cross-section, in which the head losses in the diaphragm can be analytically expressed, the direct integration of the equations of motion can be performed in the following two situations:

(a) total instantaneous closure of the maximum flow-rate

$$\left(\frac{W}{W_o}\right)^2 = \frac{Z}{P_o + R_o} + \frac{LfW_o^2}{2gF\,(P_o + R_o)^2}\left[1 - e^{\frac{2gF(K_1+K_2)f}{L}(Z-P_o-R_o)}\right] \tag{5.4}$$

with $P = K_1 Q^2$; $R = K_2\, Q_c^2$, the 0 index corresponding to the steady regime;

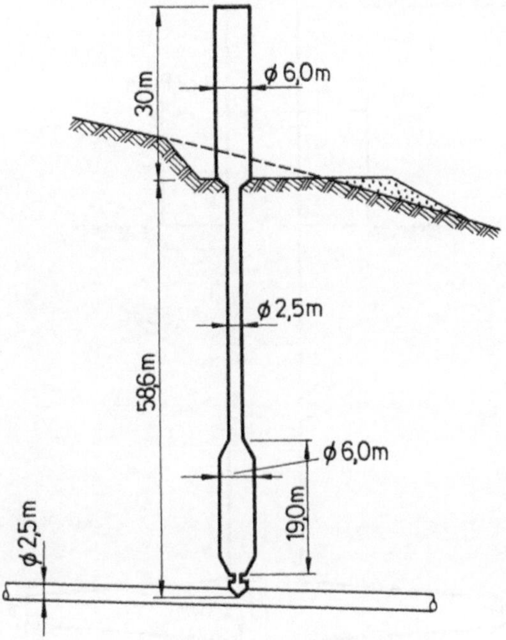

Figure 5.2. Surge tank of the Pancrazio hydroelectric power plant ($N = 35\,\text{MW}$).

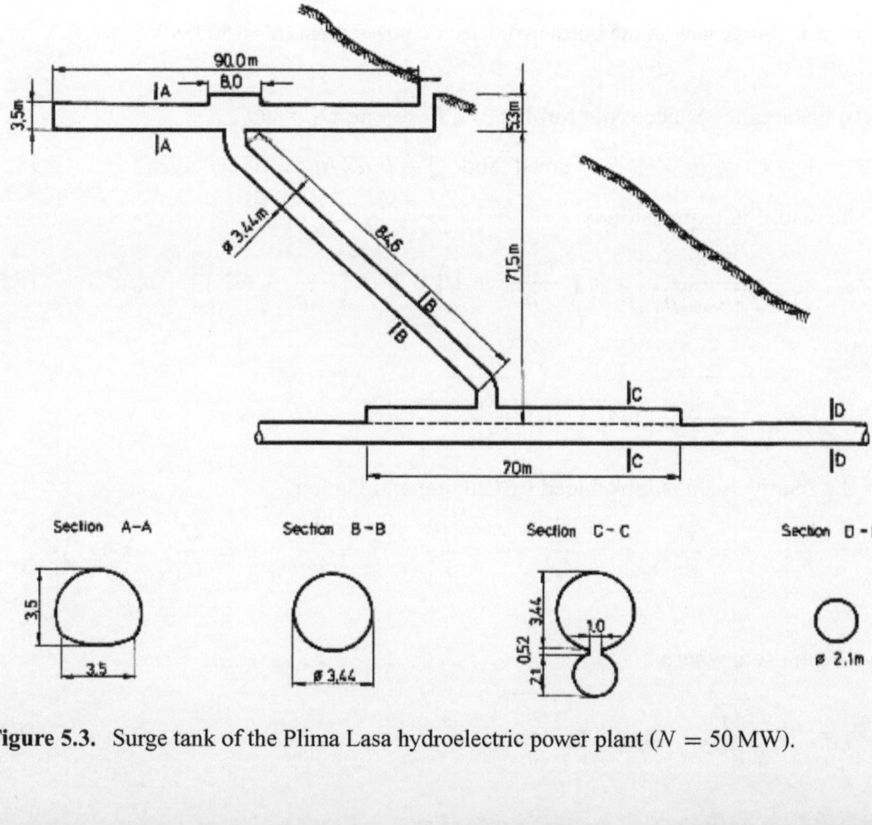

Figure 5.3. Surge tank of the Plima Lasa hydroelectric power plant ($N = 50\,\text{MW}$).

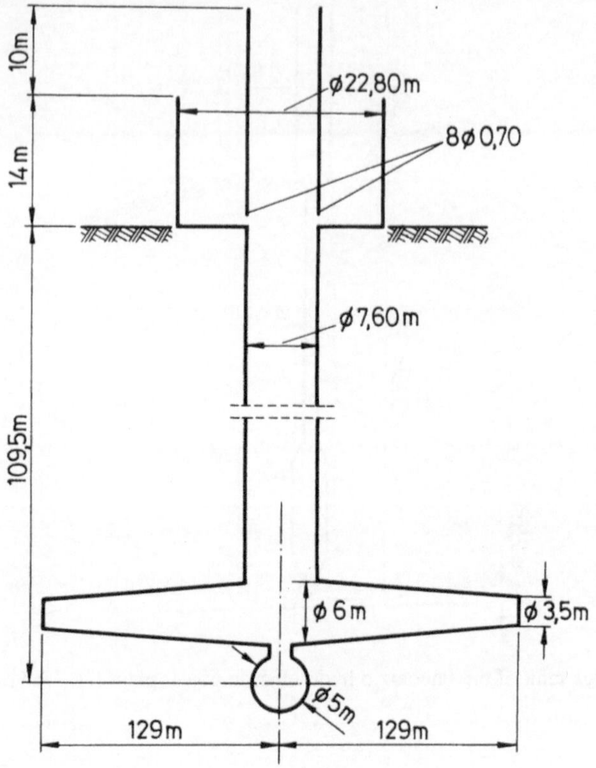

Figure 5.4. Surge tank of the Lotru hydroelectric power plant ($N = 500$ MW).

(b) instantaneous access to turbine of a flow-rate $Q_T = nQ_o$,

$$0 < n \leq 1, \text{ with } Z + R = \text{const and } Q - F\,\mathrm{d}Z/\mathrm{d}t = Q_T = \text{const} \tag{5.5}$$

The resulting expression is:

$$Z_{min} = \frac{1}{2\sqrt{-Z_{min}\,p_o}}\left[a \ln\left(\frac{n-1}{a}+1\right) - b \ln\left(\frac{n-1}{b}+1\right)\right] - p_o\,n^2 \tag{5.6}$$

where

$$a = 1 + \sqrt{\frac{-Z_{min}}{p_o}}; \quad b = 1 - \sqrt{\frac{-Z_{min}}{p_o}}$$

and the relative values introduced by Calame and Gaden

$$z = \frac{Z}{Z_*}; \quad p = \frac{P}{Z_*}; \quad r = \frac{R}{Z_*}; \quad w = \frac{W}{W_o}; \quad v = \frac{V}{V_o}; \quad u = \frac{U}{U_o}; \quad \tau = \frac{t}{T_*} \tag{5.7}$$

have been used, where

$$Z_* = W_o\sqrt{\frac{Lf}{gF}}; \quad T_* = 2\pi\sqrt{\frac{LF}{gf}} \tag{5.8}$$

represent the amplitude and period of the ideal oscillations in the tank (when the head losses are neglected). Relation (5.6) allows the calculation of one of the parameters $Z_{\min}$, p_o and n when two of them are known. Results are systematically presented in graphs [22, 35].

The pressure in the tunnel remains almost constant at the surge tank section during the first lowering of the water level in the surge tank, and formula (5.6) can consequently be used, in the following two cases:

- surge tank with diaphragm producing during the considered operations an initial depression equal to the minimum jump value;
- surge tank with a large cross-section lower chamber as compared to the well cross-section.

5.1. Computational models for mass oscillations

5.1.1. Differential equation of oscillations

In order to obtain the differential equation of oscillations in the gallery surge tank system, one should start from the system (5.1), written in the simplified form [12, 19, 24, 37]

$$\begin{aligned} \frac{L}{g}\frac{\mathrm{d}W}{\mathrm{d}t} + Z + P'W|W| &= 0 \\ Wf &= F\frac{\mathrm{d}Z}{\mathrm{d}t} + Q_{\mathrm{T}} \end{aligned} \tag{5.9}$$

where

$$P' = \left(\lambda\frac{L}{d} + \sum\zeta\right)\frac{1}{2g} \tag{5.10}$$

and from the flow law (5.3) of the form

$$N = \eta\gamma\, Q_{\mathrm{T}}(H + Z) = \eta\gamma\, Q_{\mathrm{o}}H_{\mathrm{o}} = \text{const.} \tag{5.11}$$

corresponding to the automatic operation of the plant at constant power.

Eliminating the variables Q and Q_{T} between eqns. (5.9), (5.11), we get

$$\frac{LF}{gf}\frac{\mathrm{d}^2Z}{\mathrm{d}t^2} + \frac{k}{2g}\left|\frac{\mathrm{d}Z}{\mathrm{d}t} + \frac{N}{\eta\gamma F(H+Z)}\right|\left(\frac{\mathrm{d}Z}{\mathrm{d}t} + \frac{N}{\eta\gamma F(H+Z)}\right)$$
$$- \frac{NL}{\eta g\gamma f(H+Z)^2}\frac{\mathrm{d}Z}{\mathrm{d}t} + Z = 0 \tag{5.12}$$

i.e. a non-linear second-order differential equation, in which the notation

$$k = \left(\lambda\frac{L}{d} + \sum\zeta\right)\frac{F^2}{f^2} \tag{5.13}$$

has been used.

Uniqueness of the solution is ensured by the initial conditions

$$t = 0 \rightarrow Z = Z_o; \quad \left.\frac{dZ}{dt}\right|_o = \frac{f}{F}W_o - \frac{N}{\eta\gamma F(H+Z_o)} \tag{5.14}$$

In the particular case of the total halt of the plant (sudden closing of the valve), the oscillation equation simplifies to the form

$$\frac{LF}{gf}\frac{d^2Z}{dt^2} + \frac{k}{2g}\left|\frac{dZ}{dt}\right|\frac{dZ}{dt} + Z = 0 \tag{5.15}$$

with initial conditions

$$t = 0 \rightarrow Z = Z_o; \quad \frac{dZ}{dt} = \frac{Q_o}{F}$$

5.1.2. Non-dimensional form of the oscillation equation

In the specialized literature several methods of non-dimensionalisation of oscillation equation are known. Among the best known are those proposed by Vogt, Colame and Gaden, Gardel, Morozov [12, 22].

Based on an optimisation study performed by the authors [2] on the non-dimensional form of the oscillation equation, the following relative values are suggested for the equation variables:

$$\zeta = \frac{Z}{Z_*}, \quad \tau = \frac{t}{T_*} \tag{5.16}$$

where

$$Z_* = \frac{Lf}{kF} \quad \text{and} \quad T_* = \sqrt{\frac{LF}{gf}} \tag{5.17}$$

Using these relative values, the oscillation eqn. (5.15) in the case of a total halt of the plant, takes a very simple form:

$$\frac{d^2\zeta}{d\tau^2} + \frac{1}{2}\left|\frac{d\zeta}{d\tau}\right|\frac{d\zeta}{d\tau} + \zeta = 0 \tag{5.18}$$

while the initial conditions ensuring the uniqueness of the solution become

$$\begin{aligned} &\zeta|_{\tau=0} = \zeta_o = \frac{Z_o}{Z_*} = -\frac{1}{2}\left(kW_o\sqrt{\frac{f}{gLF}}\right)^2 < 0 \\ &\left.\frac{d\zeta}{d\tau}\right|_{\tau=0} = \frac{T_*}{Z_*}\left.\frac{dZ}{dt}\right|_{t=0} = kW_o\sqrt{\frac{f}{gLF}} = \sqrt{-2\zeta_o} \end{aligned} \tag{5.19}$$

Note that these initial conditions are written as a function of a single parameter ζ_o called the *initial head loss parameter*.

Coming back to the oscillation equation in the general form (5.12), the use of the relative values (5.16) yields the following non-dimensional form:

$$\frac{d^2\zeta}{d\tau^2} + \frac{1}{2}\left|\frac{d\zeta}{d\tau} + \frac{\beta}{1+\alpha\zeta}\right|\left(\frac{d\zeta}{d\tau} + \frac{\beta}{1+\alpha\zeta}\right) - \frac{\alpha\beta}{(1+\alpha\zeta)^2}\frac{d\zeta}{d\tau} + \zeta = 0 \tag{5.20}$$

where the following additional non-dimensional parameters have been used:

$$\alpha = \frac{Z_*}{H} = \frac{FL}{kfH} < 1$$
$$\beta = \frac{Q_o}{F}\frac{T_*}{Z_*} = \frac{kQ_o}{F}\sqrt{\frac{f}{gFL}} \tag{5.21}$$

In the case of the plant start-up (sudden opening of the valve), the initial conditions to solve eqn. (5.20) take the form

$$\tau = 0 \rightarrow \zeta_o = 0; \qquad \left.\frac{d\zeta}{d\tau}\right|_o = -\frac{Q_o}{F}\frac{T_*}{Z_*} = -\beta \tag{5.22}$$

5.1.3. Integral variational formulation

This formulation represents a relatively new approach of the technical problems. The method relies on the equivalence which under some restrictive conditions can be established between a differential equation and the stationarity condition of an integral [9, 14]. The main advantage of the method consists in the opportunity it gives to admit several assumptions on the global evolution of the phenomenon, necessary to apply specific numerical techniques, e.g. the finite element method.

The study of the oscillatory flow in the gallery surge tank system, starts from Hamilton's principle [32], considering the following expressions:

- *kinetic energy of the system*

$$E_c = \frac{\rho W^2}{2}Lf + C_1 = \frac{\rho f L}{2f^2}\left(F\frac{dZ}{dt} + Q_T\right)^2 + C_1$$

- *potential energy of the system*

$$E_p = -\frac{\rho g F^2}{2}Z^2 + C_2$$

- *energy dissipated in the system*

$$E_d = -\int_{Z_o}^{Z} k\rho F\left|\frac{dZ}{dt} + \frac{Q_o H_o}{F(H+Z)}\right|\left(\frac{dZ}{dt} + \frac{Q_o H_o}{F(H+Z)}\right) dZ + C_3$$

where C_1, C_2, C_3 are arbitrary constants. Eventually, one arrives to the following non-dimensional variational formula [5]

$$\delta \int_{\tau_o}^{\tau} \frac{1}{2}\left\{\dot{\zeta}^2 - \zeta^2 + 2\int_{\dot{\zeta}_o}^{\dot{\zeta}} \frac{\beta}{1+\alpha\zeta} d\dot{\zeta} - \int_{\zeta_o}^{\zeta}\left|\dot{\zeta} + \frac{\beta}{1+\alpha\zeta}\right|\left(\dot{\zeta} + \frac{\beta}{1+\alpha\zeta}\right) d\zeta\right\} d\tau = 0 \tag{5.23}$$

where δ is the variation symbol specific to variational calculus.

In the particular case of the plant shutdown (total valve closing), the β parameter vanishes and eqn. (5.23) takes the simpler form

$$\delta \int_{\tau_o}^{\tau} \frac{1}{2}\left[\dot{\zeta}^2 - \zeta^2 - \int_{\zeta_o}^{\zeta} |\dot{\zeta}|\, \zeta \mathrm{d}\zeta\right] \mathrm{d}\tau = 0 \tag{5.24}$$

Moreover, in this case ζ can be exactly calculated by a first integration of the oscillation equation, hence eqn. (5.24) can be successively transformed under the form

$$\begin{aligned} &\delta \int_{\tau_o}^{\tau} \frac{1}{2}\left[\dot{\zeta}^2 - \zeta^2 - 2\int_{\zeta_o}^{\zeta} (1 - \zeta - \mathrm{e}^{\zeta_o - \zeta})\mathrm{d}\zeta\right] \mathrm{d}\tau = 2\delta \int_{\tau_o}^{\tau_1} (1 - \zeta - \mathrm{e}^{\zeta_o - \zeta})\mathrm{d}\tau \\ &\quad = \delta \int_{\zeta_o}^{\zeta_1} \dot{\zeta}^2 \mathrm{d}\tau = \delta \int_{\zeta_o}^{\zeta_1} \frac{\mathrm{d}\zeta}{\mathrm{d}\tau}\mathrm{d}\zeta = \delta \int_{\zeta_o}^{\zeta} q \mathrm{d}\zeta = 0 \end{aligned} \tag{5.25}$$

where q denotes the non-dimensional value of flow.

5.1.4. Finite difference formulation

The numerical methods for mass oscillations calculation, as well as the graphical methods formerly used, rely on the finite difference transcription of the equations of motion. We present a brief commentary trying to justify the optimal choice—from the point of view of results accuracy—of the finite difference model for writing these equations.

The accuracy of the employed numerical methods has been directly checked by comparing the calculation results with oscillation measurements in the surge tanks of two hydroelectric power plants [21]. We have started from the general expression of the equations of motion in finite differences, for the case of a surge tank of variable cross-section

$$\begin{aligned} &\frac{L}{g}\frac{\Delta W}{\Delta t} + Z + P = 0 \\ &Wf = F\frac{\Delta Z}{\Delta t} + Q_{\mathrm{T}} \end{aligned} \tag{5.26}$$

or

$$\begin{aligned} \Delta W &= -(Z + P)\,\frac{g\Delta t}{L} \\ \Delta Z &= \frac{f\Delta t}{F}W - \frac{\Delta t}{F}Q_{\mathrm{T}} \end{aligned} \tag{5.27}$$

where

$$P = \pm\left(\lambda\frac{L}{d} + \sum \zeta + 1\right)\frac{1}{2g}W^2 = \pm\varepsilon W^2 \tag{5.28}$$

As a first comparison method Mayer's and Pressel's method has been chosen, solving the finite difference system (5.27) by a direct calculation of quantities ΔZ and ΔW

in terms of the quantities Z and W taken at the beginning of the chosen time interval $\Delta t = t_n - t_{n-1}$. It follows

$$\begin{aligned} \Delta W_n &= -(Z_{n-1} \pm \varepsilon W_{n-1}^2) \frac{g\Delta t}{L} \\ \Delta Z_n &= \frac{f\Delta t}{F} W_{n-1} - \frac{\Delta t}{F} Q_{\mathrm{T}} \end{aligned} \tag{5.29}$$

The reduction of the time interval Δt generally leads to an increased accuracy of the method. Leiner and Streck have improved this method, by writing the equations in the form

$$\begin{aligned} \Delta W_n &= -(Z_n \pm \varepsilon W_{n-1}^2) \frac{g\Delta t}{L} \\ \Delta Z_n &= \frac{f\Delta t}{F} W_{n-1} - \frac{\Delta t}{F} Q_{\mathrm{T}} \end{aligned} \tag{5.30}$$

Results can also be obtained by direct computation.

Finally, a third comparison method, the iterative method, has been chosen using the mean values of quantities Z and W on the time interval Δt; in this case the finite differences equations take the form

$$\begin{aligned} \Delta W_n &= -(Z_m \pm \varepsilon W_m^2) \frac{g\Delta t}{L} \\ \Delta Z_n &= \frac{f\Delta t}{F} W_m - \frac{\Delta t}{F} Q_{\mathrm{T}} \end{aligned} \tag{5.31}$$

Written in this form, the equations allow numerical integration of the system either by the direct iterative method, or by an improved iterative method.

From the three analysed computation methods, which in fact are three different finite difference formulations of the same computation method, the last one, which uses the mean values of quantities Z and W on the time interval Δt, looked as the most accurate. In fact, we also have adopted this form in the numerical methods we shall present in the following.

Generally speaking, the computation methods in the case of one surge tank can also be extended to the case of several surge tanks. For the case of variable section surge tanks, it is advisable to choose and analyse the quantities ΔZ_n, which generally ensure a better convergence of the solution.

5.2. Computation of mass oscillation by analytical methods

5.2.1. Computation of hydroelectric power plants shutdown

In this case, to compute the oscillatory motion in the tunnel surge-tank system, one has to start with the oscillation equation in the non-dimensional form (5.18), with the boundary conditions (5.19), and make the change of variable [1, 2–7]

$$\zeta \to \frac{\mathrm{d}\zeta}{\mathrm{d}\tau} = \dot{\zeta}, \quad \tau \to \zeta \tag{5.32}$$

suggested by the non-explicit form of the time variable in the equation. The following form of the oscillation equation is obtained:

$$\frac{d(\dot{\zeta}^2)}{d\tau} \pm \dot{\zeta}^2 + 2\zeta = 0 \tag{5.33}$$

the + or − sign corresponding to the two directions of water flow in the tunnel, i.e. $\dot{\zeta} \geq 0, \dot{\zeta} \leq 0$, respectively. In computing the extreme levels, we should note that to a maximum level there corresponds an odd index, while to a minimum level there corresponds an even one.

The general solution of the equation system (5.33), called by some authors a "chained" equations system, takes the form

$$\dot{\zeta} = \frac{d\zeta}{d\tau} = (-1)^{j-1}\sqrt{2}\sqrt{C_j^{(-1)^j \zeta} + (-1)^j + 1}; \quad \tau \in [\tau_{j-1}, \tau_j] \tag{5.34}$$

After the computation of the integration constants using the initial conditions, one finally obtains the following solution [3]:

$$\tau \in [\tau_o, \tau_1] \rightarrow \dot{\zeta} = \sqrt{2}\sqrt{1 - \zeta - e^{\zeta_o - \zeta}} \tag{5.35}$$

$$\tau \in [\tau_{2k-1}, \tau_{2k}] \rightarrow \dot{\zeta} = -\sqrt{2}\sqrt{1 + \zeta - (1 + \zeta_{2k-1})\, e^{\zeta - \zeta_{2k-1}}} \tag{5.36}$$

$$\tau \in [\tau_{2k}, \tau_{2k+1}] \rightarrow \dot{\zeta} = \sqrt{2}\sqrt{1 - \zeta - (1 - \zeta_{2k})\, e^{\zeta_{2k} - \zeta}}, \tag{5.37}$$

The computation of the maximum jump in the surge tank is made from eqn. (5.35) and corresponds to $\dot{\zeta} = 0$, whence

$$1 - \zeta - e^{\zeta_o - \zeta} = 0 \quad \text{or} \quad \zeta_1 + \ln(1 - \zeta_1) = \zeta \tag{5.38}$$

For the remaining successive extreme levels, the following general relation applies:

$$\ln(1 - |\zeta_j|) + |\zeta_j| = \ln(1 + |\zeta_{j-1}|) - |\zeta_{j-1}|, \qquad j = 2, 3, 4, \ldots \tag{5.39}$$

where

$$|\zeta_j| = \begin{cases} \zeta_j, & \text{for odd } j, \text{ when } \zeta \geq 0 \\ -\zeta_j, & \text{for even } j, \text{ when } \zeta_j \leq 0 \end{cases}$$

In the following, we briefly present two practical procedures for calculating the solutions of these equations, respectively the succesive extreme oscillations in the surge tank: a procedure relying on some computation tables and another one relying on approximate formulae.

In the first case, the following two functions associated with eqns. (5.38) and (5.39) are tabulated

$$\begin{aligned} f_1(|\zeta|) &= \ln(1 - |\zeta|) + |\zeta|, \quad |\zeta| < 1 \\ f_2(|\zeta|) &= \ln(1 + |\zeta| - |\zeta|), \quad |\zeta| < 1 \end{aligned} \tag{5.40}$$

and the following computation stages are succesively performed: the value of ζ_o being known and taking into account the equality $f_1(|\zeta_1|) = \zeta_o$, one reads from the first table the value of ζ_1 (possibly by a linear interpolation). From the second table one reads $f_2(|\zeta_1|)$ and coming back to the first table, one reads $|\zeta_2|$. So, in general, the extreme

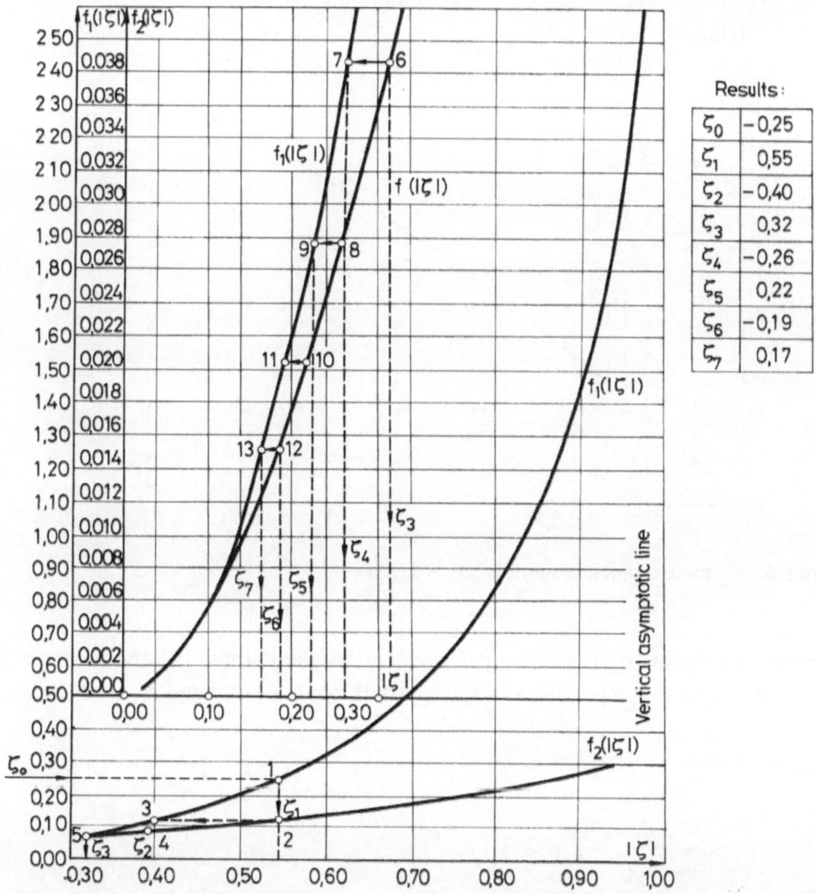

Figure 5.5. Example of determination of the successive extreme levels in a surge tank by means of functions $f_1(|\zeta|)$ and $f_2(|\zeta|)$.

level $|\zeta_{I-1}|$ being known, one reads $f_2(|\zeta_{j-1}|)$ from the second table, after which one reverts to the first table, reading $|\zeta_j|$. For an even j, $\zeta_j < 0$, and for an odd j, $\zeta_j > 0$.

By a graphical plot of functions $f_1(|\zeta|)$ and $f_2(|\zeta|)$, valid for any gallery surge-tank system, the practical application of the method becomes very easy, as can be seen in Fig. 5.5. The graphical plot in Fig. 5.6, can also be used where two successive extreme levels are connected, which follow from the solution of equation

$$|\zeta_{j+1}| + \ln(1 - |\zeta_{j+1}|) = -|\zeta_j| + \ln(1 + |\zeta_j|) \tag{5.41}$$

The analyses and processing conducted by the authors [3] with a view to computing the maximum jump have led to the following approximate formula

$$\zeta_1 = \begin{cases} 1 - \dfrac{10}{7}\dfrac{(2.1 + \zeta_o)\left(3 + \sqrt{-2\zeta_o}\right)}{\left(3 + 2\sqrt{-2\zeta_o} - \zeta_o\right)^2} & \text{for} \quad \zeta_o \geq -2.1 \\ 1 & \text{for} \quad \zeta_o < -2.1 \end{cases} \tag{5.42}$$

ensuring a satisfactory accuracy in practical applications.

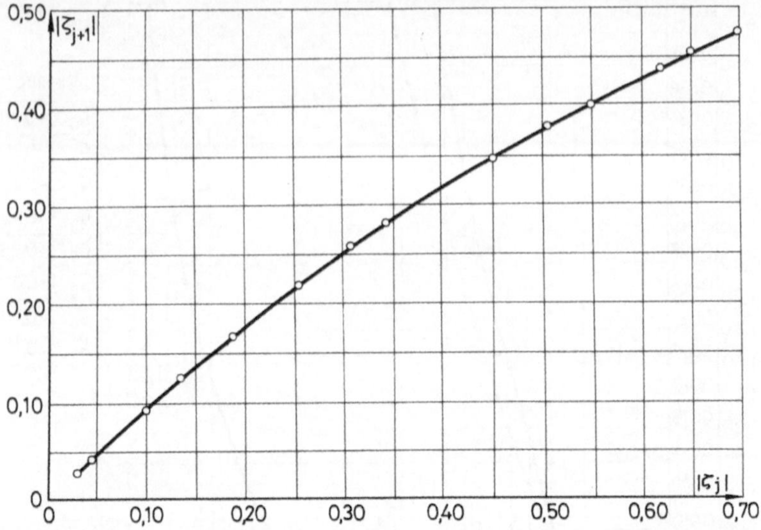

Figure 5.6. Graphical plot of function $\zeta_{j+1} = f(|\zeta_j|)$.

Finally, the following two formulae deduced by the authors [4] are also mentioned in which different extreme values of the oscillations in a surge tank are interconnected:

$$\zeta_{j+1} = -\frac{\zeta_j}{1 + \frac{2}{3}\left|\zeta_j\right|}, \quad j = 2, 3, \ldots \tag{5.43}$$

$$\zeta_j = \frac{(-1)^{j-1}\,\zeta_1}{1 + \frac{2(j-1)}{3}\zeta_1}, \quad j = 2, 3, \ldots \tag{5.44}$$

A special analysis devoted to the computation of the semiperiods of the damped oscillatory motion in a surge tank should also be mentioned [4], which led to some practical computation relations of the following form:

$$\begin{aligned} \tau_1 &= \frac{\pi}{2} + \frac{42}{100}\sqrt{-2\zeta_o} \\ \tau_k &= \pi + \frac{42}{1000\,(k-1)}\sqrt{-2\zeta_o}, \quad k > 1 \end{aligned} \tag{5.45}$$

(that adds to the analytical computation methods for the case under investigation).

Computation example

The maximum hydraulic jump produced at a sudden valve closing was computed for the experimental laboratory setup of the type: reservoir–pressure pipe–surge tank, with the following geometric and hydraulic characteristics (Fig. 5.7):

- pressure pipe: $L = 180$ m; $d = 108$ mm;
- surge tank, in three variants, of diameter: (a) $D = 0.355$ m; (b) $D = 0.255$ m; (c) $D = 0.155$ m;
- regime characterized by $Z_o = -1.60$ m.

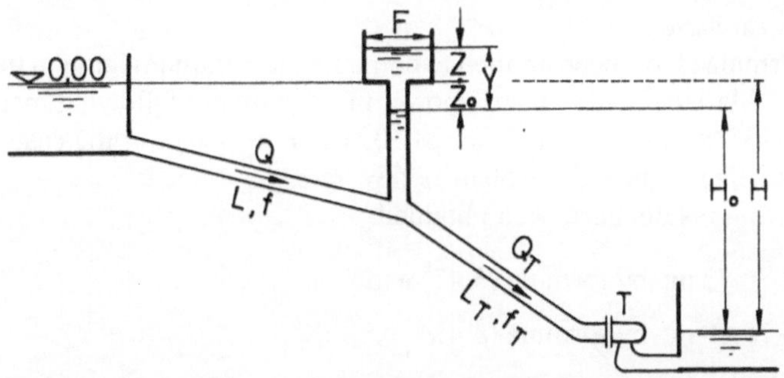

Figure 5.7. Scheme of a hydroelectric power plant with a surge tank of variable section.

By applying the analytical methods formerly presented, the following results for each variant are obtained:

(a) $Z_{max} = 0.276\,\text{m}, \quad t_1 = 42.5\,\text{s},$
(b) $Z_{max} = 0.535\,\text{m}, \quad t_1 = 26.3\,\text{s},$
(c) $Z_{max} = 1.357\,\text{m}, \quad t_1 = 11.6\,\text{s}.$

5.2.2. Computation of hydroelectric power plant start-up

The hydraulic study of this case has practical importance in the computation of the minimum hydraulic jump in the surge tank, which during operation must not exceed certain imposed limits.

Starting from the general equations of oscillatory motion in the tunnel–surge tank system, written in the non-dimensional form

$$\frac{d^2\zeta}{d\tau^2} + \frac{1}{2}\left(\frac{d\zeta}{d\tau} + \frac{\beta}{1+\alpha\zeta}\right)^2 - \frac{\alpha\beta}{(1+\alpha\zeta)^2}\frac{d\zeta}{d\tau} + \zeta = 0$$
$$\tau = 0 \rightarrow \zeta_o = 0, \quad \frac{d\zeta}{d\tau} = -\beta \tag{5.46}$$

and using special computation procedures [8], there have been obtained approximate computation methods for the minimum jump in the tank, that led to satisfactory results in practice.

In this respect, we mention two formulae ensuring an improved computational accuracy [8, 9]:

$$\zeta_{min} = \zeta_f - \frac{\sqrt{-2\zeta_f}}{1 + 0.35\sqrt{-2\zeta_f} - \zeta_f} \tag{5.47}$$

where the f index designates the final hydraulic regime in the system and

$$\zeta_{min} = -\frac{\beta\left(1 + \dfrac{\beta\pi}{4}\right)}{(1+\beta)(1-\alpha\beta) - \left(\dfrac{\beta}{2}\right)^3} \tag{5.48}$$

ensuring an improved accuracy [8, 11].

Computation example
The above formulae have been applied to compute the minimum jump in the surge tank of the Somes–Mariselu hydroelectric power plant, with the following main characteristics: $L = 8130\,\text{m}$; $d = 4.4\,\text{m}$; $F = 28.30\,\text{m}^2$; $H = 470\,\text{m}$ in the case of a sudden start-up of the plant with a $Q_f = 50\,\text{m}^3/\text{s}$ flow rate.

The following results have been obtained:

$$Z_{\min} = -70.72\,\text{m}, \text{ by formula (5.47) and}$$

$$Z_{\min} = -75.09\,\text{m}, \text{ by formula (5.48).}$$

5.2.3. Surge tanks currently used in practice

Among the surge tanks currently applied in practice, we shall present the type of tank with diaphragm and chambers [10, 22–24].

Surge tanks with diaphragm
The local hydraulic resistance at the tank basis (in the diaphragm) is most usually chosen so that at a sudden shutdown of the plant the head loss in the diaphragm be approximately equal to the overpressure given by the maximum hydraulic jump in the surge tank.

For the surge tank of constant section, this condition leads to

$$\zeta_d^* = \frac{Z_{\max}}{W_o^2/2g}\frac{F^2}{f^2} \tag{5.49}$$

where ζ_d^* denotes the optimum value of the diaphragm head loss ζ_d.

On the other hand, at the sudden start-up of the plant, the diaphragm must not induce any depression in the pressure tunnel, which reverts to the condition that the head loss in the diaphragm be less than the piezometric height of the gallery at the tank basis, a condition currently satisfied in practice. These conditions generally lead to the practical use of diaphragms that ensure asymmetrical hydraulic resistances.

Returning to the computation of the maximum jump in the surge tank at the sudden halt of the plant, one finds that this problem must be solved at the same time and together with the optimum choice of the hydraulic characteristics of diaphragms. With this aim, the oscillation equation in the tunnel–surge tank system is considered in the form

$$\frac{LF}{gf}\frac{\mathrm{d}^2Z}{\mathrm{d}t^2} + \frac{k(1+\eta)}{2g}\left|\frac{\mathrm{d}Z}{\mathrm{d}t}\right|\frac{\mathrm{d}Z}{\mathrm{d}t} + Z = 0, \quad \eta = \frac{\zeta_d}{k} > 0 \tag{5.50}$$

Introducing the notation $k_d = k(1+\eta)$, the following non-dimensional form of the oscillation equation is obtained identical to the one encountered for the simple tank

$$\frac{\mathrm{d}^2\zeta_d}{\mathrm{d}\tau^2} + \frac{1}{2}\left|\frac{\mathrm{d}\zeta_d}{\mathrm{d}\tau}\right|\frac{\mathrm{d}\zeta_d}{\mathrm{d}\tau} + \zeta_d = 0 \tag{5.51}$$

where $\zeta_d = Z/Z_{*d}$ and $Z_{*d} = Z_*/(1+\eta)$, but with different initial conditions

$$\tau = 0 \rightarrow \zeta_d = \zeta_{do}, \quad \frac{\mathrm{d}\zeta_d}{\mathrm{d}\tau} = \sqrt{-2(1+\eta)\zeta_{do}} \tag{5.52}$$

In this case, the logarithmic equation of the maximum jump ($\zeta_{d\ell} = \zeta_{d\,max}$) takes the form

$$\zeta_{d\ell} + \ln(1 + \zeta_{d\ell}) = \zeta_{do} + \ln(1 + \eta\zeta_{do}) = \delta_o < 0 \tag{5.53}$$

with the additional condition

$$1 + \eta\zeta_{do} > 0 \rightarrow \zeta_d < \frac{k}{2}\left(-1 + \sqrt{1 - \frac{4FL}{kfZ_o}}\right) = \zeta_{d\,lim} \tag{5.54}$$

In what follows, the computation scheme is similar to the one used for the simple surge tank, with corresponding adaptations: instead of the initial parameter ζ_o, one takes δ_o, and instead of the reference quantity Z_*, one considers Z_{*d}.

Eventually, the computation formula of the maximum jump [6] reads

$$\zeta_{max\,d} = \frac{Z_{max}}{Z_{*d}} = \begin{cases} 1 - \dfrac{10}{7}\dfrac{(2.1 + \delta_o)\left(3 + \sqrt{-2\delta_o}\right)}{\left(3 + 2\sqrt{-2\delta_o} - \delta_o\right)^2} & \text{for} \quad \delta_o \geq -2.1 \\ 1 & \text{for} \quad \delta_o < -2.1 \end{cases} \tag{5.55}$$

and contains two unknown quantities Z_{max} and η.

Considering the practical case of the optimum diaphragm, we have an additional relation

$$\zeta_d^* = \frac{Z_{max}}{W_o^2/2g}\frac{F^2}{f^2} \rightarrow \eta_* = -\frac{Z_{max}}{Z_o} \tag{5.56}$$

and the problem reduces to the solution of the algebraic system (5.55) and (5.56).

Computation example

The maximum jump in surge tank of the Somes–Mariselu hydroelectric power plant has been computed in a modified computation variant with respect to the surge tank diaphragm, assuming the following basic data:

$$L = 8130\,\text{m}, \quad d = 4.4\,\text{m}, \quad F = 28.3\,\text{m}^2, \quad \eta = 0.886$$

and considering the sudden shutdown of the plant at the $N = 100$ MW regime.

The resulting maximum jump was $Z_{max} = 22.5$ m, when selecting an optimum diaphragm with $\eta_* = 8.12$.

Surge tank with chambers

We shall consider the case of the surge tank with upper chamber and present a computation method of the maximum jump after a sudden closing of the valve.

Assuming the surge tank well, as well as the upper chamber of constant cross-section, the following non-dimensional values are introduced:

- *for the surge tank well*

$$Z_{*p} = \frac{LF_p}{kf}, \quad T_{*p} = \sqrt{\frac{LF_p}{gf}}, \quad \zeta_p = \frac{Z}{Z_{*p}}, \quad \tau_p = \frac{t}{T_{*p}} \tag{5.57}$$

- *for the upper chamber*

$$Z_{*c} = \frac{LF_c}{kf}, \quad T_{*c} = \sqrt{\frac{LF_c}{gf}}, \quad \zeta_c = \frac{Z}{Z_{*c}}, \quad \tau_c = \frac{t}{T_{*c}} \tag{5.58}$$

Taking into account the equations expressing the water lift speed in the tank

- *for the surge tank well*

$$\frac{d\zeta_p}{d\tau_p} = \sqrt{2}\sqrt{1 - \zeta_p - e^{\zeta_{po} - \zeta}} \tag{5.59}$$

- *for the connection section*, located at level Z_A, $\zeta_{pA} = Z_A/Z_{*p}$ respectively

$$\frac{d\zeta_p}{d\tau_p} = \sqrt{2}\sqrt{1 - \zeta_{pA} - e^{\zeta_{po} - \zeta_{pA}}} = \sqrt{\varphi}\,\frac{d\zeta_c}{d\tau_c} \tag{5.60}$$

where $\varphi = F_p/F_c < 1$;

- *for the upper chamber*

$$\frac{d\zeta_c}{d\tau_c} = \sqrt{2}\sqrt{1 - \zeta_c + \left(\frac{1}{\varphi} - 1 - \frac{1}{\varphi}e^{\zeta_{po} - \zeta_{pA}}\right)e^{\frac{1}{\varphi}\zeta_{pA} - \zeta_c}} \tag{5.61}$$

we obtain the equation of the maximum jump in the upper chamber, of the form [7]

$$\zeta_{c\,max} + \ln|1 - \zeta_{c\,max}| = \frac{1}{\varphi}\zeta_{pA} + \ln\left|1 + \frac{1}{\varphi}e^{\zeta_{po} - \zeta_{pA}} - \frac{1}{\varphi}\right| = \alpha_o \tag{5.62}$$

similar to the case of a simple surge tank. We note that, while for the case of a simple surge tank we always had $\zeta_{max} < 1$, in the case of the tank with upper chamber $\zeta_{max} > 1$ can also occur.

For the case $\zeta_{c\,max} < 1$, one can use the computation scheme of the simple surge tank, by applying the approximate formula

$$\zeta_{c\,max} = \begin{cases} 1 & \text{for} \quad \alpha_o < -2.1 \\ 1 - \dfrac{10}{7}\dfrac{(2.1 + \alpha_o)\left(3 + \sqrt{-2\alpha_o}\right)}{\left(3 + 2\sqrt{-2\alpha_o} - \alpha_o\right)} & \text{for} \quad \alpha_o \geq -2.1 \end{cases} \tag{5.63}$$

while in the case $\zeta_{c\,max} > 1$, corresponding to

$$1 + \frac{1}{\varphi}e^{\zeta_{po} - \zeta_{pA}} - \frac{1}{\varphi} < 0 \tag{5.63'}$$

eqn. (5.62) can be solved by direct trial. For more simplicity, we recommend tabulation or graphical plotting of the function

$$f_3(\zeta_{c\,max}) = \zeta_{c\,max} + \ln(\zeta_{c\,max} - 1), \quad \zeta_{c\,max} < 1 \tag{5.64}$$

Computation example

This example concerns the surge tank with upper chamber of the Bicaz hydroelectric power plant in Romania, in the computation variant without diaphragm. Considering the complete halt of the plant at a maximum flow $Q_o = 178\,\text{m}^3/\text{s}$, the following basic data have been used:

$$L = 4649\,\text{m}, \quad d = 7\,\text{m}, \quad F_p = 530\,\text{m}^2, \quad F_c = 855\,\text{m}^2, \quad Z_A = 0$$

For the maximum jump in the upper chamber, the value $Z_{max} = 16.80$ m is obtained [20].

5.3. Computation of mass oscillation by numerical methods

When designing the hydraulic dimension of surge tanks at extreme (maximum and minimum) oscillation values the following flow manoeuvres at the station are usually considered: complete instantaneous shutdown of the maximum flow and instantaneous opening of n% of the maximum flow. Generally speaking, n varies between 20 and 80.

This section presents a general method for automatic numerical computation of the extreme oscillations in a surge tank or air chamber and in a system of two surge tanks, respectively for different flow manoeuvres at the station [11, 16, 23, 24, 27, 35].

5.3.1. Systems with a single surge tank or air chamber

Surge tank

For the numerical integration of the general differential equation system of the oscillatory water motion in the gallery–surge tank system computations start from these equations written in the form (Fig. 5.7):

$$\begin{aligned} &\frac{L}{g}\frac{\mathrm{d}W}{\mathrm{d}t} + Z + P'W|W| + R'V_s|V_s| = 0 \\ &Wf = F\frac{\mathrm{d}Z}{\mathrm{d}t} + Q_T \\ &Q_T = \Phi(t) \end{aligned} \tag{5.65}$$

and adopt the following system of finite differences and mean values:

$$\begin{aligned} &\Delta t = t_{i+1} - t_i, \quad t_m = t_i + \frac{\Delta t}{2}, \quad Z_m = Z_i + \frac{\Delta Z}{2} \\ &W_m = W_i + \frac{\Delta W}{2}, \quad F_m = \left|\frac{\text{Vol}_{i+1} - \text{Vol}_i}{\Delta Z}\right|, \quad Q_{Tm} = \frac{Q_{Ti} + Q_{Ti+1}}{2} \\ &V_{sm} = \frac{F_m}{F_s}\frac{\Delta Z}{\Delta t}, \quad f_m = \frac{L}{\sum \frac{L_f}{f_i}} \end{aligned} \tag{5.66}$$

where Vol denotes the instantaneous volume of water in the tank against an arbitrary reference level.

The system (5.65) is written in finite differences

$$\begin{cases} \dfrac{L}{g}\dfrac{\Delta W}{\Delta t} + Z_m + P' W_m\,|W_m| + R' V_{sm}\,|V_{sm}| = 0 \\ W_m f_m = F_m \dfrac{\Delta Z}{\Delta t} + Q_{Tm} \\ Q_{Tm} = \Phi(t_m) \end{cases} \tag{5.67}$$

and further processed to read [26]

$$
\begin{aligned}
\Delta W &= 2\left(\frac{F_{\mathrm{m}}}{f_{\mathrm{m}}}\frac{\Delta Z}{\Delta t} - W_i + \frac{Q_{\mathrm{Tm}}}{f_{\mathrm{m}}}\right) \\
\Delta Z &= -2\left[Z_i + \frac{L}{g}\frac{\Delta W}{\Delta t} + P'\left|W_i + \frac{\Delta W}{2}\right|\left(W_i + \frac{\Delta W}{2}\right) + R'\left|V_{\mathrm{sm}}\right|V_{\mathrm{sm}}\right] \qquad (5.68) \\
Q_{\mathrm{Tm}} &= \Phi(t_{\mathrm{m}})
\end{aligned}
$$

Using a simplified form for the algebraic system (5.68),

$$\Delta W = \varphi_1(\Delta Z), \quad \Delta Z = \varphi_2(\Delta W), \quad Q_{\mathrm{Tm}} = \Phi(t_{\mathrm{m}})$$

and notations ΔZ_{p}, ΔZ_{c} for the purpose and computed value of ΔZ respectively, and Δ_i, δ_i as simplified notations for ΔZ, ΔW, the following computation algorithm can be applied:

Stage I

$$\Delta Z_{\mathrm{p}} = \Delta_1, \quad \Delta W = \varphi_1(\Delta_1) = \delta_1, \quad \Delta Z_{\mathrm{c}} = \varphi_2(\delta_1)$$

$$\left|\Delta Z_{\mathrm{p}} - \Delta Z_{\mathrm{c}}\right| = |\varepsilon_1| < \varepsilon$$

Stage II

$$
\begin{aligned}
\Delta Z_{\mathrm{p}} &= 2\Delta_1 = \Delta_2, \quad \Delta W = \varphi_1(\Delta_2) = \delta_2 \\
\Delta Z_{\mathrm{c}} &= \varphi_2(\delta_2), \quad \left|\Delta Z_{\mathrm{p}} - \Delta Z_{\mathrm{c}}\right| = |\varepsilon_2| < \varepsilon
\end{aligned}
$$

Stage III

$$\Delta Z_{\mathrm{p}} = \Delta_1 - \left(\frac{\Delta_1 - \Delta_2}{\varepsilon_1 - \varepsilon_2}\right)\varepsilon_1 = \Delta_3$$

$$\Delta W = \varphi_1(\Delta_3) = \delta_3, \quad \Delta Z_{\mathrm{c}} = \varphi_2(\delta_3), \quad \left|\Delta Z_{\mathrm{p}} - \Delta Z_{\mathrm{c}}\right| = |\varepsilon_3| < \varepsilon$$

During the iteration process over any arbitrary time step Δt, for the first proposed value of increment ΔZ one should assume its value at the previous computation step, while its value at the first time step Δt one should assume for ΔZ to be any arbitrary value consistent with the physical situation. The computation is considered to be completed for each computation step if the accuracy condition $|\Delta Z_{\mathrm{p}} - \Delta Z_{\mathrm{c}}| < \varepsilon$ is met. One can then proceed to the following computation step. If not, the computations are resumed from the IIIrd stage, considering the already performed IInd and IIIrd stages as Ist and IInd stages.

The numerical solution of the problem has proved to be stable and rapidly convergent, the accuracy of the results being very good for time steps ΔT equal to a fraction of the period T_* of the ideal oscillations

$$\Delta t = T_*/n, \quad \text{with} \quad n = 50\text{–}100$$

Air chamber

The computation of the air chamber in a hydraulic surge system assuming a slowly varied motion, has been treated in the specialised literature by various methods. For instance, Haindl [18] presents several analytical methods, whose results are similar to the ones encountered in the hydraulics of surge tanks. Escande [13] suggests an analytical method on a case, while Schlag [31] exposes a numerical method applicable by tabulated computations [29]. The numerical method presented by the authors of this book is similar to the one used for surge tanks. This method starts from the general equations of motion written in the form (3.3) and uses the mean values of the variables over the time interval Δt [25].

For the numerical integration of the system (3.3) the following finite difference form is used:

$$\begin{aligned}
&\frac{L}{gf}\frac{\Delta Q}{\Delta t} - (H_{\mathrm{m}} + Z_{\mathrm{m}}) + P' Q_{\mathrm{m}}|Q_{\mathrm{m}}| + M' Q_{\mathrm{Rm}}|Q_{\mathrm{Rm}}| = 0 \\
&Q_{\mathrm{Pm}} = F\frac{\Delta Z}{\Delta t} + Q_{\mathrm{m}} \\
&(H_{\mathrm{m}} + h_{\mathrm{a}})V_{\mathrm{m}} = K \\
&Q_{\mathrm{Pm}} = Q(t_{\mathrm{m}})
\end{aligned} \tag{5.69}$$

Further, from these equations, the expressions of the increments ΔZ and ΔQ are deduced as

$$\begin{aligned}
\Delta Z = 2\Bigg[&\frac{L}{gf}\frac{\Delta Q}{\Delta t} + P'\left(Q_i + \frac{\Delta Q}{2}\right)\left|Q_i + \frac{\Delta Q}{2}\right| + M'\left(F\frac{\Delta Z}{\Delta t}\right)\left|F\frac{\Delta Z}{\Delta t}\right| \\
&- \left(\frac{K}{V_i - F\frac{\Delta Z}{2}} + Z_i - h_{\mathrm{a}}\right)\Bigg] \\
\Delta Q = 2&\left(Q_{\mathrm{Pm}} - F\frac{\Delta Z}{\Delta t} - Q_i\right)
\end{aligned} \tag{5.70}$$

Using the simplified form of the above algebraic system

$$\begin{aligned}
\Delta Q &= \Phi(\Delta Z) \\
\Delta Z &= \Psi(\Delta Z, \Delta Q)
\end{aligned} \tag{5.71}$$

one can then apply the three-stage computation algorithm for surge tanks.

A similar numerical method applied to the computation of the air chamber in the system of a hydroelectric power plant is presented in reference [33].

5.3.2. Hydraulic systems with several surge tanks

We consider the system of a hydroelectric power plant with two surge tanks in the system (Fig. 5.8), and treat it identically to the hydraulic scheme with two serial surge tanks located along the pressure tunnel (Fig. 5.8a). Any other hydraulic power plant scheme can be treated in a similar manner [24, 28].

The numerical integration of the oscillating motion equations in the considered system is performed from the general differential equations (Fig. 5.8a)

$$\begin{aligned}
&\frac{L_1}{g}\frac{\mathrm{d}W_1}{\mathrm{d}t} + Z_1 + P_1' W_1\,|W_1| + R_1' V_{\mathrm{s1}}\,|V_{\mathrm{s1}}| = 0 \\
&\frac{L_2}{g}\frac{\mathrm{d}W_2}{\mathrm{d}t} + Z_2 - Z_1 + P_1' W_2\,|W_2| + R_2' V_{\mathrm{s2}}\,|V_{\mathrm{s2}}| = 0 \\
&W_1 f_1 = F_1\frac{\mathrm{d}Z_1}{\mathrm{d}t} + W_2 f_2, \quad W_2 f_2 = F_2\frac{\mathrm{d}Z_2}{\mathrm{d}t} + Q_{\mathrm{T}}, \quad Q_{\mathrm{T}} = \Phi(t)
\end{aligned} \tag{5.72}$$

using a finite difference system and mean values similar to (5.66).

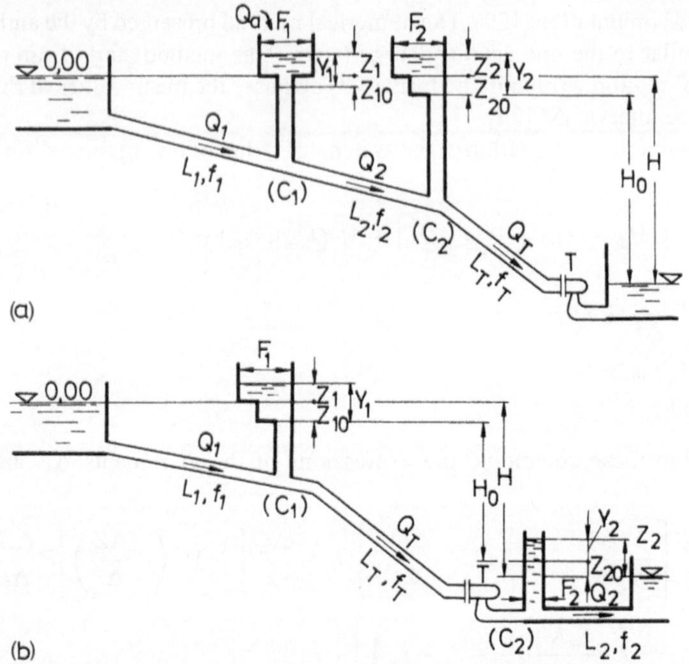

Figure 5.8. The scheme of a hydroelectric power plant with two surge tanks of variable cross-section.

To compute the extreme oscillations in the considered tank system, the eqns. (5.72) are written in finite differences and become

$$
\begin{aligned}
&(a)\quad \frac{L_1}{g}\frac{\mathrm{d}W_1}{\mathrm{d}t} + Z_{1\mathrm{m}} + P_1' W_{1\mathrm{m}}\,|W_{1\mathrm{m}}| + R_1' V_{\mathrm{s}1}\,|V_{\mathrm{s}1}| = 0\\
&(b)\quad \frac{L_2}{g}\frac{\mathrm{d}W_2}{\mathrm{d}t} + Z_{2\mathrm{m}} - Z_{1\mathrm{m}} + P_2' W_{2\mathrm{m}}\,|W_{2\mathrm{m}}| + R_2' V_{\mathrm{s}2}\,|V_{\mathrm{s}2}| = 0\\
&(c)\quad W_{1\mathrm{m}} f_1 = F_{1\mathrm{m}}\frac{\Delta Z_1}{\Delta t} + W_{2\mathrm{m}} f_2 \qquad\qquad (5.73)\\
&(d)\quad W_{2\mathrm{m}} f_2 = F_{2\mathrm{m}}\frac{\Delta Z_2}{\Delta t} + Q_{\mathrm{Tm}}\\
&(e)\quad Q_{\mathrm{Tm}} = \Phi(t_{\mathrm{m}})
\end{aligned}
$$

Under a simplified form, the algebraic system (5.73) becomes

$$
\begin{aligned}
&(a)\quad \varphi_1(\Delta Z_1, \Delta W_1) = 0\\
&(b)\quad \varphi_2(\Delta Z_1, \Delta Z_2, \Delta W_2) = 0\\
&(c)\quad \varphi_3(\Delta Z_1, \Delta W_1, \Delta W_2) = 0 \qquad\qquad (5.74)\\
&(d)\quad \varphi_4(\Delta Z_2, \Delta W_2) = 0\\
&(e)\quad Q_{\mathrm{Tm}} = \Phi(t_{\mathrm{m}})
\end{aligned}
$$

Solving the equation system (5.74c, d) with respect to ΔW_1 and ΔW_2, as well as eqns. (5.74a, b) with respect to ΔZ_1 and ΔZ_2, the following relations are obtained

$$\begin{aligned} \Delta W_1 &= \psi_1(\Delta Z_1, \Delta Z_2) \\ \Delta W_2 &= \psi_2(\Delta Z_2) \\ \Delta Z_1 &= \psi_3(\Delta W_1) \\ \Delta Z_2 &= \psi_4(\Delta W_2, \Delta Z_1) \\ \Delta Q_{\mathrm{Tm}} &= \Phi(t_{\mathrm{m}}) \end{aligned} \tag{5.75}$$

where functions ψ_1, ψ_2, ψ_3, ψ_4 shall be subsequently defined.

Further, in order to solve the algebraic system (5.75) with respect to a time step Δt, the following computation algorithm is applied, in stages:

Stage I

$$\begin{aligned} &\Delta Z_1^{\mathrm{p}} = \Delta_1', \quad \Delta Z_2^{\mathrm{p}} = \Delta_1'' \\ &\Delta W_1 = \psi_1\left(\Delta_1', \Delta_1''\right) = \delta_1', \quad \Delta W_2 = \psi_2\left(\Delta_1''\right) = \delta_2'' \\ &\Delta Z_1^{\mathrm{c}} = \psi_3\left(\delta_1'\right), \quad \Delta Z_2^{\mathrm{c}} = \psi_4\left(\delta_1'', \Delta_1'\right) \\ &\left|\Delta Z_1^{\mathrm{p}} - \Delta Z_1^{\mathrm{c}}\right| = \left|\varepsilon_1'\right| < \varepsilon, \quad \left|\Delta Z_2^{\mathrm{p}} - \Delta Z_2^{\mathrm{c}}\right| = \left|\varepsilon_1''\right| < \varepsilon \end{aligned}$$

Stage II

$$\begin{aligned} &\Delta Z_1^{\mathrm{p}} = 2\Delta_1' = \Delta_2', \quad \Delta Z_2^{\mathrm{p}} = 2\Delta_1'' = \Delta_2'' \\ &\Delta W_1 = \psi_1\left(\Delta_2', \Delta_2''\right) = \delta_2', \quad \Delta W_2 = \psi_2\left(\Delta_2''\right) = \delta_2'' \\ &\Delta Z_1^{\mathrm{c}} = \psi_3(\delta_2'), \quad \Delta Z_2^{\mathrm{c}} = \psi_4\left(\delta_2'', \Delta_2'\right) \\ &\left|\Delta Z_1^{\mathrm{p}} - \Delta Z_1^{\mathrm{c}}\right| = \left|\varepsilon_2'\right| < \varepsilon, \quad \left|\Delta Z_2^{\mathrm{p}} - \Delta Z_2^{\mathrm{c}}\right| = \left|\varepsilon_2''\right| < \varepsilon \end{aligned}$$

Stage III

$$\begin{aligned} &\Delta Z_1^{\mathrm{p}} + \Delta_1' + \left(\frac{\Delta_2' - \Delta_1'}{\varepsilon_1' - \varepsilon_2'}\right)\varepsilon_1' = \Delta_3' \\ &\Delta Z_2^{\mathrm{p}} + \Delta_1'' + \left(\frac{\Delta_2'' - \Delta_1'''}{\varepsilon_1'' - \varepsilon_2''}\right)\varepsilon_1'' = \Delta_3'' \\ &\Delta W_1 = \psi_1\left(\Delta_3', \Delta_3''\right) = \delta_3', \quad \Delta W_2 = \psi_2\left(\Delta_3''\right) = \delta_3'' \\ &\Delta Z_1^{\mathrm{c}} = \psi_3\left(\delta_3'\right), \quad \Delta Z_2^{\mathrm{c}} = \psi_1\left(\delta_3'', \Delta_3'\right) \\ &\left|\Delta Z_1^{\mathrm{p}} - \Delta Z_1^{\mathrm{c}}\right| = \left|\varepsilon_3'\right| < \varepsilon, \quad \left|\Delta Z_2^{\mathrm{p}} - \Delta Z_2^{\mathrm{c}}\right| = \left|\varepsilon_3''\right| < \varepsilon \end{aligned}$$

We shall make more explicit the above computation scheme: eqns. (c) and (d) of system (5.73) can be written as

$$\begin{aligned} \left(W_{1i} + \frac{\Delta W_1}{2}\right) f_1 &= F_{1\mathrm{m}}\frac{\Delta Z_1}{\Delta t} + F_{2\mathrm{m}}\frac{\Delta Z_2}{\Delta t} + Q_{\mathrm{Tm}} \\ \left(W_{2i} + \frac{\Delta W_2}{2}\right) f_2 &= F_{2\mathrm{m}}\frac{\Delta Z_2}{\Delta t} + Q_{\mathrm{Tm}} \end{aligned} \tag{5.76}$$

or

$$\begin{aligned} &f_1\Delta W_1 - b\Delta Z_1 - c\Delta Z_2 + a - d = 0 \\ &f_2\Delta W_2 - c\Delta Z_2 + e - d = 0 \end{aligned} \tag{5.77}$$

where the following notation

$$a = 2f_1 W_{1i}, \quad b = 2\frac{F_{1\mathrm{m}}}{\Delta t}, \quad c = 2\frac{F_{2\mathrm{m}}}{\Delta t}, \quad d = 2Q_{\mathrm{Tm}}, \quad e = 2f_2 W_{2i} \tag{5.78}$$

has been used. By solving the system (5.77), we get

$$\Delta W_1 = \frac{b\Delta Z_1 + c\Delta Z_2 + d - a}{f_1}, \qquad \Delta W_2 = \frac{c\Delta Z_2 + d - e}{f_2} \tag{5.79}$$

Furthermore, developing equations (a) and (b) from system (5.73), one can write

$$\begin{aligned} &\frac{L_1}{g}\frac{\Delta W_1}{\Delta t} + Z_{1i} + \frac{\Delta Z_1}{2} + P_1' \left| W_{1i} + \frac{\Delta W_1}{2} \right| \left(W_{1i} + \frac{\Delta W_1}{2} \right) \\ &\quad + R_1' \left| \frac{F_{1\mathrm{m}}}{F_{s1}} \frac{\Delta Z_1}{\Delta t} \right| \left(\frac{F_{1\mathrm{m}}}{F_{s1}} \frac{\Delta Z_1}{\Delta t} \right) = 0 \\ &\frac{L_2}{g}\frac{\Delta W_2}{\Delta t} + Z_{2i} - Z_{1i} + \frac{\Delta Z_2}{2} - \frac{\Delta Z_1}{2} P_2' \left| W_{2i} + \frac{\Delta W_2}{2} \right| \left(W_{2i} + \frac{\Delta W_2}{2} \right) \\ &\quad + R_2' \left| \frac{F_{2\mathrm{m}}}{F_{s2}} \frac{\Delta Z_2}{\Delta t} \right| \left(\frac{F_{2\mathrm{m}}}{F_{s2}} \frac{\Delta Z_2}{\Delta t} \right) = 0 \end{aligned} \tag{5.80}$$

or

$$\begin{aligned} &A_1\Delta W_1 + 2Z_{1i} + \Delta Z_1 + C_1 \left|2W_{1i} + \Delta W_1\right| (2W_{1i} + \Delta W_1) + B_1 \left|\Delta Z_1\right| \Delta Z_1 = 0 \\ &A_2\Delta W_2 + D + \Delta Z_2 - \Delta Z_1 + C_2 \left|2W_{2i} + \Delta W_2\right| (2W_{2i} + \Delta W_2) \\ &\quad + B_2 \left|\Delta Z_2\right| \Delta Z_2 = 0 \end{aligned} \tag{5.81}$$

where the following notation

$$\begin{aligned} &A_1 = \frac{2L_1}{g\Delta t}, \quad A_2 = \frac{2L_2}{g\Delta t} \\ &B_1 = 2\left(\frac{F_{1\mathrm{m}}}{F_{s1}}\right)^2 \frac{R_1'}{(\Delta t)^2}, \quad B_2 = 2\left(\frac{F_{2\mathrm{m}}}{F_{s2}}\right)^2 \frac{R_2'}{(\Delta t)^2} \\ &C_1 = \frac{P_1'}{2}, \quad C_2 = \frac{P_2'}{2}, \quad D = 2\,(Z_{2i} - Z_{1i}) \end{aligned} \tag{5.82}$$

has been used.

By solving eqns. (5.81) and taking into account relations (5.79), we get

$$\begin{aligned} &\Delta Z_1 = -\frac{A_1\Delta W_1 + 2Z_{1i} + C_1 \left|2W_{1i} + \Delta W_1\right| (2W_{1i} + \Delta W_1)}{1 + B_1 \left| \dfrac{a - e + f_1\Delta W_1 - f_2\Delta W_2}{b} \right|} \\ &\Delta Z_2 = \frac{\Delta Z_1 - A_2\Delta W_2 - D - C_2 \left|2W_{2i} + \Delta W_2\right| (2W_{2i} + \Delta W_2)}{1 + B_2 \left| \dfrac{e - d + f_2\Delta W_2}{c} \right|} \end{aligned} \tag{5.83}$$

Thus, eqns. (5.79) and (5.83) make explicit the algebraic system (5.75) written in a symbolic form.

During the iteration process over an arbitrary time step Δt, one should assume for the first proposed values of $\Delta Z_1, \Delta Z_2$ their values at the previous computation step, while at the first step, one should take for $\Delta Z_1, \Delta Z_2$ any arbitrary set of acceptable values which are compatible with the physics of the phenomenon. The iterative computation step by step is performed according to the provisions made for the case of one surge tank. The computation method can be easily written in an algorithmic language and used on a mean capacity computer. Finally, mention should be made also of some other numerical methods for computing the extreme oscillations in surge tanks and systems of surge tanks [1, 30].

5.4. Computation examples for hydroelectric power plants and pumping stations

5.4.1. The hydroelectric system on Bistrita river

The general project of the hydroelectric system on Bistrita river consists of the main hydroelectric power plant Bicaz, followed by another 12 smaller hydrostations downstream the river up to the town of Bacau. The "Izvorul Muntelui" dam, of the hydroelectric power plant Bicaz is a concrete cavity dam with the following main characteristics: total height 127 m, length of dam crest 435 m, concrete volume 1.625 million m^3, reservoir 1230 million m^3 water. High water discharges are evacuated through a four-gate weir 11.5 m long each, endowed with sector type valves and through four bottom galleries, 2.50 m diameter.

In the hydroelectric scheme of the Bicaz hydroelectric power plant one can mention (Fig. 5.9): the pressure tunnel of length $L = 4649$ m and diameter $d = 7$ m, the cylindrical surge tank ($D_1 = 26$ m) with upper chamber ($D_2 = 33$ m) and hydraulic resistance (diaphragm) at the basis, two penstocks of length $L_T = 157$ m and diameter $D_T - 4$ and the power station, equipped with six Francis type vertical turbines, with a total rated power of 210 MW (Fig. 5.10), using water fall $H = 149$ m and a flow-rate $Q_T = 178$ m^3/s. The annual mean electric power output of the plant is of 392 GWh.

Along the Bistrita river, downstream from the Bicaz plant, 12 hydroelectric power stations of mean head and power are located, two of the dam station type and 10 of

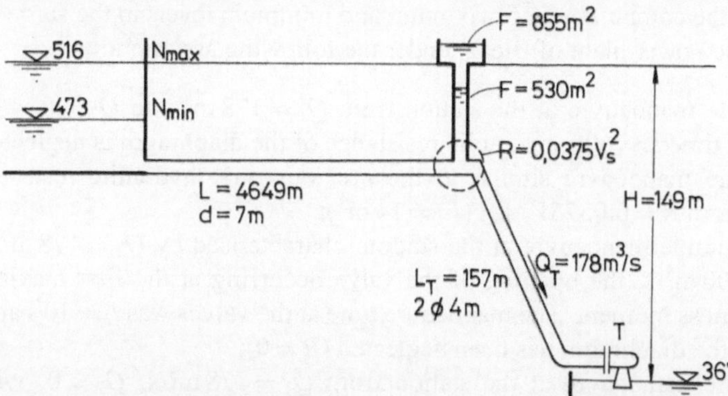

Figure 5.9. Scheme of the Bicaz hydroelectric power plant.

Figure 5.10. Bicaz hydroelectric power plant: engine hall.

the deviation hydroelectric power plant type. These stations have a head of 15–26 m, flow rates of 80–200 m^3/s, and are equipped with vertical Kaplan type turbines with installed capacity of 11–44 MW, total installed capacity of 244 MW and mean annual power output of 750 GWh. Coordination, control and checks of the entire operation of the hydroelectric power plants along the Bistrita river are automatic and distance performed by a dispatch centre located in the city of Piatra Neamt.

The Bicaz reservoir, as well as the downstream lakes, allow regulation and control of the Bistrita river and its tributaries satisfying drinking and industrial water needs of neighbouring towns and contributing at the same time to the harmony of tourist and leisure landscape in the zone.

We shall present the results of a hydraulic study performed by numerical methods concerning the computation of maximum and minimum levels in the surge tank of the hydroelectric power plant of Bicaz, under the following assumptions:

(a) flow-rate manoeuvre at the station from $Q_1 = 178\,m^3/s$ to $Q_2 = 0$, with a valve closing time $t_i = 60$ s; hydraulic resistance of the diaphragm is neglected ($R = 0$);
(b) flow-rate manoeuvre similar to the previous one, hydraulic resistance of the diaphragm $R = 0.0375 V_s|V_s|$ ($F_s = 14\,m^2$);
(c) unfavourable manoeuvre at the station, characterised by $Q_1 = 178\,m^3/s$, $Q_2 = 0$, $Q_3 = 100\,m^3/s$, the opening of the valve occurring at the first maximum water unevenness moment. The manoeuvre time at the valves was $t_i = 10$ s and the head loss in the diaphragm has been neglected ($R = 0$);
(d) flow-rate manoeuvre at the station from $Q_1 = 178\,m^3/s$, $Q_2 = 0$, with a valve closing time $t_i = 10$ s, $R = 0$ and considering a law chamber at surge tank $F_3 = 2000\,m^2$.

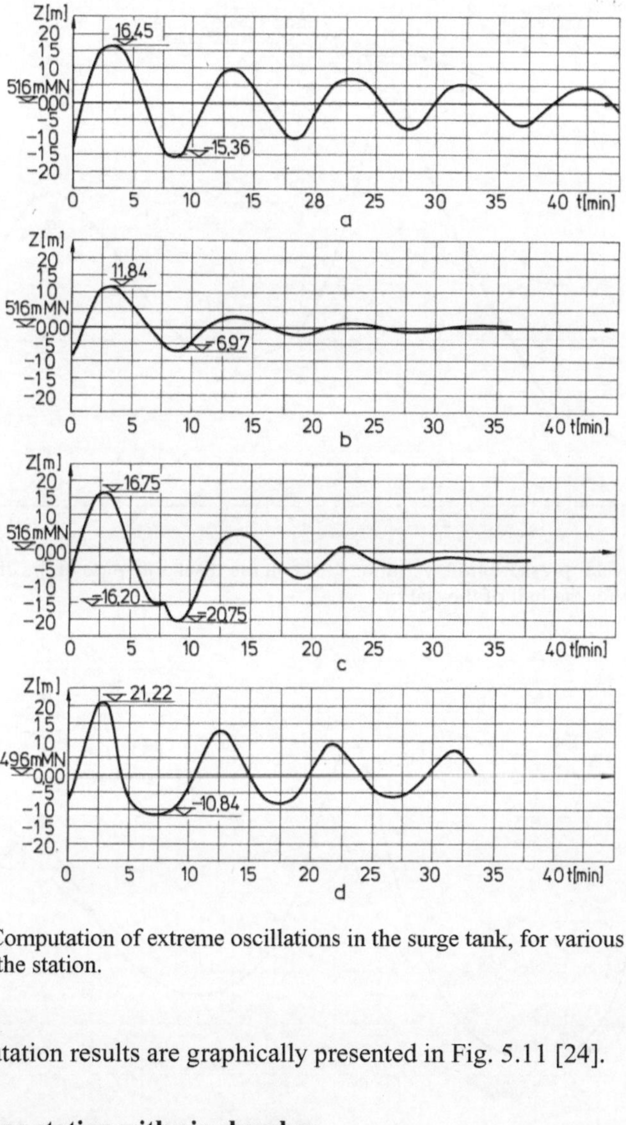

Figure 5.11. Computation of extreme oscillations in the surge tank, for various flow-rate manoeuvres at the station.

The computation results are graphically presented in Fig. 5.11 [24].

5.4.2. Pumping station with air chamber

We consider the case of a pumping station with the following main characteristics: $H_G = 41$ m; $Q_P = 9\,\text{m}^3/\text{s}$; $L = 760$ m; $d = 1900$ mm and we investigate the water-hammer protection solution of the station by means of an air chamber located near the station.

By the numerical computation method relying on the assumption of a slowly varying motion, several variants have been analysed and the retained optimum solution consists of an air chamber with the following characteristics: $V_t = 60\,\text{m}^3$ (total volume of the air chamber), $V_{air} = 25\,\text{m}^3$ (volume of the air cushion at the regime pressure); $d = 700$ mm (diameter of the connection pipe with a hydraulic resistance $R = 0.55Q_R^2$). Results of the numerical computations graphically presented in Figs. 5.12 and 5.13 exhibit time variation curves of some hydraulic parameters of the water motion in the system and justify the selected solution [25].

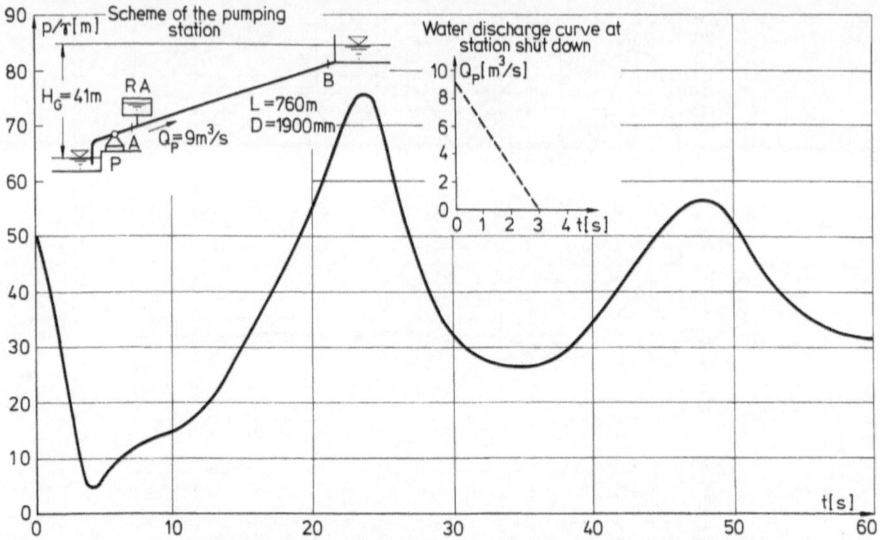

Figure 5.12. Water pressure time variation curve in the discharge pipe in the chamber cross-section, after the halt of the station.

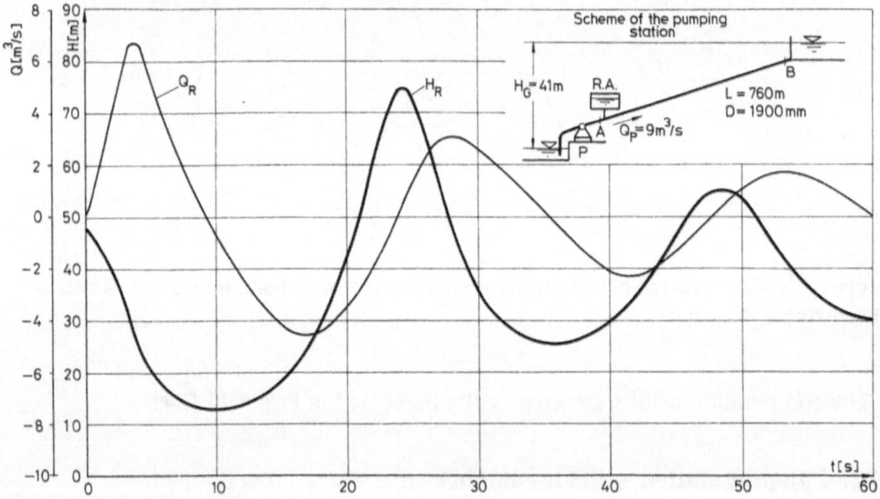

Figure 5.13. Variation curves of the air cushion pressure (H_R) and of the air chamber conduit discharge exchanges (Q_R).

5.4.3. I.C.H. Laboratory set-up

Hydraulic computations by numerical methods have been performed for the experimental laboratory installation in the variant I-A, described in Chapter 2 (Fig. 2.4), representing the model of a diversion hydroelectric power plant with surge tank. Maximum and minimum hydraulic jumps in the tank have been computed for various sudden manoeuvres of valve closing and opening.

Part of the results of numerical computations is presented under graphical form in Fig. 5.14 and allows a direct comparison with experimental measurements presented in

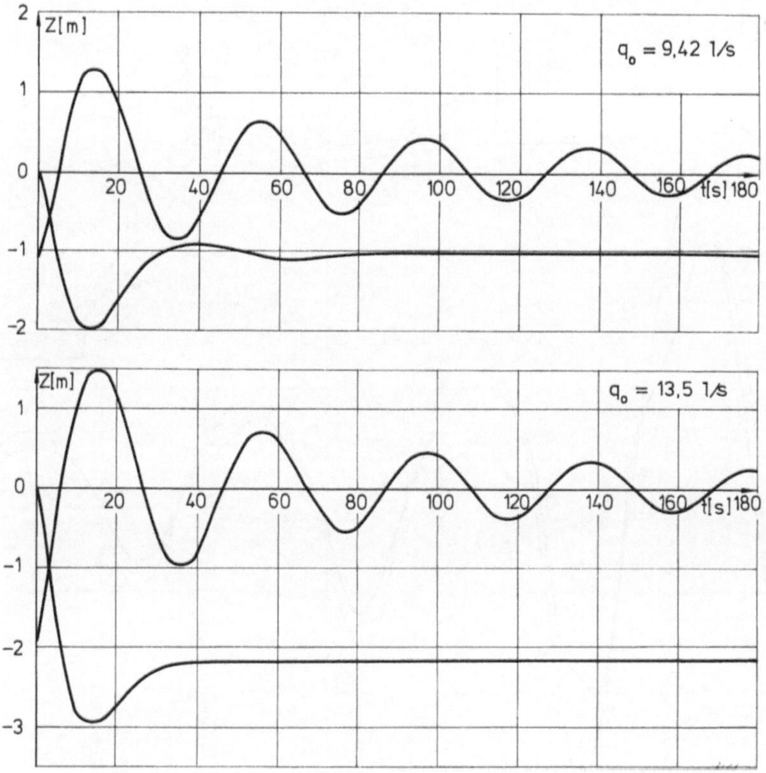

Figure 5.14. Hydraulic scheme I-A. Computation of water level oscillations in the surge tank at a sudden closing and opening of the valve, for two flow regimes.

Chapter 7 (Fig. 7.1). As usually, this comparison reminds us that computation methods rely on mathematical models involving the selection of some coefficients which appear in the motion equations and introducing a certain approximation degree (e.g. roughness coefficients of pipes, tunnels or channels, etc) [24].

5.4.4. Hydroelectric power plant with two surge tanks

We consider the case of a hydroelectric power plant with two surge tanks located in-line on the pressure tunnel with the following main characteristics:

$$L_1 = 4000\,\text{m}, \quad f_1 = 40\,\text{m}^2, \quad L_2 = 2000\,\text{m}, \quad f_2 = 40\,\text{m}^2,$$

$$F_1' = 200\,\text{m}^2, \quad F_1'' = 400\,\text{m}^2, \quad F_2' = 300\,\text{m}^2, \quad F_2'' = 600\,\text{m}^2,$$

$$L_T = 170\,\text{m}, \quad f_{T'} = 25\,\text{m}^2, \quad Q_T = 150\,\text{m}^3/\text{s}, \quad H = 200\,\text{m}.$$

and investigate the oscillatory motion of water in the system, by applying the previously exhibited numerical method. Head losses in the tanks diaphragms have been taken of the form $R_1 = R_2 = 0.1 V_{s1}|V_{s1}| = 0.1 V_{s2}|V_{s2}|$; $F_{s1} = F_{s2} = 20\,\text{m}^2$ and one has computed maximum jumps in the tanks at a flow manoeuvre at the station from $Q_1 = 150\,\text{m}^3/\text{s}$ to $Q_2 = 0$, with valve closing time $t_i = 60\,\text{s}$.

Results of numerical computations for the considered variant are also presented, in graphical form, in Fig. 5.15 [27].

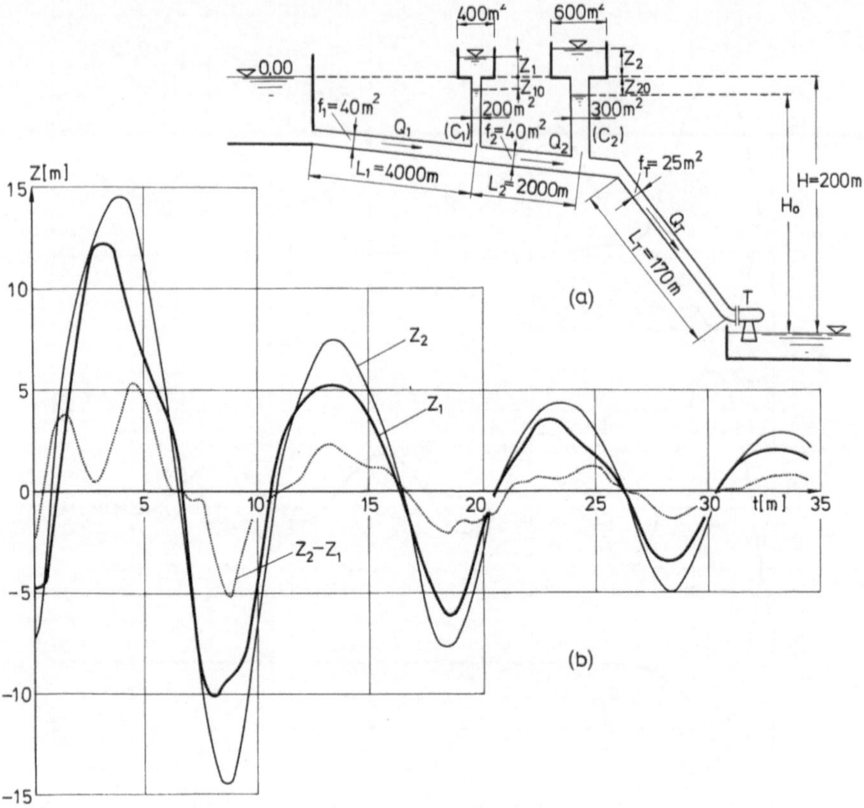

Figure 5.15. Hydroelectric power plant with two surge tanks with upper chamber: (a) power plant scheme; (b) maximum oscillations computed for the total halt of the plant.

References

1. Anderson, A., Robbie, J.: Effect of Equation Formulation on the Prediction of Mass Oscillations in Closed Conduits and Surge Tanks. *Proceedings of the 3rd International Conference on Pressure Surges*, Canterbury, England, 1980.
2. Arsenie, D.: O Forma Rationala a Ecuatiilor Nivelului Apei in Castele de Echilibru si unele Consecinte Care Rezulta din Aceasta (A Rational Form of the Equation of the Oscillations of Water Level in Simple Surge Tanks and some Consequences). *Hidrotehnica*, 11 (1971).
3. Arsenie, D.: Determinarea Analitica a Nivelurilor Extreme in Castele de Echilibru Simple, in Cazul Inchiderii Rapide si Totale a Obturatorului (Analytic Determination of the Extreme Levels in Simple Surge Tanks in the Case of a Rapid and Total Closing of the Shutter). *Hidrotehnica*, 8 (1972).
4. Arsenie, D., Asupra Legii Cvasioscilatiilor in Masa Generate de Inchiderea Brusca si Totala a Obturatorului, in Cazul Aductiunilor sub Presiune Prevazute cu Castel de Echilibru Simplu (On the Law of the Mass Quasi-Oscillations Generated by the Sudden and Total Closing of the Shutter in the Case of Surge Feeds Endowed with a Simple Surge Tank). *Hidrotehnica*, 12 (1973).
5. Arsenie, D.: Studiul Oscilatiilor in Masa ce Apar in Sistemul Galerie de Aductiune–Castel de Echilibru la Schimbarea Brusca a Puterii Centralei Hidroelectrice, avind la Baza Principiul lui Hamilton (The Study of the Mass Oscillations Occurring in the Feed Gallery–Surge Tank System at the Sudden Change of the Hydroelectric Power Plant on the Basis of the Hamilton Principle). *Studii si Cercetari de Mecanica Aplicata*, 4 (1973).

6. Arsenie, D.: Saltul Maxim in Castele de Echilibru cu Diafragma (The Maximum Jump in the Surge Tanks with Diaphragm). *Hidrotehnica*, 4 (1978).
7. Arsenie, D.: Determinarea Saltului Maxim in Castele de Echilibru cu Camera Superioara (The Determination of the Maximum Jump in the Surge Tanks with Upper Chamber). *Hidrotehnica*, 7 (1978).
8. Arsenie, D.: O Formula Imbunatatita pentru Saltul Minim in Castele de Echilibru Simple (An improved formula for the jump in surge tanks). *Hidrotehnica*, 2 (1985).
9. Arsenie, M., Arsenie, D.: Contributii Variationale asupra Pierderilor de Energie in Curgerea Laminara a Lichidelor Newtoniene in Conducte Rectilinii de Sectiune Circulara (Variation Considerations on the Energy Losses in Laminar Flows of Newtonian Liquids in Rectilinear Pipes of Circular Cross-Section). *Studii si Cercetari de Mecanica Aplicata*, 4 (1968).
10. Bernhart, H.: The Dependence of Pressure Wave Transmission through Surge Tanks on the Valve Closure Time. *Proceedings of 2nd Conference on Pressure Surges*, London, England, 1976.
11. Cioc, D., Popescu, M., Stematiu, D.: *Calcul Automatique des Oscillations dans une Chambre d'Equilibre a Section Variable*. I.S.C.H., Vol. XVII, Bucharest, 1968 (in Romanian).
12. Escande, L.: *Méthodes Nouvelles pour le Calcul des Cheminées d'Équilibre*. Dunod, Paris, 1950.
13. Escande, L.: *Calcul Rapide des Réservoirs d'Air de Stations de Pompage*. International Commission on Irrigation and Drainage, Annual Bulletin, 1963.
14. Forray, M.: *Calculul Variational in Stiinta si Tehnica* (Variation Calculus in Science and Technique—translation from English). Ed. Tehnica, Bucharest, 1975.
15. Finzi, D.: Impianto Plima Lasa. *L'Energia Elettrica*, 7 (1957).
16. Forrest, J., Robbie, J.: Mass Oscillation Prediction—a Comparative Study of Mass Surge and Waterhammer Methods. *Proceedings of 3rd Conference on Pressure Surges*, Canterbury, England, 1980.
17. Gardel, A., Rechsteiner, G.: *Les Pertes de Charge dans les Branchements en TE des Conduites à Section Circulaire*. Ecole polytechnique fédérale de Lausane, Communication du Laboratoire d'Hydraulique, no. 23, 1971.
18. Haindl, K.: *Air Chamber and its Modifications as Anti-Waterhammer Device*. Prague, 1968.
19. Ionescu, G.: *Introduction in Hydraulics*. Ed. Tehnice, Bucharest, 1977 (in Romanian).
20. Maruta, A.: Surse de Aproximatie in Calculul Analitic si in Studiile pe Model ale Castelelor de Echilibru (Sources of Approximation in the Analytic Computation and in Model Studies of the Surge Tanks). *Metodologia, Hidrologia si Gospodarirea Apelor*, 2 (1960).
21. Naber, G.: Mass Oscillations in Surge Tanks. *Water Power and Dam Construction*, 6 (1964).
22. Popescu, M.: *Probleme Actuale in Domeniul Hidraulicii Castelelor de Echilibru* (Actual Problems in Hydraulics of Surge Tanks). I.S.C.H., Vol. XXI, Bucharest, 1969, 225 pp.
23. Popescu, M.: Contributii la Calculul Hidraulic al Castelelor de Echilibru (Contributions to the Hydraulic Computation of the Surge Tanks). Ph. D. Thesis, Institutul de Constructii, Bucharest, 1970.
24. Popescu, M.: *Metode Noi pentru Calculul Hidraulic al Castelelor de Echilibru* (New Methods for the Hydraulic Calculation of the Surge Tanks). I.S.C.H., Vol. XXII, Bucharest, 1970, 74 pp.
25. Popescu, M., Zarea, S.: Calculul Rezervoarelor de Aer din Sistemele Hidraulice sub Presiune in Ipoteza Miscarii Lent Variabile (The Computation of the Air Chambers of the Surge Hydraulic Systems in the Hypothesis of the Slowly Varying Motion). *Studii si Cercetari de Mecanica Aplicata*, 4 (1977) 502–512.
26. Popescu, M., Novotny, P.: Calculul Numeric Automat al Oscilatiilor din Castelul de Echilibru de Sectiune Variabila (The Automatic Computation of the Oscillations in Surge Tanks with Variable Section). *Hidrotehnica*, 7 (1969) 324–328.
27. Popescu, M., Novotny, P., Stere, C.: Model Matematic pentru Calculul Oscilatiilor Maxime din Castele si din Sisteme de Castele de Echilibru (Mathematical Model for the Computations of Maximal Oscillations in Surge Tanks and Systems of Surge Tanks). *Studii si Cercetari de Mecanica Aplicata*, 30 (1971) 469–486.

28. Popescu, M., Stere, C.: Calculul Numeric Automat al Oscilatiilor dintr-un Sistem de Doua Castele de Echilibru Situate pe Galeria de Aductiune a Unei Uzine Hidroelectrice (The Automatic Numerical Computation of the Oscillations in a System of Two Surge Tanks Located on the Feeding Gallery of a Hydroelectric Power Plant). *Hidrotehnica*, 9 (1970), 465–472.
29. Popescu, M.: O Statie de Pompare cu Castel de Echilibru (Pumping Station Provided with a Surge Tank). *Hidrotehnica*, 6 (1972) 308–311.
30. Robbie, J., Robson, F.: Computer Aided Design of Surge Chambers in Pumped Storage Conduit Systems. *Proceedings of 1st International Conference on Pressure Surges*, Canterbury, England, 1980.
31. Schlag, A.: *Méthode Approchée de Calcul d'une Cloche à Air pour une Installation de Pompage*. Centre Belge d'Etudes et de Documentation des Eaux, 228, 1962.
32. Serrin, J.: Mathematical Principles of Classical Fluid Mechanics. In: *Handbuch der Physik*, Springer, Berlin, 1959.
33. Svee, R.: Surge Chamber with an Enclosed, Compressed Air-Cushion. *Proceedings of 1st International Conference on Pressure Surges*, Canterbury, England, 1972.
34. Thirriot, C.: Des Méfaits de l'Épanouissement Supérieur d'une Cheminée d'Équilibre lors d'Ouverture a Niveau Haut. *XII Convergo di Idraulica e Costruzioni Idrauliche*, Bari, Italia, 1970.
35. Zicman, B.: Méthodes Nouvelles pour le Calcul des Cheminées d'Équilibre. *La Houille Blanche*, 5 (1953).
36. Impianto Idroelettrico di S. Pancrazio sul Torrente Valsura. *L'Energia Elettrica*, 1 (1958).
37. Wylie, E., Streeter, V.: *Fluid Transients in Systems*. Library Technical University of Nova Scotia, 1993.

Chapter 6

Hydraulic Stability of Systems Endowed with Surge Tanks

The first studies concerning the oscillation stability of the water level in surge tanks belong to D. Thoma (1910), who proved that a surge tank in a hydroelectric power plant with automatic operation is stable only if the cross-section F of the tank is larger than a certain limiting value. If this condition is not fulfilled there appears the phenomenon of hydraulic instability, consisting either in maintained water oscillations in the tank, or in their amplification. In current engineering practice, the technical phrase "stable hydraulic operation of hydroelectric power plants with surge tanks" is usually abridged to "hydraulic stability of surge tanks". Technical literature in the field quotes the case of the Heinbach hydroelectric power plant (1904) provided with a surge tank, and exhibiting an unstable hydraulic operation.

The hydraulic instability can be explained on a surge tank scheme, by simple physical relations using, besides eqn. (5.1), a regulating equation of the form

$$N = \eta_o H_o Q_o = \eta(H + Z)Q_T \tag{6.1}$$

For the general case, we have three equations: the dynamic equation, the continuity equation and the regulation equation.

For the case of surge tanks systems, the dynamic and continuity equations split in several equations. In the study of the oscillations stability one should try to reduce the above equations to equations of the form

$$\frac{d^2Z}{dt^2} + \Phi(t)\frac{dZ}{dt} + \Psi(t)Z = 0 \tag{6.2}$$

or of the form

$$A_n\frac{d^nZ}{dt^n} + A_{n-1}\frac{d^{n-1}Z}{dt^{n-1}} + \cdots + A_1Z + A_o = 0 \tag{6.3}$$

and to discuss these equations [15].

Before proceeding to the analysis of small and large oscillations, several remarks on the regulation equation should be made. An arbitrary regulating equation $\Phi(Q_T, Z) = 0$ can be represented as a graph $Q_T = Q_T(Z)$. For instance, $Q_T = \text{constant}$ corresponds to line II in Fig. 6.1, and $Q_T = Q_o\sqrt{(H + Z)/Z}$ (regulation by opening the constant speed valve) corresponds to an increasing function of Z.

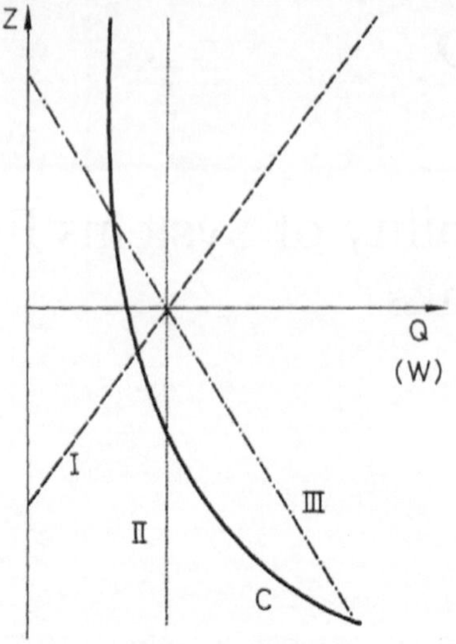

Figure 6.1. Characteristic regulating curves. Curve C corresponds to a regulation at a constant power.

In this case, taking for Q_T an approximate value

$$Q_T = Q_o\sqrt{\frac{H+Z}{H}} \cong Q_o\left(1+\frac{Z}{2H}\right) \tag{6.4}$$

the function is represented by an arbitrary straight line (e.g. straight line I in Fig. 6.1). On the other hand, the regulating equation at constant speed

$$Q_T = \frac{\eta_o Q_o H_o}{\eta(H+Z)} \tag{6.5}$$

is represented by a continuous falling curve, e.g. the C curve in the same figure.

For the case $Q_T =$ constant, Frank and Schuller have shown that all oscillations are stable, if the head losses in the pressure tunnel [15] are accounted for. Also, all regulating rising characteristics generate stable oscillations for any cross-section of the tank. Instability appears only for falling regulating characteristics (the straight line III or curve C in Fig. 6.1). As a consequence, the exact form of the regulating equation is of great importance in a correct stability study [6, 7, 23]. The validity of the incompressible fluid model used in the hydraulic stability study of surge tanks is discussed in Appendix 1.

6.1. Small and large oscillations of the water level in surge tanks. Generalities

6.1.1. The case of small oscillations

For the study of small oscillations, we start with the differential equations of motion, written in the form

$$\frac{L}{g}\frac{\mathrm{d}W}{\mathrm{d}t} + Z + P'W^2 = 0 \tag{6.6}$$

$$Wf = F\frac{\mathrm{d}Z}{\mathrm{d}t} + Q_T \tag{6.7}$$

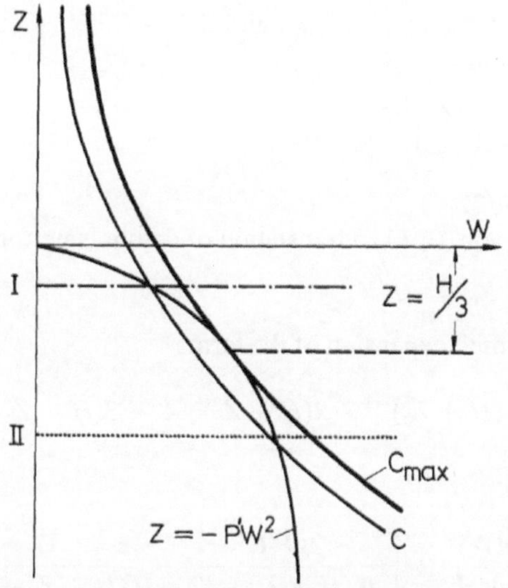

Figure 6.2. Graph of the $Z = -P'W^2$ curve and of the characteristic regulating curve C.

$$Q_T(H + Z) = Q_o H_o = C = \frac{N}{\eta g} \tag{6.8}$$

where N is the power of the station (assumed constant) after a sudden variation, while η and F are constant [15, 17].

By differentiating eqns. (6.7) and (6.8), one obtains

$$\frac{dQ_T}{dt} = -\frac{C}{(H+Z)^2}\frac{dZ}{dt} = f\frac{dW}{dt} - F\frac{d^2Z}{dt^2} \tag{6.9}$$

whence

$$\frac{dW}{dt} = \frac{F}{f}\frac{d^2Z}{dt^2} - \frac{C}{f(H+Z)^2}\frac{dZ}{dt} \tag{6.10}$$

Introducing this relation in eqn. (6.6) and expanding the $P'W^2$ term yields, after some calculations,

$$\frac{d^2Z}{dt^2} + \frac{gP'F}{fL}\left(\frac{dZ}{dt}\right)^2 + \left[\frac{2gP'C}{fL(H+Z)} - \frac{C}{F(H+Z)^2}\right]\frac{dZ}{dt} - \frac{gf}{LF}Z = -\frac{P'C^2g}{fLF(H+Z)^2} \tag{6.11}$$

The equation corresponds to an oscillatory motion. The oscillation data are found by making $Z = Z_o$ and $dZ/dt = 0$, $d^2Z/dt^2 = 0$, i.e. the conditions of the movement in steady regime. The position of the oscillation datum is graphically illustrated in Fig. 6.2 by the intersection points of the $Z = P'W^2$ curve with the characteristic regulating curve C. Hence, two oscillation data are possible, I and II. There exists a curve $C = C_{max}$, corresponding to the maximum power, for which the data I and II coincide.

The tangent point of curves $C = C_{max}$ and $Z = -P'W^2$ is found by differentiating eqn. (6.8) with respect to Z

$$\frac{\partial C}{\partial Z} = \frac{f(H+Z)}{2(P'Z)^{1/2}} + \left(\frac{Z}{P'}\right)^{1/2} f = 0 \tag{6.12}$$

which yields $Z = -H/3$.

In order to discuss eqn. (6.11) a translation of datum is performed.

$$Z = Z_o + Y \quad \text{by} \quad Z_o = -P'W_o^2$$

Considering the binomial expansion of the term

$$\frac{1}{(H+Z_o+Y)^2} = (H+Z_o)^{-2} - 2(H+Z_o)^{-3}Y + 3(H+Z_o)^{-4}Y^2 + \cdots$$

eqn. (6.11) becomes

$$\frac{d^2Y}{dt^2} + \frac{gP'F}{fL}\left(\frac{dY}{dt}\right)^2 + \left[\frac{2gP'C}{fL(H+Z_o+Y)} - \frac{C}{f(H+Z_o+Y)^2}\right]\frac{dY}{dt} + \left[\frac{gf}{LF} - \frac{gP'C^2}{LFf(H+Z_o)^2}\left(\frac{2}{H+Z_o} - \frac{3Y}{(H+Z_o)^2} + \cdots\right)\right]Y = 0 \tag{6.13}$$

or, in a symbolized form,

$$\frac{d^2Y}{dt} + \Phi(t)\frac{dY}{dt} + \Psi(t) = 0 \tag{6.14}$$

In the case of small oscillations the terms containing Y^2 and $(dY/dt)^2$ can be neglected as small infinities of higher order with respect to Y and (dY/dt), and eqn. (6.13) reduces to

$$\frac{d^2Y}{dt^2} + \left[\frac{2gP'C}{fL(H+Z_o)} - \frac{C}{F(H+Z_o)^2}\right]\frac{dY}{dt} + \frac{gf}{LF}\frac{H+3Z_o}{H+Z_o}Y = 0 \tag{6.15}$$

which is of the form

$$\frac{d^2Y}{dt^2} + 2a\frac{dY}{dt} + bY = 0 \tag{6.16}$$

As it is well known, for $b > 0$ and $a^2 < b$, the oscillation is periodic, more precisely

$a > 0$ damped oscillation;
$a = 0$ purely sinusoidal oscillation;
$a < 0$ maintained oscillation;
$a^2 > b$ non-periodic oscillation, namely:
$a > \sqrt{b}$ damped oscillation;
$-a > \sqrt{b}$ maintained oscillation.

Consequently, in order for the oscillations to be damped, the following conditions must be simultaneously fulfilled:

$$a > 0 \quad \text{and} \quad b \geq 0 \tag{6.17}$$

Condition $b > 0$ implies

$$-Z_o \leq \frac{H}{3} \tag{6.18}$$

which in practice is always fulfilled, while condition $a \geq 0$ yields

$$F_{\min} \geq \frac{W_o^2}{2g} \frac{Lf}{P'W_o^2(H - P'W_o^2)} = \frac{Lf}{2gP'H_o} = F_{\mathrm{Th}} \tag{6.19}$$

This is Thoma's classical condition for the stability of small oscillations. F_{Th} is called the Thoma's minimum section. Instead of the inequality (6.19), the inequality

$$F_{\min} = nF_{\mathrm{Th}} \tag{6.20}$$

is usually written, where $n \geq 1$, n being called the *safety coefficient* of stability. It is worth noting that this condition must be met for any cross-section of the surge tank; consequently, if the tank has several chambers, it must also be met by the cross-section of the well.

An interesting case in studying hydraulic stability of surge tanks appears when assuming a smaller tank cross-section than Thoma's limit cross-section and computing the limit amplitude of maintained oscillations when the plant is set on constant power [27].

In a first approximation, we assume that oscillations eventually reach a sinusoidal form, i.e.

$$Y = Z + P_o = Y_1 \sin \frac{2\pi t}{T_*} \tag{6.21}$$

We seek for such a solution of the differential equations of motion, not for the entire evolution of the phenomenon, but only for a half-period, i.e. starting from one of the maximum water levels in the surge tank. The head loss in the tank diaphragm shall also be taken into account.

In this case, the equation becomes

$$Q_{\mathrm{T}}(H + Z + R) = Q_o(H - P_o) = Q_o H_o \tag{6.22}$$

whence

$$\frac{Q_{\mathrm{T}}}{Q_o} = \frac{H_o}{H_o + Y + R} = \frac{H_o}{H_o + M} \tag{6.23}$$

Non-dimensional parameters of the form below (5.7) are adopted

$$\begin{aligned} v &= \frac{Q_c}{Q_o}, & u &= \frac{Q_{\mathrm{T}}}{Q_o}, & w &= \frac{Q}{Q_o} \\ p &= \frac{P}{Z_*} = p_o w^2, & r &= \frac{R}{Z_*} = r_o v^2, & h_o &= \frac{H_o}{Z_*} \\ y &= \frac{Y}{Z_*}, & m &= \frac{M}{Z_*}, & \tau &= \frac{t}{T_*} \end{aligned} \tag{6.24}$$

Thus, eqns. (5.1), (6.21) and (6.23) become

$$\frac{1}{2\pi}\frac{\mathrm{d}w}{\mathrm{d}\tau} + y + p_o(w^2 - 1) + r_o v^2 = 0 \tag{6.25}$$

$$w = u + v \tag{6.26}$$

$$y = y_\ell \sin 2\pi\tau \tag{6.27}$$

$$u = \frac{h_o}{h_o + m} = 1 - \frac{m}{h_o} + \frac{m^2}{h_o^2} - \cdots \tag{6.28}$$

If the oscillations are small with respect to H_o, the terms of the order higher than $(m/h_o)^2$ can be neglected, yielding

$$u = 1 - \frac{m}{h_o} + \left(\frac{m}{h_o}\right)^2 = 1 - \frac{y}{h_o} - \frac{r}{h_o} + \frac{y^2}{h_o^2} + \frac{r^2}{h_o^2} + \frac{2ry}{h_o^2} \tag{6.29}$$

Integrating eqn. (6.25) from $\tau_o = -1/2$ to $\tau_\ell = 1/2$, and neglecting powers of y_ℓ higher than two, we get

$$\frac{2}{3\pi}\frac{p_o}{h_o}\left(2r_o \pm \frac{1}{h_o}\right)y_\ell^2 \pm \left(\frac{p_o + r_o}{4} - \frac{p_o r_o}{2h_o} \mp \frac{3p_o}{4h_o^2}\right)y_\ell + \frac{1 - 2p_o h_o}{\pi h_o} = 0 \tag{6.30}$$

The + and − signs alternate function of the integration being made from the maximum or minimum water level in the tank. Equation (6.30) can be further simplified to

$$\frac{4}{3\pi}\frac{p_o r_o}{h_o}y_\ell^2 \pm \left(\frac{p_o + r_o}{4} - \frac{p_o r_o}{2h_o}\right)y_\ell + \frac{1 - 2p_o h_o}{\pi h_o} = 0 \tag{6.31}$$

This approximation is generally possible because the term $(3p_o/4h_o^2)$ is generally small with respect to $[(p_o + r_o/4) - (p_o r_o/2h_o)]$ and $2r_o$ is generally large with respect to $(1/h_o)$.

If this last condition was not fulfilled, the coefficient of y_ℓ^2 would be small and eqn. (6.31) would yield

$$y_\ell = \frac{1 - 2p_o r_o}{\pi h_o\left(\frac{p_o + r_o}{4} - \frac{p_o r_o}{2h_o}\right)} \tag{6.32}$$

The solution of eqn. (6.31) leads to B. Zieman's formula:

$$y_\ell = \frac{\pm A \pm B}{\frac{8}{3\pi}\frac{p_o r_o}{h_o}} \tag{6.33}$$

where

$$A = \frac{p_o + r_o}{4} - \frac{r_o p_o}{2h_o}$$
$$B = \sqrt{\left(\frac{p_o + r_o}{4} - \frac{r_o p_o}{2h_o}\right)^2 - \frac{16}{3\pi^2}\frac{p_o r_o}{h_o^2}(1 - 2p_o r_o)} \tag{6.34}$$

From formula (6.33) the smallest value of y_ℓ should always be selected.

As is known, the expression $(1 - 2p_o h_o)$ enters Thoma's condition written in non-dimensional terms. If this condition is fulfilled $(1 - 2p_o h_o = 0)$, eqn. (6.33) yields $y_\ell = 0$.

In the particular cases in which p_o or r_o can be neglected, we get

- for $p_o = 0 \quad y_\ell = \dfrac{1.27}{r_o h_o}$, $\quad$ (6.35)
- for $r_o = 0 \quad y_\ell = 1.27\dfrac{1 - 2p_o r_o}{p_o r_o}$ $\quad$ (6.36)

A similar study, performed by L. Escande [8], led to the following formulae for computing y_ℓ

$$y_\ell = 1.18\frac{1 - 2p_o r_o}{r_o(h_o - 2p_o)} \tag{6.37}$$

$$y_\ell = \frac{1.18}{r_o}\frac{(1 - 2p_o r_o)\sqrt{h_o^2 - 1}}{(h_o - 2p_o)^2} \tag{6.38}$$

Considering again the above particular cases, formula (6.37) yields

- for $p_o = 0 \quad y_\ell = \dfrac{1.18}{r_o h_o}$ (6.39)
- for $r_o = 0 \quad y_\ell \to \infty$ (6.40)

while from formula (6.38) we deduce

- for $p_o = 0 \quad y_\ell = \dfrac{1.18}{r_o}\sqrt{\dfrac{1}{h_o^2} - \dfrac{1}{h_o^4}}$ (6.41)
- for $r_o = 0 \quad y_\ell \to \infty$ (6.42)

Consequently, for $r_o \to 0$, formula (6.32) still yields finite results, while formulae (6.37) and (6.38) indicate an unbounded rise of y_ℓ.

6.1.2. The case of large oscillations

In the study of large oscillations, functions $\Phi(t)$ and $\Psi(t)$ of eqn. (6.14) appear as elaborate expressions depending on time, on the amplitude Z of oscillations and on $\mathrm{d}Z/\mathrm{d}t$. The equation of oscillations is a second-order non-linear differential equation with variable coefficients. Among the classical approaches of the problem, we mention Schuller and Karas' method and Jaeger's method [15].

Schuller and Karas' method simplifies the problem by replacing functions $\Phi(t)$ and $\Psi(t)$ with constant mean values $\Phi_m(t)$ and $\Psi_m(t)$, chosen in an appropriate way. As a consequence, eqn. (6.14) changes into an equation of the form (6.16). The conditions, in order that this equation describes stable oscillations, are of the form

$$\Phi_m(t) > 0 \quad \text{and} \quad \Psi_m(t) > 0 \tag{6.43}$$

The use of the method is nevertheless limited by the fact that functions $\Phi(t)$ and $\Psi(t)$ are strongly variable with time, the choice of the mean values $\Phi_m(t)$ and $\Psi_m(t)$ being consequently difficult in practice.

In Jaeger's method, one assumes

$$F = nF_{Th} \quad \text{and} \quad Y = Z_* \sin\frac{2\pi t}{T_*} \tag{6.44}$$

One computes functions $\Phi(t)$ and $\Phi_m(t)$

$$\Phi(t) = \frac{2P'gC}{fL(H+Z_o+Y)} - \frac{C}{F(H+Z_o+Y)^2} + \frac{P'gF}{Lf}\frac{dY}{dt}$$

$$= \frac{2P'gC}{Lf(H+Z_o)}\left[\frac{1}{1+\frac{Y}{H+Z_o}} - \frac{1}{n\left(1+\frac{Y}{H+Z_o}\right)} - \frac{1}{2}\cos\frac{2\pi t}{T_*}\right] \tag{6.45}$$

$$\cong \frac{2P'gC}{Lf(H+Z_o)}\left[1-\frac{1}{n}-\left(1-\frac{2}{n}\right)\frac{Y}{H+Z_o} - \frac{1}{2}\cos\frac{2\pi t}{T_*}\right]$$

$$\Phi_m(t) = \frac{\int_o^t \Phi(t)\left(\frac{dY}{dt}\right)^2 dt}{\int_o^t \left(\frac{dY}{dt}\right)^2 dt} \tag{6.46}$$

under the condition that $Y/H + Z_o$ is sufficiently small.

One obtains the following condition for the oscillations that reach the stability limit [15]

$$n_* = \frac{F}{F_{Th}} = \left(1 + 0.482\frac{Z_*}{H+Z_o}\right) \tag{6.47}$$

Comparison of the two methods leads to quite different practical results. For instance, Schuller obtains the result $n_* = 2$, while Karas suggests $n_* > 2$–3, these results being independent of the ratio $Z_*/(H+Z_o)$.

Other factors influencing the stability of oscillations

Besides the factors entering the water motion equation written in the form (6.6)–(6.8), the stability of oscillations in surge tanks is influenced by the following factors:

- the hydraulic resistance in the tank diaphragm;
- the kinetic term of the water motion in the gallery, in the section downstream the tank ($W'^2/2g = P'W'^2$);
- the form of the output curve of the turbines;
- the longitudinal head loss in the penstock;
- the parallel operation of several electric power stations connected to the same network.

Mention is made of the existence of certain formulae obtained by analytical methods, according to the influence of each factor on the limit section of the surge tank. However, their value is only theoretical [10, 25].

The safety coefficient for the stability of oscillations is usually defined by the ratio

$$n = \frac{F}{F_{Th}} \tag{6.48}$$

yielding a constant value, for any characteristic of the hydroelectric power plant and its type of connection to the network. On the optimum value of this coefficient, a lot of suggestions have been given by various authors, ranging from $n = 1$ to $n = 4$–6. Last time, the suggestion $n = 1.5$–1.8 seems to have been admitted.

But such a complex phenomenon as the stability of water motion in the gallery–tank system, depending on a multitude of factors, cannot be characterized by the single ratio "n" of two areas (the effective and the limit areas of the surge tank). For instance,

n can be easily influenced by the choice of head loss coefficient on the pressure gallery. In fact, if the same tank is calculated for two values, C_1 and C_2 one obtains two values n_1 and n_2 whose ratio is

$$\frac{n_1}{n_2} = \left(\frac{c_2}{c_1}\right)^2 \tag{6.49}$$

A question of practical interest is to determine the oscillations damping with time in the surge tank, characterized for instance by the ratio of two consecutive amplitudes of the oscillation.

In the case of small oscillations one can find a relation between the oscillations damping and the safety coefficient [15], of the form (Fig. 6.3)

$$\Delta = e^{2\pi p_{o1}\left(\frac{1}{\sqrt{n}}-\sqrt{n}\right)T} \tag{6.50}$$

where Δ is the ratio of two consecutive oscillations and

$$T = \frac{1}{\sqrt{1 - p_{o1}^2 \frac{(1+n)^2}{n}}}, \quad p_{o1} = \frac{p_o}{\sqrt{n}} \tag{6.51}$$

noticing that only relative quantities of the form (6.24) have been used.

We see that the phenomenon of oscillation damping is influenced in a quite complicated manner by the safety coefficient $n = F/F_{Th}$, even for the case in which a single factor (the head loss in the water conveyance gallery) that influences the oscillations damping has been taken into account. According to certain measurements of the quantity Δ performed on existent installations, the safety coefficient n appears to be considerably lower than the one computed with Thoma's formula, especially when the plants work in partial regime. In other cases, investigations and computations have led to more optimistic conclusions [15].

Finally, we mention another approach of the problem of hydraulic stability of surge tanks, which takes into account the influence of the non-permanent initial working regime of the hydroelectric power plant as well. In this case, as a consequence of some successive unfavourable manoeuvres of the turbines start-up/shutdown there can occur the situation of an irreversible decrease in the water level in the tank (described by the author as "collapse"), with very serious consequences on the hydroelectric power plant. In order to avoid the occurrence of such a phenomenon during operation, computations have yielded a new value of the *limit cross-section of the surge tank*, which is recommended to be taken into account in modern hydraulic computations involved with the design of surge tanks [26].

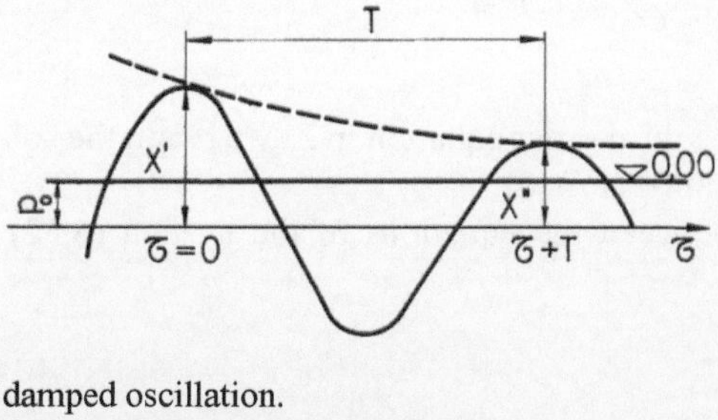

Figure 6.3. Graph of damped oscillation.

6.2. Methods for analysis of hydraulic stability of surge tanks

6.2.1. The case of a single surge tank

After the general considerations, the problem of hydraulic stability of surge tanks will be approached more extensively by using some significant results obtained recently.

Analytical methods approach the problem of hydraulic stability of surge tanks starting from the differential equations of the water motion in the gallery–tank system, with small and large oscillations dealt separately, based on simplifying assumptions. The system of differential equations of motion, corresponding to a given value of the power, admits a constant solution describing a steady regime of the plant operation. The problem consists in the study of its stability.

The problem of motion stability generally involves two aspects:

- the study of the stability of the linearised equation system, which ensures a good behaviour of the physical system for small initial deviations that would correspond in the hydroelectric power plants operation, to manoeuvres that would not entail large variations of power (small oscillations);
- determination of the attraction domain around the stationary point and analysis of the points associated with usual manoeuvres, in order to establish whether they are included in this domain.

We shall present two methods for analysing the hydraulic stability of surge tanks based on Liapunov's function, illustrating their practical applicability [2–4, 12, 22].

(a) Consider the case of a hydroelectric power plant with a surge tank of constant cross-section with diaphragm at the basis ($R \neq 0$), working at constant power in a separated network [12, 22] (Fig. 5.1).

The differential equations of the oscillatory water motion in the system are

$$\begin{aligned} &\frac{L}{g}\frac{\mathrm{d}W}{\mathrm{d}t} + Z + P'W^2 + R'\left|\frac{\mathrm{d}Z}{\mathrm{d}t}\right|\frac{\mathrm{d}Z}{\mathrm{d}t} = 0 \\ &F\frac{\mathrm{d}Z}{\mathrm{d}t} + \frac{N_2}{\eta g(H+Z)} - fW = 0, \quad \eta g(H+Z)Q_{\mathrm{T}} = N_2 \end{aligned} \tag{6.52}$$

with the usual notation (we assume $W|W| = W^2$).

The characteristics of Z_o, W_o of the stationary point (corresponding to the regime situation) are computed by solving the algebraic system

$$\frac{N_2}{\eta g(H+Z)} = fW, \quad Z = -P'W^2 \tag{6.53}$$

This system leads to a third order equation in Z; we retain the solution of this equation situated in the interval $(-H/3, 0)$.

Eliminating W between the equations of the system (6.52) and performing the change of variable

$$Y = Z - Z_o \tag{6.54}$$

we get

$$F\frac{\mathrm{d}^2Y}{\mathrm{d}t^2} - \frac{N_2\frac{\mathrm{d}Y}{\mathrm{d}t}}{\eta g(H_o+Y)^2} + \frac{fg}{L}\left[Y + Z_o + \frac{P'}{f^2}\left(F\frac{\mathrm{d}Y}{\mathrm{d}t} + \frac{N^2}{\eta g(H_o+Y)}\right)^2 + R'\frac{\mathrm{d}Y}{\mathrm{d}t}\left|\frac{\mathrm{d}Y}{\mathrm{d}t}\right|\right] = 0 \tag{6.55}$$

where $H_o = H + Z_o$ denotes the net head.

We select Liapunov's function of the form

$$V(Y,V) = \frac{1}{2}FV^2 + \frac{gf}{2L}Y^2\left(1 - \frac{2P'N_2^2}{H_o^2 f^2 g^2 \eta^2 (H_o+Y)}\right) \tag{6.56}$$

and we compute its derivative based on the system[1]

$$W(Y,V) = -V^2\left[\frac{fgR'}{L}|V| + \frac{gP'F^2}{fL}V - \frac{N_2}{\eta g(H_o+Y)^2} + \frac{2P'FN_2}{\eta LF(H_o+Y)}\right] \tag{6.57}$$

where $V = \mathrm{d}Y/\mathrm{d}t$.

According to the Liapunov's theorem, the stability condition to large oscillations is $V(Y,V) > 0$ for $Y^2 + V^2 \neq 0$, with the condition that the derivative of this function related to the system be negative [11].

The first condition leads to the inequalities (known from the case of small oscillations)

$$F > F_{\mathrm{Th}} = \frac{1}{2g}\frac{fL}{P'H_o} \quad \text{and} \quad -Z_o < \frac{H}{3} \tag{6.58}$$

In order to analyse the second condition, the coefficient below is introduced

$$\alpha = \frac{R}{P} = \frac{R'\left(\frac{\mathrm{d}Z}{\mathrm{d}t}\right)^2}{P'W^2} \quad \text{for } Q_{\mathrm{T}} = 0$$

or

$$\alpha = \frac{f^2R'}{F^2P'} \tag{6.59}$$

If $\alpha > 1$, the condition imposed to the derivative, by virtue of the system, reduces to

$$\frac{2P'F}{Lf} > \frac{1}{g(H_o+Y)}$$

hence

$$H_o + Y > \frac{Lf}{2P'Fg} = \frac{F_{\mathrm{Th}}H_o}{F} \tag{6.60}$$

or

$$\frac{F}{F_{\mathrm{Th}}} > \frac{H_o}{H_o+Y} \tag{6.61}$$

[1] $W(Y,V) = \frac{\partial v(Y,V)}{\partial Y}\cdot\frac{\mathrm{d}Y}{\mathrm{d}t} + \frac{\partial v(Y,V)}{\partial V}\cdot\frac{\mathrm{d}V}{\mathrm{d}t}$, where $\frac{\mathrm{d}Y}{\mathrm{d}t} = V$ and $\frac{\mathrm{d}Y}{\mathrm{d}t}$ is computed from eqn. (6.55).

In this case, from the point of view of the Liapunov function the situation is similar to the one of the linearised system; besides Thoma's condition, the above inequality gives information about the maximum allowed levels.

If $\alpha < 1$, a case[2] currently met in practice, the condition on the derivative by virtue of the system leads to the inequality

$$\frac{gP'F^2}{fL}(\alpha|V| + V) - \frac{N_2}{\eta g(H_o + Y)^2} + \frac{2P'FN_2}{\eta Lf(H_o + Y)} > 0 \tag{6.62}$$

which, in the most unfavourable situation, becomes

$$\frac{gP'F^2}{fL}(1 - \alpha)V > \frac{N_2}{\eta g(H_o + Y)^2} - \frac{2P'FN_2}{\eta Lf(H_o + Y)} \tag{6.63}$$

or

$$V > \frac{fLN_2}{\eta gP'F^2(1 - \alpha)(H_o + Y)}\left[\frac{1}{g(H_o + Y)} - \frac{2P'F}{Lf}\right] \tag{6.64}$$

According to a theorem of Lasalle, in order to find the attraction domain around the stationary point, we seek the highest level of the function $V(Y, V) = C$, situated in the region in which the derivative by virtue of the system is negative.

To this end, we build the curve (Φ)

$$V = \frac{fLQ_oH_o}{P'F^2(1 - \alpha)(H_o + Y)}\left[\frac{1}{g(H_o + Y)} - \frac{2P'F}{Lf}\right] \tag{6.65}$$

and the family of curves (ψ_c)

$$FV^2 + \frac{fg}{L}\left(1 + \frac{2Z_o}{H_o + Y}\right)Y^2 = C \tag{6.66}$$

and retain *the interior domain of the curve* (ψ_c), corresponding to the largest value of C, situated in the region above the curve (Φ).

Knowing the *attraction domain*, we check whether the points (Y, V), corresponding to the most unfavourable manoeuvre during the working of the plant, belong to this domain. At any rate we shall take into account the points corresponding to the manoeuvre of the total start of the plant ($N_1 = 0$, $N_2 = N_{max}$)

$$\left(-Z_o, -\frac{N_{max}}{\eta gHF}\right), \quad (Z_{min} - Z_o, 0), \quad (Z_{max} - Z_o, 0) \tag{6.67}$$

which, in fact, are initial conditions for eqn. (6.55). The analysis is performed for several values of the area F of the tank section. The curves (Φ) and (Ψ_c) can be plotted by using a computer.

For the study of hydraulic stability of surge tanks against small oscillations there also are general formulae, e.g. the formula of Gardel [10], taking into account all factors influencing this phenomenon; but the value of these formulae is only theoretical, due to the simplifying assumptions used for their deduction.

---

[2] For the case $\alpha = 0$, corresponding to $R = 0$ (surge tank without diaphragm) we recall the Jaeger's stability condition for large oscillations presented in the previous section under relation (6.47).

(b) Consider the case of a hydroelectric power plant with a simple surge tank ($R = 0$), working at constant power in a separated network (Fig. 5.1). Starting from the oscillations equation [4]

$$\frac{d^2\zeta}{d\tau^2} + \frac{1}{2}\left(\frac{d\zeta}{d\tau} + \frac{\beta}{1+\alpha\zeta}\right)^2 - \frac{\alpha\beta}{(1+\alpha\zeta)^2}\frac{d\zeta}{d\tau} + \zeta = 0 \tag{6.68}$$

with the boundary conditions

$$\tau = 0 \rightarrow \zeta = 0, \quad \frac{d\zeta}{d\tau} = -\beta = -\frac{T_*}{Z_*}\frac{dZ}{dt}\bigg|_{t=0} \tag{6.69}$$

we get for the regime situation the relation

$$\frac{1}{2}\frac{\beta^2}{(1+\alpha\zeta_f)^2} + \zeta_f = 0 \tag{6.70}$$

Instead of the function $\zeta(\tau)$ one usually takes the function $\psi(\tau)$ obtained by the substitution $\psi = \zeta - \zeta_f$; then the equation becomes

$$\frac{d^2\psi}{d\tau^2} + \frac{1}{2}\left(\frac{d\psi}{d\tau} + \frac{\beta}{1+\alpha\zeta_d+\alpha\psi}\right)^2 - \frac{\alpha\beta}{(1+\alpha\zeta_f+\alpha\zeta)^2}\frac{d\psi}{d\tau} + \psi + \zeta_f = 0 \tag{6.71}$$

This equation can be written in the following matrix form, which makes evident the linear and non-linear part of the equation, justifying the name of perturbed linear system

$$\frac{dX}{d\tau} = AX + F(X) \tag{6.72}$$

where

$$X = \begin{bmatrix}\psi\\ \xi\end{bmatrix}, \quad \frac{dX}{d\tau} = \begin{bmatrix}\dfrac{d\psi}{d\tau}\\[2ex] \dfrac{d\xi}{d\tau}\end{bmatrix}$$

$$A = \begin{bmatrix} 0 & 1 \\ \dfrac{\alpha\beta}{(1+\alpha\zeta_f)^3} - 1 & \dfrac{\beta}{1+\alpha\zeta_f}\left(\dfrac{\alpha}{1+\alpha\zeta_f} - 1\right)\end{bmatrix} \tag{6.73}$$

$$F(X) = \begin{bmatrix}0\\ R(\psi,\xi)\end{bmatrix}$$

and the following notation has been used:

$$\xi = \frac{d\psi}{d\tau}$$

$$R(\psi,\xi) = -\frac{1}{2}\xi^2 + \frac{\alpha\beta}{(1+\alpha\zeta_f)(1+\alpha\zeta_f+\alpha\zeta)}\left[1 - \frac{\alpha(2+2\alpha\zeta_f+\alpha\psi)}{(1+\alpha\zeta_f)(1+\alpha\zeta_f+\alpha\psi)}\right]\psi\xi$$
$$- \frac{\alpha^2\beta^2(3+3\alpha\zeta_f+2\alpha\psi)}{2(1+\alpha\zeta_f)^3(1+\alpha\zeta_f+\alpha\psi)^2}\psi^2 \tag{6.74}$$

For the study of small oscillations we neglect function $F(X)$, and choose the following Liapunov function

$$v(\psi,\xi) = \frac{a+b+c}{d} \geq 0 \tag{6.75}$$

where

$$\begin{aligned}
a &= \left[1 - \frac{\alpha\beta^2}{(1+\alpha\zeta_f)^3}\right]\left[2 - \frac{\alpha\beta^2}{(1+\alpha\zeta_f)^2}\right]\psi^2 \\
b &= \left[\frac{\beta}{1+\alpha\zeta_f}\left(1 - \frac{\alpha}{1+\alpha\zeta_f}\right)\psi + \zeta\right]^2 \\
c &= \left[1 - \frac{\alpha\beta^2}{(1+\alpha\zeta_f)^3}\right]\xi^2 \\
d &= 2\frac{\beta}{1+\alpha\zeta_f}\left(1 - \frac{\alpha}{1+\alpha\zeta_f}\right)\left[1 - \frac{\alpha\beta^2}{(1+\alpha\zeta_f)^2}\right]
\end{aligned} \tag{6.76}$$

and its derivative by virtue of the system is computed

$$\frac{\mathrm{d}v}{\mathrm{d}\tau} = \frac{\partial v}{\partial \psi}\frac{\mathrm{d}\psi}{\mathrm{d}\tau} + \frac{\partial v}{\partial \xi}\frac{\mathrm{d}\xi}{\mathrm{d}\tau} = -(\psi^2 + \xi^2) \leq 0 \tag{6.77}$$

The conditions for stability against small oscillations are obtained from the condition $v(\psi,\xi) \geq 0$, yielding

$$\begin{aligned}
&1 + \alpha\zeta_f > 0, \qquad \frac{1}{2}\frac{\beta^2}{(1+\alpha\zeta_f)^2} = \zeta_f < \frac{1+\alpha\zeta_f}{2\alpha} \\
&\frac{\alpha}{1+\alpha\zeta_f} < 1
\end{aligned} \tag{6.78}$$

or, finally,

$$-Z_f < \frac{1}{3}H \quad \text{and} \quad F > F_{Th} = \frac{Lf}{2gP'(H+Z_f)} \tag{6.79}$$

representing well-known conditions which are regained in this way.

We note that, in fact, the above derivative of the Liapunov function expresses the total hydraulic energy of the oscillatory system, as seen from the relations

$$\begin{aligned}
E_p &= \frac{1}{2}\rho g F (Z - Z_f)^2 \\
E_c &= \frac{1}{2}\frac{\rho L F^2}{f}\left(\frac{\mathrm{d}Z}{\mathrm{d}t}\right)^2
\end{aligned} \tag{6.80}$$

hence,

$$\frac{E_p + E_c}{E_*} = \varepsilon = \psi^2 + \xi^2 \tag{6.81}$$

where

$$E_* = \frac{1}{2}\rho g F Z_*^2$$

denotes the reference energy.

Using the theory of quadratic forms [5], one can write

$$\frac{v}{\lambda_1} \leq -(\psi^2 + \xi^2) \leq -\frac{v}{\lambda_2} \tag{6.82}$$

where λ_1, λ_2 are the characteristic roots

$$\lambda_{1,2} = \frac{\left[2 - \dfrac{\alpha\beta^2}{(1+\alpha\zeta_f)^3}\right] + \dfrac{\beta^2}{1+\alpha\zeta_f}\left(1 - \dfrac{\alpha}{1+\alpha\zeta_f}\right)^2}{4\dfrac{\beta}{1+\alpha\zeta_f}\left(1 - \dfrac{\alpha}{1+\alpha\zeta_k}\right)\left[1 - \dfrac{\alpha\beta^2}{(1+\alpha\zeta_f)^3}\right]}$$
$$\times \left[1 \pm \sqrt{\frac{\left(1 - \dfrac{\alpha}{1+\alpha\zeta_f}\right)^2 + \dfrac{\alpha^2\beta^2}{(1+\alpha\zeta_f)^4}}{\left[2 - \dfrac{\alpha\beta^2}{(1+\alpha\zeta_f)^3}\right]^2 + \dfrac{\beta^2}{(1+\alpha\zeta_f)^2}\left(1 - \dfrac{\alpha}{1+\alpha\zeta_f}\right)^2}}\,\right] \tag{6.83}$$

On the other hand, writing the variation with time of the oscillatory system energy [4], we get

$$\frac{v_o e^{-\frac{\tau}{\lambda_1}}}{\lambda_2} \leq \varepsilon \leq \frac{v_o e^{-\frac{\tau}{\lambda_2}}}{\lambda_2} \tag{6.84}$$

where $v_o(\psi, \xi)$ denotes the initial moment value of function $v(\psi, \xi)$.

For the moments τ_j and τ_{j+2} at which two extreme levels of the same sense ($\tau_j = \tau_{j+2} = 0$) are reached, one obtains the relations

$$\begin{aligned} \frac{v_o e^{-\frac{\tau_j}{\lambda_1}}}{\lambda_2} &\leq \psi_j^2 \leq \frac{v_o e^{-\frac{\tau_j}{\lambda_2}}}{\lambda_1} \\ \frac{v_o e^{-\frac{\tau_{j+2}}{\lambda_1}}}{\lambda_2} &\leq \psi_{j+2}^2 \leq \frac{v_o e^{-\frac{\tau_{j+2}}{\lambda_2}}}{\lambda_1} \end{aligned} \tag{6.85}$$

Taking into account relation (6.84) which at the initial moment has the form

$$\frac{v_o}{\lambda_2} \leq \zeta_f^2 + \beta^2 \leq \frac{v_o}{\lambda_1} \tag{6.86}$$

we get the inequality

$$\delta_{min} = \frac{1}{2}\left(\frac{\tau_{j+2}}{\lambda_2} - \frac{\tau_j}{\lambda_1}\right) \leq \delta_j \leq \frac{1}{2}\left(\frac{\tau_{j+2}}{\lambda_1} - \frac{\tau_j}{\lambda_2}\right) = \delta_{max} \tag{6.87}$$

where the known concept of logarithmic decrement

$$\delta_j = \ln\frac{\psi_j}{\psi_{j+2}} \tag{6.88}$$

has been used. With respect to this result one should note that the expression

$$\delta_j = 2\pi \frac{\beta_{\mathrm{Th}}\left(\sqrt{n} - \dfrac{1}{\sqrt{n}}\right)}{\sqrt{1 - \beta_{\mathrm{Th}}^2 \dfrac{(n+1)^2}{n}}} \tag{6.89}$$

obtained by Jaeger and applied in relation (6.50), laying outside the interval given by relation (6.87), explains its approximate value, as pointed out in [15].

For the study of hydraulic stability to large oscillations, we choose as Liapunov function the energy of the oscillatory system itself [4]

$$v(\psi, \xi) = \psi^2 + \xi^2 \geq 0 \tag{6.90}$$

and compute the derivative of this function by virtue of the system

$$\frac{\mathrm{d}v}{\mathrm{d}\tau} = -\xi^2 \left[\xi + \frac{2\beta}{1 + \alpha\zeta_{\mathrm{f}} + \alpha\psi} - \frac{2\alpha\beta}{(1 + \alpha\zeta_{\mathrm{f}} + \alpha\psi)^2}\right] \leq 0 \tag{6.91}$$

Writing the condition for which the derivative of Liapunov's function satisfies the relation (6.91) for the most unfavourable power manoeuvre ($N_1 = 0, N_2 = N_{\max}$), one obtains the inequality

$$\xi_{\mathrm{o}} = -\beta > \frac{2\alpha\beta}{(1 + \alpha\zeta_{\mathrm{f}} + \alpha\psi)^2} - \frac{2\beta}{1 + \alpha\zeta_{\mathrm{f}} + \alpha\psi} \tag{6.92}$$

whence

$$\alpha \leq \frac{1}{2} \quad \text{and} \quad 1 + \alpha\zeta_{\mathrm{f}} + \alpha\psi < 1 + \sqrt{1 - 2\alpha} \tag{6.93}$$

These relations, expressed by means of non-dimensional quantities, finally lead to the following condition of stability to large oscillations

$$n = \frac{F}{F_{\mathrm{Th}}} > 4 \frac{H + Z_{\mathrm{f}}}{H + \sqrt{H^2 - 8Z_{\min}^2}} \tag{6.94}$$

yielding an alternative to the formerly quoted Jaeger's relation (6.47).

Computation example
We considered the case of the Somes–Mariselu hydroelectric power plant and we used this case to test the hydraulic stability when operating at large oscillations. We assumed the cross-section of the surge tank ($F = 28.3\,\mathrm{m}^2$) satisfied relation (6.94).

The topological method allows very interesting stability investigations, considering the operation the hydraulic system as a non-linear oscillator. For example, we analyse the case of a surge tank, starting from the system of differential equations of motion

$$\begin{aligned} &\frac{L}{g}\frac{\mathrm{d}W}{\mathrm{d}t} + Z + P'W^2 = 0 \\ &FW = FV + Q_{\mathrm{T}} \\ &Q_{\mathrm{T}}(H + Z) = K \end{aligned} \tag{6.95}$$

Considering $x = W$ and $y = Z$, we obtain the following dynamical system of the characteristic curves [24]:

$$\frac{dx}{X} = \frac{dy}{Y} = \frac{dt}{T} \begin{cases} X = -F(H+y)(y+P'x^2) \\ Y = \dfrac{L}{g}[fx(H+y) - K] \\ T = \dfrac{Lf}{g}(H+y) \end{cases} \tag{6.96}$$

The topological study of this system leads to the following conclusions [24]:

- the system admits a characteristic $T = 0$ which is the line $y = -H$; this defines the domain of operation $y > -H$;
- the parabola $y = P'x^2$ is connected to the equilibrium position (if it exists); its position is independent of the used power K;
- the regulating curve is an equilateral hyperbola $fx(H + y) = K$, whose asymptotic lines are the boundary line $y = -H$ and the y-axis;
- these last two curves generally intersect at two points: the focal point A and the node B (Fig. 6.4); the system is stable or unstable according to the integral curve position in Fig. 6.5.

From the point of view of stability, only point A is a stable point of equilibrium, under the condition

$$F > \frac{Lf}{2gP'(H + y_A)} \tag{6.97}$$

When the power K increases, point A approaches C, the point at which the parabola and the regulating curve corresponding to a maximum power are tangent. By writing the

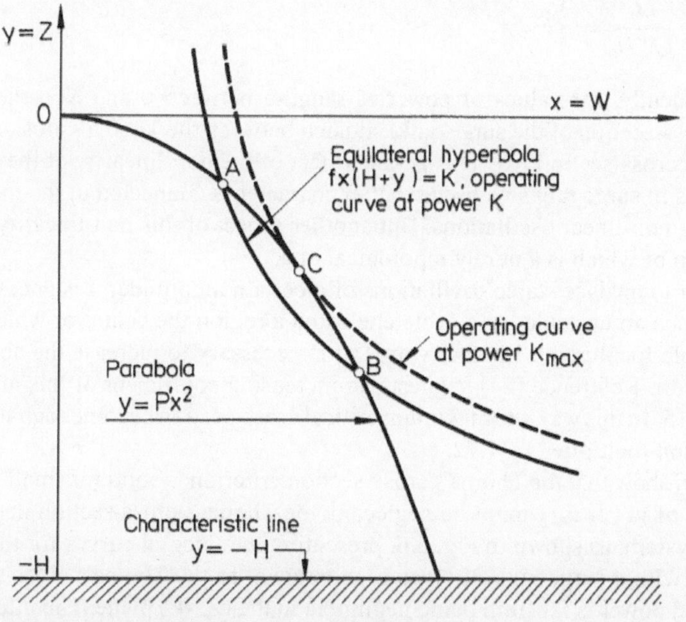

Figure 6.4. Graphical representation of the curves X, Y and T.

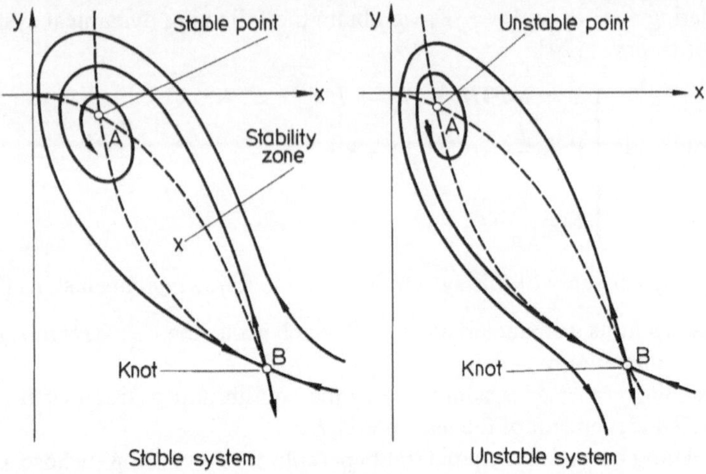

Figure 6.5. Analysis of the stability of the system by means of integral curves.

condition for which the equation satisfied by the abscissa of the intersections points of the two curves has a double root, one obtains the maximum value of K and the minimum ordinate of point A, which in this case coincides with C

$$K_M = 0.38fH\sqrt{\frac{H}{P'}}, \quad Y_C = -\frac{H}{3} \tag{6.98}$$

When points A and C coincide, the system reaches the instability boundary, and the maximum critical cross-section of the surge tank is

$$F_M = \frac{3}{4}\frac{Lf}{gP'H} = \frac{3}{2}F_{Th} \tag{6.99}$$

Consequently, for values of power K ranging between 0 and K_M, there exists a critical cross-section of the surge tank, situated between the Thoma's cross-section and the critical cross-section F_M. This is one aspect of the non-linearity of the problem of oscillations in surge tanks whose essential character is connected to the more general problem of non-linear oscillations. But another aspect of this non-linearity exists, the explanation of which is a purely topological one.

In order to analyse stable oscillations of a certain amplitude, it is necessary to put into evidence an unstable limit cycle enclosing a region the centre of which becomes then a stable focal point. In other words, it is necessary to increase the above critical cross-section. Reference [24] suggests an increasing coefficient of this cross-section equal to 1.15. In this way, the maximum critical cross-section becomes equal to Thoma's cross-section multiplied by 1.72.

One can show that the Thoma's cross-section criterion is valid for small values of K and hence of y_A. If y_A cannot be neglected, the Thoma's cross-section determines an unstable system, as shown in Fig. 6.6, presenting the integral curves for the operation of a plant with a surge tank of cross-section equal to the Thoma's cross-section, but whose used power is far from being negligible in the (Z, W) plane. The integral curves of the same system in the (W, Q_T) plane have also been presented.

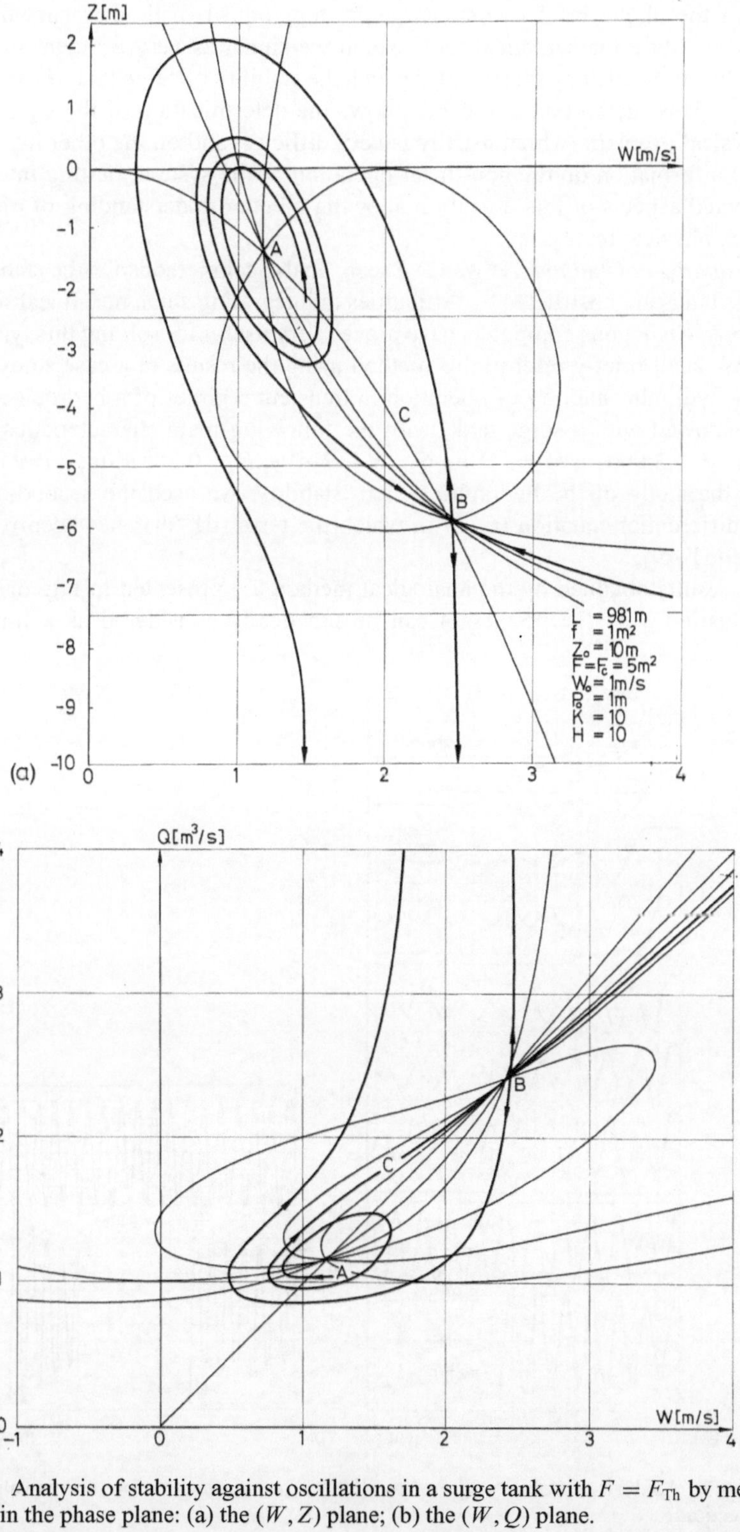

Figure 6.6. Analysis of stability against oscillations in a surge tank with $F = F_{\mathrm{Th}}$ by means of trajectories in the phase plane: (a) the (W, Z) plane; (b) the (W, Q) plane.

From the above brief commentary, the topological method is shown to be an extremely subtle mathematical tool, which we signal as very efficient in the study of non-linear oscillatory systems, for which the stability of surge tanks is an important example. This method on one hand, allows the determination of the singularities of the physical problem (which usually is very difficult) and on the other hand, yields a general information on the non-linear operation of the system, putting into evidence unexpected aspects of this operation allowing a better understanding of often a very complex physical reality.

The analogical method. As we have seen, analytical methods for the stability study of surge tanks are obstructed by difficulties in integrating some non-linear differential equations. Analogous computers have proved very useful for solving this type of problems. We shall briefly refer to this method using the results of a case study. We shall analyse hydraulic stability of operation in transient regimes of a hydroelectric power plant endowed with a surge tank, with the following main characteristics [16]: $f = 7.07\,\text{m}^2$; $L = 2{,}000\,\text{m}$; $H = 175\,\text{m}$; $C = 3.439\,\text{m}^4/\text{s}$; $P' = 0.383\,\text{s}^2/\text{m}$; $g = 9.81\,\text{m/s}^2$.

For the study of the hydraulic system stability, we used the second-order non-linear differential equation (6.13) in which the term $(\text{d}Y/\text{d}t)^2$ has been replaced by $(\text{d}Y/\text{d}t)|\text{d}Y/\text{d}t|$.

The results obtained by the analogical method are presented in Fig. 6.7. We note that situation VII ($F = 15.62\,\text{m}^2$) can be practically considered as a limit for the

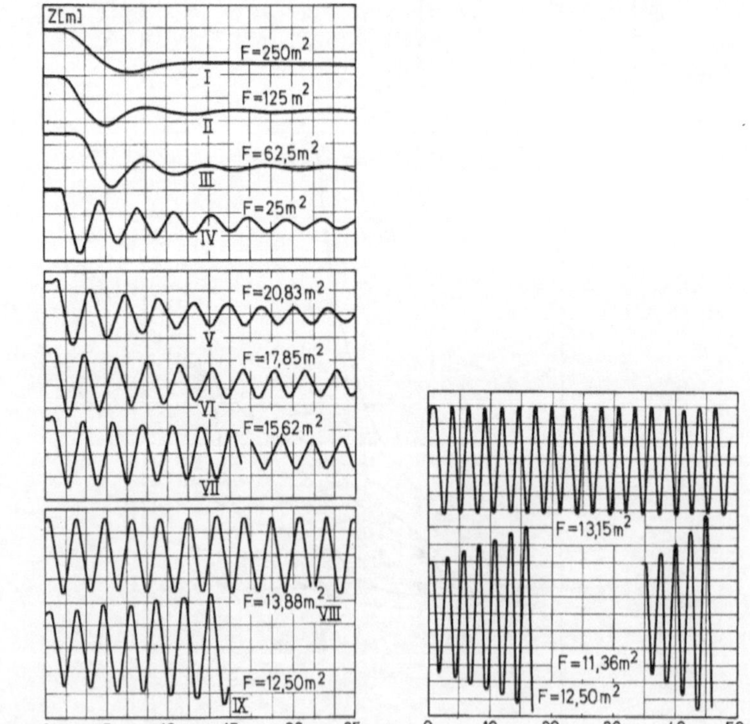

Figure 6.7. Analysis of stability against oscillations in a surge tank for various values of the cross-section F.

damping of oscillations. The computation of the Thoma's cross-sections for situations VI and VII yields the following stability safety coefficients: $n_{VI} = 1.64$ and $n_{VII} = 1.44$, sufficiently well framed into the prescription concerning stability against small perturbations $n = 1.5$–1.8.

The use of analogous electronic computers is limited as they are generally designed to solve a restricted number of differential equation types, which in analysed case requires the use of a special analogical computer.

6.2.2. The case of two surge tanks in the system

The study of the stability of a two-surge-tank system implies much greater difficulties than in the case of a single surge tank. Famous specialists in hydraulics as L. Escande and Ch. Jaeger [9, 14] have contributed to solve this problem. In reference [17] these methods have been widely referred to, being so far considered as the only ones giving an acceptable mathematical solution of the problem.

Consider the case of a hydroelectric power plant endowed with a system of two surge tanks of constant cross-section with bottom diaphragms (R_1, $R_2 \neq 0$) located on both sides of the turbines (Fig. 6.8b), or located on the water conveyance gallery (Fig. 6.8a). We assume the plant operates at constant power, in a separated power network.

We shall analyse the hydraulic scheme in Fig. 6.8b. As it is well-known, in modern underground hydroelectric power plants, the tailrace of which may be pressurised under certain conditions, besides the surge tank located on the feeding gallery, another surge

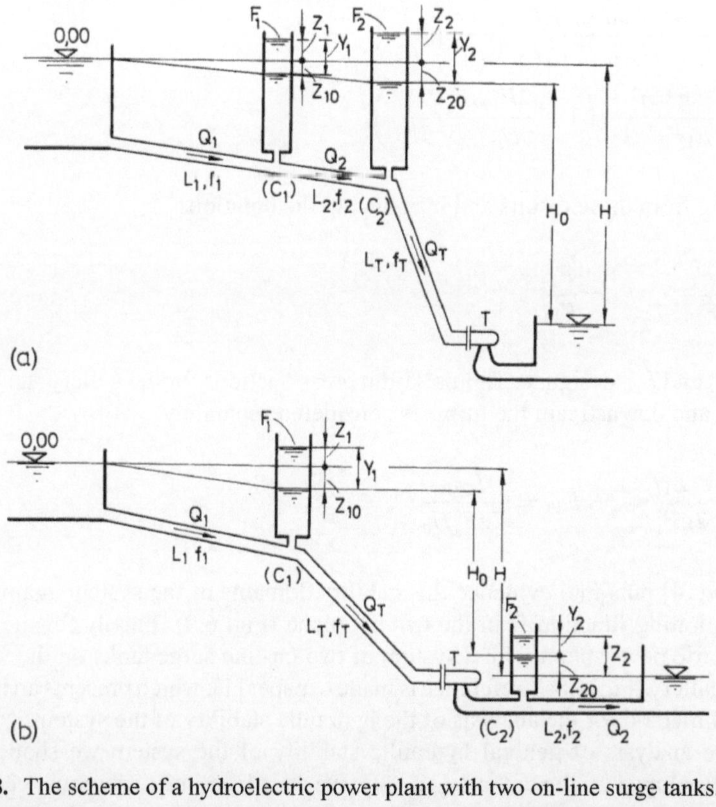

Figure 6.8. The scheme of a hydroelectric power plant with two on-line surge tanks.

tank is designed on the tailrace, just downstream the turbine. Besides the protection to overpressure of the tailrace during some flow manoeuvres at the station this contributes to a general improvement of the hydraulic operation of the plant at transient regimes. The study of the stability of oscillations by analytical methods in such a system of surge tanks has proved to be very difficult. For the case of small oscillations the results obtained by Escande and Huron [9] should be mentioned.

The necessary and sufficient condition of stability against small oscillations of this system (deduced by means of the Routh–Hurwith criterion) appears as

$$\begin{aligned} &\alpha_1 > 0, \quad \alpha_2 > 0, \quad \alpha_3 > 0, \quad \alpha_4 > 0 \\ &\Gamma = \alpha_3^2 - \alpha_1(\alpha_2\alpha_3 - \alpha_1\alpha_4) < 0 \end{aligned} \tag{6.100}$$

where

$$\begin{aligned} \alpha_1 &= \frac{2g}{Q_o}\left(P_{10}\frac{f_1}{L_1} + P_{20}\frac{f_2}{L_2}\right) - \frac{Q_o}{H_o}\left(\frac{1}{F_1} + \frac{1}{F_2}\right) \\ \alpha_2 &= 4g^2\frac{P_{10}P_{20}}{Q_o^2}\frac{f_1f_2}{L_1L_2} - \frac{2g}{H_o}\left(\frac{1}{F_1} + \frac{1}{F_2}\right)\left(P_{10}\frac{f_1}{L_1} + P_{20}\frac{f_2}{L_2}\right) + g\left(\frac{f_1}{L_1f_1} + \frac{f_2}{L_2f_2}\right) \\ \alpha_3 &= \frac{2g^2}{Q_o}\frac{f_1f_2}{L_1L_2}\left(\frac{P_{10}}{F_2} + \frac{P_{20}}{F_1}\right) - \frac{gQ_o}{F_1F_2H_o}\left(\frac{f_2}{L_2} + \frac{f_1}{L_1}\right) \\ &\quad - \frac{4g^2P_{10}P_{20}}{H_oQ_o}\frac{f_1f_2}{L_1L_2}\left(\frac{1}{F_1} + \frac{1}{F_2}\right) \\ \alpha_4 &= \frac{g^2f_1f_2}{F_1F_2L_1L_2}\left(1 - \frac{2P_{10}}{H_o} - \frac{2P_{20}}{H_o}\right) \end{aligned} \tag{6.101}$$

Starting from these results and introducing the notations

$$n_1 = \frac{F_1}{F_{10}}, \quad n_2 = \frac{F_2}{F_{20}} \tag{6.102}$$

where F_{10} and F_{20} designate Thoma's limit cross-sections for the gallery–tank systems upstream and downstream the turbines considered separately

$$F_{10} = \frac{L_1f_1}{2gP_1'H_o}, \quad F_{20} = \frac{L_2f_2}{2gP_2'H_o} \tag{6.103}$$

Jaeger [14] puts into evidence the stability domains of the system against oscillations by plotting function Γ in the (n_1, n_2) plane (Fig. 6.9). Finally for the case of a hydroelectric power plant with a system of two on-line surge tanks on the water conveyance gallery (Fig. 6.8a), reference is made to paper [1], which presents an interesting analytical method for the analysis of the hydraulic stability of the system.

For the analysis of general hydraulic stability of the system we shall present a method developed by the authors [12, 22]. We start from the differential equations of

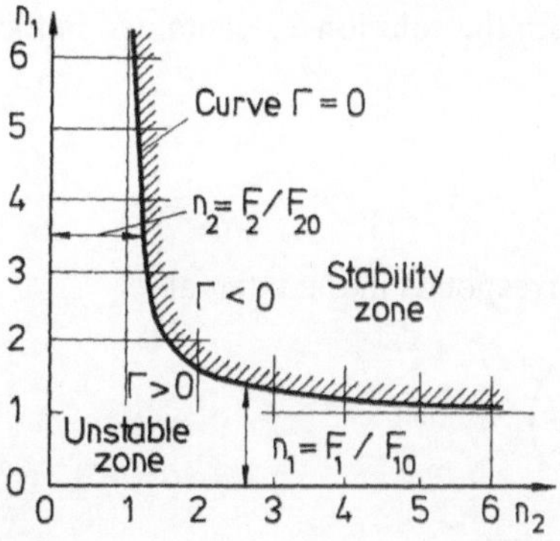

Figure 6.9. Graph of function Γ.

the oscillatory water motion in the system, written in the form

$$\begin{aligned}
&\frac{L_1}{g}\frac{dW_1}{dt} + Z_1 + P_1'W_1^2 + R_1'\left|\frac{dZ_1}{dt}\right|\frac{dZ_1}{dt} = 0\\
&F_1\frac{dZ_1}{dt} + \frac{N_2}{\eta g(H + Z_1 - Z_2)} - f_1W_1 = 0\\
&\frac{L_2}{g}\frac{dW_2}{dt} + Z_2 + P_2'W_2^2 + R_2'\left|\frac{dZ_2}{dt}\right|\frac{dZ_2}{dt} = 0\\
&F_2\frac{dZ_2}{dt} + \frac{N_2}{\eta g(H + Z_1 - Z_2)} + f_2W_2 = 0\\
&\eta g(H + Z_1 - Z_2)Q_T = N_2
\end{aligned} \tag{6.104}$$

We shall go through the computation and analysis stages of hydraulic stability adopted in the case of a single surge tank.

The characteristics $Z_1^o, W_1^o, Z_2^o, W_2^o$ of the stationary point are computed by solving the algebraic system

$$\frac{N_2}{\eta g(H + Z_1 - Z_2)} = f_1W_1 = f_2W_2; \quad Z_1 = -P_1'W_1^2; \quad Z_2 = P_2'W_2^2 \tag{6.105}$$

By making in the first equation the substitution

$$Z_2 = -\frac{P_2'f_1^2}{P_1'f_2^2}Z_1 \tag{6.106}$$

we get

$$\frac{N_2^2}{\eta^2 g^2\left(H + Z_1 + \frac{P_2'f_1^2}{P_1'f_2^2}Z_1\right)^2} = f_1^2W_1^2 = -\frac{f_1^2}{P_1'}Z_1 \tag{6.107}$$

or

$$\left(H + \frac{P_1'f_2^2 + P_2'f_1^2}{P_1'f_2^2}\right)^2 Z_1 + \frac{P_1'N_2^2}{\eta^2 g^2 f_1^2} = 0 \tag{6.108}$$

From eqn. (6.108) we retain the solution Z_1° contained in the interval

$$\left(-\frac{H}{3}\frac{P_1' f_2^2}{P_1' f_2^2 + P_2' f_1^2}, 0\right)$$

to which a solution Z_2° corresponds in the interval

$$\left(0, -\frac{H}{3}\frac{P_2' f_1^2}{P_1' f_2^2 + P_2' f_1^2}\right)$$

Using the changes of functions

$$Y_1 = Z_1 - Z_1^{\circ} \quad \text{and} \quad Y_2 = Z_2 - Z_2^{\circ} \tag{6.109}$$

and a successive elimination of W_1 and W_2, we obtain

$$\begin{aligned}
&F_1\frac{\mathrm{d}^2 Y_1}{\mathrm{d}t^2} + \frac{N_2}{\eta g(H_o + Y_1 - Y_2)^2}\left(\frac{\mathrm{d}Y_2}{\mathrm{d}t} - \frac{\mathrm{d}Y_1}{\mathrm{d}t}\right) + \frac{f_1 g}{L_1}\left\{Z_1^{\circ} + Y_1 + \frac{P_1'}{f_1^2}\right.\\
&\left.\times\left[F_1\frac{\mathrm{d}Y_1}{\mathrm{d}t} + \frac{N_2}{\eta g(H + Y_1 - Y_2)}\right]^2 + R_1'\left|\frac{\mathrm{d}Y_1}{\mathrm{d}t}\right|\frac{\mathrm{d}Y_1}{\mathrm{d}t}\right\} = 0\\
&F_2\frac{\mathrm{d}^2 Y_2}{\mathrm{d}t^2} + \frac{N_2}{\eta g(H_o + Y_1 - Y_2)^2}\left(\frac{\mathrm{d}Y_1}{\mathrm{d}t} - \frac{\mathrm{d}Y_2}{\mathrm{d}t}\right) + \frac{f_2 g}{L_2}\left\{Z_2^{\circ} + Y_2 + \frac{P_2'}{f_2^2}\right.\\
&\left.\times\left[F_2\frac{\mathrm{d}Y_2}{\mathrm{d}t} + \frac{N_2}{\eta g(H + Y_1 - Y_2)}\right]^2 + R_2'\left|\frac{\mathrm{d}Y_2}{\mathrm{d}t}\right|\frac{\mathrm{d}Y_2}{\mathrm{d}t}\right\} = 0
\end{aligned} \tag{6.110}$$

where $H_o = H + Z_1^{\circ} - Z_2^{\circ}$ denotes the net head.

For the following computations, a new change of function

$$Y_{12} = Y_1 - Y_2 \quad \text{and} \quad Z_{12} = Y_2 + \frac{\dfrac{f_1^2}{P_1'}}{\dfrac{f_1^2}{P_1'} + \dfrac{f_2^2}{P_2'}} Y_{12} \tag{6.111}$$

yields

$$\frac{\mathrm{d}Y_{12}}{\mathrm{d}t} = V_1 - V_2 \quad \text{and} \quad \frac{\mathrm{d}Z_{12}}{\mathrm{d}t} = \frac{1}{\dfrac{f_1^2}{P_1'} + \dfrac{f_1^2}{P_2'}}\left(\frac{f_1^2}{P_1'}V_1 + \frac{f_2^2}{P_2'}V_2\right) \tag{6.112}$$

where

$$V_1 = \frac{\mathrm{d}Y_1}{\mathrm{d}t}, \quad V_2 = \frac{\mathrm{d}Y_2}{\mathrm{d}t}$$

As a consequence, the system (6.104) takes the new form

$$\frac{dV_1}{dt} = \frac{N_2}{\eta g F_1 (H_o + Y_{12})^2}(V_1 - V_2) - \frac{f_1 g}{L_1 f_1}\left\{ Z_1^o + \frac{\dfrac{f_2^2}{P_2'}}{\dfrac{f_1^2}{P_1'} + \dfrac{f_2^2}{P_2'}} Y_{12} + Z_{12} + \frac{P_1'}{f_1^2}\left[F_1 V_1 + \frac{N_2}{\eta g (H_o + Y_{12})}\right]^2 + R_1' V_1 |V_1| \right\}$$

$$\frac{dV_2}{dt} = \frac{N_2}{\eta g F_2 (H_o + Y_{12})^2}(V_1 - V_2) - \frac{f_2 g}{L_2 f_2}\left\{ Z_2^o - \frac{\dfrac{f_1^2}{P_1'}}{\dfrac{f_1^2}{P_1'} + \dfrac{f_2^2}{P_2'}} Y_{12} + Z_{12} - \frac{P_2'}{f_2^2}\left[F_2 V_2 - \frac{N_2}{\eta g (H_o + Y_{12})}\right]^2 + R_2' V_2 |V_2| \right\} \tag{6.113}$$

We introduce the coefficients

$$\alpha_1 = \frac{f_1^2 R_1'}{F_1^2 P_1'}, \quad \alpha_2 = \frac{f_2^2 R_2'}{F_2^2 P_2'} \tag{6.114}$$

and we choose the Liapunov function

$$2V(Y_{12}, Z_{12}, V_1, V_2) = \frac{f_1 L_1 F_1}{P_1'} V_1^2 + \frac{f_2 L_2 F_2}{P_2'} V_2^2 + g \frac{\dfrac{f_1^2}{P_1'}\dfrac{f_2^2}{P_2'}}{\dfrac{f_1^2}{P_1'} + \dfrac{f_2^2}{P_2'}} Y_{12}^2 + g\left(\frac{f_1^2}{P_1'} + \frac{f_2^2}{P_2'}\right) Z_{12}^2 - \frac{2N_2^2}{\eta^2 g}\int_o^{Y_{12}} \left[\frac{1}{H_o^2} - \frac{\eta}{(H_o + Y_{12})^2}\right] dY \tag{6.115}$$

which, after computations, becomes

$$V(Y_{12}, Z_{12}, V_1, V_2) = \frac{f_1 L_1 F_1}{2P_1'} V_1^2 + \frac{f_2 L_2 F_2}{2P_2'} V_2^2 + g\left[\frac{f_1^2 f_2^2}{2(P_2' f_1^2 + P_1' f_2^2)} - \frac{N_2^2}{\eta^2 g^2 H_o^2 (H_o + Y_{12})}\right] Y_{12}^2 + \frac{1}{2} g \frac{P_2' f_1^2 + P_1' f_2^2}{P_1' P_2'} Z_{12}^2 \tag{6.116}$$

Its derivative by virtue of the system is—$\boldsymbol{W}(Y_{12}, Z_{12}, V_1, V_2)$, where

$$\begin{aligned}\boldsymbol{W}(Y_{12}, Z_{12}, V_1, V_2) = &\left[\frac{2F_1N_2}{\eta(H_o+Y_{12})} - \frac{F_1L_1N_2}{\eta g P'_1(H_o+Y_{12})} + gF_1^2(\alpha_1|V_1|+V_1)\right]V_1^2\\ &+\frac{N_2}{\eta g(H_o+Y_{12})^2}\left(\frac{f_1L_1}{P'_1}+\frac{f_2L_2}{P'_2}\right)V_1V_2\\ &+\left[\frac{2F_2N_2}{\eta(H_o+Y_{12})} - \frac{F_2L_2N_2}{\eta g P'_2(H_o+Y_{12})} + gF_2^2(\alpha_2|V_2|+V_2)\right]V_2^2\end{aligned} \tag{6.117}$$

In the case of small oscillations (corresponding to the linearised system), the Liapunov function takes the form

$$\begin{aligned}\boldsymbol{V}_{\mathrm{L}}(Y_{12}, Z_{12}, V_1, V_2) = &\frac{f_1L_1F_1}{2P'_1}V_1^2 + \frac{f_2L_2F_2}{2P'_2}V_2^2\\ &+g\left[\frac{f_1^2f_2^2}{2(P'_2f_1^2+P'_1f_2^2)} - \frac{N_2^2}{\eta^2g^2H_o^3}\right] + \frac{1}{2}g\frac{P'_2f_1^2+P'_1f_2^2}{P'_1P'_2}Z_{12}^2\end{aligned} \tag{6.118}$$

and its derivative by virtue of the system is $-\boldsymbol{W}_{\mathrm{L}}(V_1, V_2)$, where

$$\begin{aligned}\boldsymbol{W}_{\mathrm{L}}(V_1, V_2) = &\left(\frac{2F_1N_2}{\eta H_o} - \frac{f_2L_1N_2}{\eta g P'_1 H_o^2}\right) + \frac{N_2}{\eta g H_o^2}\left(\frac{f_1L_1}{P'_1}+\frac{f_2L_2}{P'_2}\right)V_1V_2\\ &+\left(\frac{2F_2N_2}{\eta H_o} - \frac{f_2L_2N_2}{\eta g P'_2H_o^2}\right)V_2^2\end{aligned} \tag{6.119}$$

Hence, we deduce the following conditions of stability against small oscillations

$$\begin{aligned}&F_1 > \frac{f_1L_1}{2gP'_1H_o}, \quad F_2 > \frac{f_2L_2}{2gP'_2H_o}\\ &\left(2F_1 - \frac{f_1L_1}{gP'_1H_o}\right)\left(2F_2 - \frac{f_2L_2}{gP'_2H_o}\right) - \frac{1}{4g^2H_o^2}\left(\frac{f_1L_1}{P'_1}+\frac{f_2L_2}{P'_2}\right)^2 > 0\end{aligned} \tag{6.120}$$

The last condition implies

$$H_o > \frac{1}{4g}\left[\frac{f_1L_1}{P'_1F_1} + \frac{f_2L_2}{P'_2F_2} + \sqrt{\left(\frac{f_1L_1}{P'_1F_1}+\frac{f_2L_2}{P'_2F_2}\right)^2 + \frac{1}{F_1F_2}\left(\frac{f_1L_1}{P'_1}-\frac{f_2L_2}{P'_2}\right)^2}\right] \tag{6.121}$$

or

$$H_o > \frac{1}{4g}\left[\frac{f_1L_1}{P'_1F_1} + \frac{f_2L_2}{P'_2F_2} + \sqrt{\left(\frac{1}{F_1}+\frac{1}{F_2}\right)\left(\frac{L_1^2f_1^2}{P_1'^2F_1}+\frac{L_2^2f_2^2}{P_2'^2F_2}\right)}\right] \tag{6.122}$$

This inequality yields a sufficient condition for the stability of small oscillations and can replace the formerly mentioned conditions (6.100), (6.101). Introducing the quantities F_{10} and F_{20}, condition (6.122) becomes

$$\left[\left(\frac{F_{10}}{F_1}+\frac{F_{20}}{F_2}\right)+\sqrt{\left(\frac{1}{F_1}+\frac{1}{F_2}\right)\left(\frac{F_{10}^2}{F_1}+\frac{F_{20}^2}{F_2}\right)}\right]<2 \tag{6.123}$$

When the difference between F_{10} and F_{20} is small, the sufficient condition for the stability of the system against small oscillations becomes even simpler

$$\frac{1}{F_1}+\frac{1}{F_2}<\frac{1}{\max(F_{10},F_{20})} \tag{6.124}$$

This condition is similar in form to the Escande's condition [9]

$$\frac{1}{F_1}+\frac{1}{F_2}<\frac{1}{F_{10}}+\frac{1}{F_{20}} \tag{6.125}$$

which is necessary but not sufficient for the stability of the system against small oscillations. Simplified conditions can also be obtained separately for the parameters F_1 and F_2.

In fact, using the obvious algebraic inequality $ab \geq -1/2\,(a^2+b^2)$, relation (6.119) yields the inequality

$$\begin{aligned} W_L(V_1,V_2) > & \left[\frac{2F_1N_1}{H_o}-\frac{f_1L_1N_1}{\eta g P_1' H_o^2}-\frac{N_2}{2\eta g H_o^2}\left(\frac{f_1L_1}{P_1'}+\frac{f_2L_2}{P_2'}\right)\right]V_1^2 \\ & +\left[\frac{2F_2N_2}{\eta H_o}-\frac{f_2L_2N_2}{\eta g P_2' H_o^2}-\frac{N_2}{2\eta g H_o^2}\left(\frac{f_1L_1}{P_1'}+\frac{f_2L_2}{P_2'}\right)\right]V_2^2 \end{aligned} \tag{6.126}$$

implying for F_1 and F_2 separately the simple conditions

$$\begin{aligned} F_1 &> \frac{f_1L_1}{2P_1'gH_o}+\frac{1}{4gH_o}\left(\frac{f_1L_1}{P_1'}+\frac{f_2L_2}{P_2'}\right) \\ F_2 &> \frac{f_2L_2}{2P_2'gH_o}+\frac{1}{4gH_o}\left(\frac{f_1L_1}{P_1'}+\frac{f_2L_2}{P_2'}\right) \end{aligned} \tag{6.127}$$

or

$$\begin{aligned} F_1 &> F_{10}+\frac{1}{2}(F_{10}+F_{20}) \\ F_2 &> F_{20}+\frac{1}{2}(F_{10}+F_{20}) \end{aligned} \tag{6.128}$$

For the analysis of general hydraulic stability of the system (the case of large oscillations) we consider the case $\alpha_1 < 1;\ \alpha_2 < 1$, which occurs most frequently in practice.

In this case,

$$
\begin{aligned}
W(Y_{12}, Z_{12}, V_1, V_2) = {} & \left[\frac{2F_1N_1}{\eta(H_o + Y_{12})} - \frac{f_1L_1N_2}{\eta g P_1'(H_o + Y_{12})^2} + (1-\alpha_1)gF_1^2V_1\right]V_1^2 \\
& - \frac{N_2}{2\eta g(H_o + Y_{12})^2}\left(\frac{f_1L_1}{P_1'} + \frac{f_2L_2}{P_2'}\right)(V_1^2 + V_2^2) \\
& + \left[\frac{2F_2N_2}{\eta(H_o + Y_{12})} - \frac{f_2L_2N_2}{\eta g P_2'(H_o + Y_{12})^2} + (1-\alpha_2)gF_2^2V_2\right]V_2^2
\end{aligned}
\tag{6.129}
$$

yielding the conditions

$$
\begin{aligned}
& 2F_1(H_o + Y_{12}) > \frac{f_1L_1}{P_1'g} + \frac{1}{2g}\left(\frac{f_1L_1}{P_1'} + \frac{f_2L_2}{P_2'}\right) \\
& 2F_2(H_o + Y_{12}) > \frac{f_2L_2}{P_2'g} + \frac{1}{2g}\left(\frac{f_1L_1}{P_1'} + \frac{f_2L_2}{P_2'}\right) \\
& (1-\alpha_1)gF_1^2V_1 + \frac{2F_1N_1}{\eta(H_o + Y_{12})} - \frac{N_2}{2\eta g(H_o + Y_{12})^2}\left(\frac{3f_1L_1}{P_1'} + \frac{f_2L_2}{P_2'}\right) > 0 \\
& (1-\alpha_2)gF_2^2V_2 + \frac{2F_2N_2}{\eta(H_o + Y_{12})} - \frac{N_2}{2\eta g(H_o + Y_{12})^2}\left(\frac{f_1L_1}{P_1'} + \frac{3f_2L_2}{P_2'}\right) > 0
\end{aligned}
\tag{6.130}
$$

or

$$
\begin{aligned}
& V_1 > \frac{N_2}{\eta g(1-\alpha_1)(H_o + Y_{12})F_1^2}\left[\frac{1}{2g(H_o + Y_{12})}\left(\frac{3f_1L_1}{P_1'} + \frac{f_2L_2}{P_2'}\right) - 2F_1\right] \\
& V_2 > \frac{N_2}{\eta g(1-\alpha_2)(H_o + Y_{12})F_2^2}\left[\frac{1}{2g(H_o + Y_{12})}\left(\frac{f_1L_1}{P_1'} + \frac{3f_2L_2}{P_2'}\right) - 2F_2\right]
\end{aligned}
\tag{6.131}
$$

which allow the plotting of curves (Φ_1) and (Φ_2) corresponding to the curve (Φ) from the case of a single surge tank. For the analysis of the domain of attraction we shall use the following functions given by relation (6.116)

$$
\begin{aligned}
& V_1(Y_{12}, V_1) = \frac{f_1L_1F_1}{2P_1'}V_1^2 + g\left[\frac{f_1^2f_2^2}{2(P_2'f_1^2 + P_1'f_2^2)} - \frac{N_2^2}{\eta^2g^2H_o^2(H_o + Y_{12})}\right]Y_{12}^2 \\
& V_2(Y_{12}, V_2) = \frac{f_2L_2F_2}{2P_2'}V_2^2 + g\left[\frac{f_1^2f_2^2}{2(P_2'f_1^2 + P_1'f_2^2)} - \frac{N_2^2}{\eta^2g^2H_o^2(H_o + Y_{12})}\right]Y_{12}^2
\end{aligned}
\tag{6.132}
$$

which allows the plotting of the curves family (ψ_{c1}) and ψ_{c2})

$$\frac{1}{2}F_1V_1^2 + \frac{gf_1}{2L_1}\left(\frac{1}{1+\dfrac{P_2'/f_2^2}{P_1'/f_1^2}} - \frac{2P_1'Q_o^2}{(H_o+Y_{12})f_1^2}\right)Y_{12}^2 = C_1$$

$$\frac{1}{2}F_2V_2^2 + \frac{gf_2}{2L_2}\left(\frac{1}{1+\dfrac{P_1'/f_1^2}{P_2'/f_2^2}} - \frac{2P_2'Q_o^2}{(H_o+Y_{12})f_2^2}\right)Y_{12}^2 = C_2 \tag{6.133}$$

The analysis of the general hydraulic stability of the system will be performed successively by means of the curves (Φ_1), (ψ_{c1}) and (Φ_2), (ψ_{c2}), by means of a method similar to the one used in the case of a single surge tank. In the analysis of the hydraulic scheme in Fig. 6.8a, one should start from the system of differential equations

$$\begin{aligned}
&\frac{L_1}{g}\frac{dW_1}{dt} + Z_1 + P_1'W_1^2 + R_1'\left|\frac{dZ_1}{dt}\right|\frac{dZ_1}{dt} = 0\\
&\frac{L_2}{g}\frac{dW_2}{dt} + Z_2 - Z_1 + P_2'W_2^2 + R_2'\left|\frac{dZ_2}{dt}\right|\frac{dZ_2}{dt} = 0\\
&F_1\frac{dZ_1}{dt} + W_2f_2 = W_1f_1\\
&F_2\frac{dZ_2}{dt} - W_2f_2 = -\frac{N_2}{\eta g(H_2+Z_2)}
\end{aligned} \tag{6.134}$$

and develop a method similar to the above one. The computation method presented here allows the optimum preliminary choice of hydraulic parameters for the design of surge tanks, which can subsequently be checked by numerical methods.

6.3. Numerical methods for analysis of hydraulic stability of hydroelectric power plants protected by surge tanks

The presentation of the analytical methods of hydraulic computation showed that the usual methods of investigation of the surge tanks stability rely on the classification of oscillations in small and large, without associating them with the equivalent concept of amplitude; they correspond inaccurately to the omission of some terms in the differential equations of motion. We also emphasize that the relation defining the stability safety coefficient has an approximate character and consequently must be cautiously used. The available formulae for the computation of extreme oscillations or of the stability of small and large oscillations are approximate due to the simplifying assumptions adopted in order to overcome the mathematical difficulties of the problem. With respect to the formerly mentioned analytical or graphical methods, the numerical methods recently developed have greatly increased the opportunities to understand and control the hydraulic behaviour in transient regimes of hydroelectric power plants protected by surge tanks.

Thus, the simpler numerical methods from two or three decades ago have been improved with the extensive use of electronic computers, allowing the treatment of the complete equations of motion as well as the adoptions of computation assumptions much closer to practical situations. In the following a general numerical method for analysing the general hydraulic stability in operation of hydroelectric power plants endowed with one or two in-line tanks is presented [13, 18, 20, 21].

6.3.1. Systems with a single surge tank

We analyse the case of a hydroelectric power plant provided with a surge tank of variable cross-section, operating automatically at constant power in a separate network (Fig. 5.7). The general equations of the oscillatory water motion in the gallery–tank system, following a power manoeuvre at the station from N_1 to N_2, in usual notation read

$$\frac{L}{g}\frac{\mathrm{d}W}{\mathrm{d}t} + Z + P + E + R = 0$$

$$Wf = F\frac{\mathrm{d}Z}{\mathrm{d}t} + Q_{\mathrm{T}} \tag{6.135}$$

$$\eta f(H + Z - K_{\mathrm{T}}Q_{\mathrm{T}}^2) = N_2$$

where[1]

$$P = \frac{\lambda L}{d}\frac{W\,|W|}{2g} = P'W\,|W|\,; \quad E = \mu\frac{W\,|W|}{2g} = E'W\,|W|$$

$$R = R'V_{\mathrm{s}}\,|V_{\mathrm{s}}|\,; \quad F = F(Z); \quad \eta = \eta(N, H) \tag{6.136}$$

In order to simplify notation, we denote

$$P + E = P'W\,|W| + E'W\,|W| = P^*W\,|W|$$

and adopt a finite difference and mean value system similar to the one used in the previous chapter

$$\Delta t = t_{i+1} - t_i; \quad t_{\mathrm{m}} = t_i + \frac{\Delta t}{2}; \quad Z_{\mathrm{m}} = Z_i + \frac{\Delta Z}{2}$$

$$F_{\mathrm{m}} = \left|\frac{\mathrm{Vol}_{i+1} - \mathrm{Vol}_i}{\Delta Z}\right|; \quad W_{\mathrm{m}} = W_i + \frac{\Delta W}{2}$$

$$V_{\mathrm{sm}} = \frac{F_{\mathrm{m}}}{F_{\mathrm{s}}}\frac{\Delta Z}{\Delta t}; \quad Q_{\mathrm{T}} = \frac{Q_{\mathrm{T},i} + Q_{\mathrm{T},i+1}}{2} \tag{6.137}$$

$$f_{\mathrm{m}} = \frac{L}{\sum_1^n \frac{L_j}{f_j}}$$

[1] The coefficient μ depends on the form of the connection of the surge tank with feed pipe and varies with time as a function of Q_{c}/Q^* where $Q^* = \max(Q, Q_{\mathrm{T}})$; in [6, 10] several solutions of the tank–gallery connection are studied and the values of this coefficient are determined experimentally.

As seen in the previous chapters, the use of the mean variable values on the time interval Δt has proved to ensure a very good computation accuracy. Thus, system (6.135) written in finite differences becomes

$$\begin{aligned}
&\frac{L}{g}\frac{\Delta W}{\Delta t} + Z_{\text{m}} + P^{*} W_{\text{m}} \left| W_{\text{m}} \right| + R' V_{\text{sm}} \left| V_{\text{sm}} \right| = 0 \\
&W_{\text{m}} f_{\text{m}} = F_{\text{m}} \frac{\Delta Z}{\Delta t} + Q_{\text{Tm}} \\
&\eta g (H + Z_{\text{m}} - K_{\text{T}} Q_{\text{T}}^2) Q_{\text{Tm}} = N_2
\end{aligned} \tag{6.138}$$

Substituting the Q_{Tm} term from the second equation in the third one and performing the calculations, we can write system (6.138) in the form

$$\begin{aligned}
&(\Delta W)^3 + a(\Delta W)^2 + b(\Delta W) + c = 0 \\
&\Delta Z = -2 \left[\frac{L}{g}\frac{\Delta W}{\Delta t} + Z_i + R' V_{\text{sm}} \left| V_{\text{sm}} \right| + P^{*} \left| W_i + \frac{\Delta W}{2} \right| \left(W_i + \frac{\Delta W}{2} \right) \right]
\end{aligned} \tag{6.139}$$

where the following notation has been used

$$\begin{aligned}
a &= 6 \left(W_i - \frac{W_{\text{m}}}{f_{\text{m}}} \frac{\Delta Z}{\Delta t} \right) \\
b &= 12 \left(W_i - \frac{F_{\text{m}}}{f_{\text{m}}} \frac{\Delta Z}{\Delta t} \right)^2 - 4 \left(\frac{H + Z_i + \dfrac{\Delta Z}{2}}{K_{\text{T}} f_{\text{m}}^2} \right) \\
c &= 8 \left[\left(W_i - \frac{F_{\text{m}}}{f_{\text{m}}} \frac{\Delta Z}{\Delta t} \right)^2 + \frac{N_2}{\eta g K_{\text{T}} f_{\text{m}}^3} - \frac{\left(W_i - \dfrac{F_{\text{m}}}{f_{\text{m}}} \dfrac{\Delta Z}{\Delta t} \right) \left(H + Z_i + \dfrac{\Delta Z}{2} \right)}{K_{\text{T}} f_{\text{m}}^2} \right]
\end{aligned} \tag{6.140}$$

The third-order equation from system (6.139) can be solved by using Cardan's formulae (if it allows for a single real root) or by an iterative procedure (if it allows for three real roots). In the last case, the root that suits better the physical phenomenon is retained.

For the case in which the influence of the plant connection to a power system is taken into account, the following regulating relation

$$\eta g (H + Z - K_{\text{T}} Q_{\text{T}}^2) Q_{\text{T}} = N_2 \left[1 + \frac{3}{2} \frac{Y}{H_{\text{o}}} (1 - K) \right] \tag{6.141}$$

can be admitted, and the system of equations of motion can be similarly dealt with. We denoted by K the ratio between the power of the station and the total power in the network.

Using a simplified form of the algebraic system (6.139)

$$\begin{aligned}
\Delta W &= \Phi(\Delta Z) \\
\Delta Z &= \Psi(\Delta W, \Delta Z)
\end{aligned} \tag{6.142}$$

and the notation ΔZ_p for the proposed value for ΔZ; ΔZ_c for the computed value for ΔZ; $\Delta_i \delta_i$ as a simplified notation for ΔZ, ΔW, we use the following computation algorithm

Stage I

$$\Delta Z_p = \Delta_1; \quad \Delta W = \Phi(\Delta_1) = \delta_1;$$
$$\Delta Z_c = \Psi(\delta_1, \Delta_1); \quad |\Delta Z_p - \Delta Z_c| = |\varepsilon_1| < \varepsilon$$

Stage II

$$\Delta Z_p = 2\Delta_1 = \Delta_2; \quad \Delta W = \Phi(\Delta_2) = \delta_2;$$
$$\Delta Z_c = \Psi(\delta_2, \Delta_2); \quad |\Delta Z_p - \Delta Z_c| = |\varepsilon_2| < \varepsilon$$

Stage III

$$\Delta Z_p = \Delta_1 - \left(\frac{\Delta_1 - \Delta_2}{\varepsilon_1 - \varepsilon_2}\right)\varepsilon_1 = \Delta_3; \quad \Delta W = \Phi(\Delta_3) = \delta_3;$$
$$\Delta Z_c = \Psi(\delta_3, \Delta_3); \quad |\Delta Z_p - \Delta Z_c| = |\varepsilon_3| < \varepsilon$$

In the iteration process, one takes as a first selected value for the increment ΔZ its value at the previous computation step, and at the first computation step an arbitrary value, consistent with the physical phenomenon. If the accuracy condition $|\Delta Z_p - \Delta Z_c| < \varepsilon$ is satisfied, the corresponding computation is considered to be finished and one passes to the following computation step. If not, the computation is resumed from step III, considering the previously carried through steps II and III, as steps I and II respectively.

The numerical solution of the problem has proved to be stable and rapidly convergent. The accuracy of results is very good for time steps Δt equal to a fraction, $1/n = 1/50 \ldots 1/100$, of the period of ideal oscillations in the tank T^*.

6.3.2. Systems with several surge tanks

Let us consider the case of a hydroelectric power plant protected by two surge tanks of variable cross-section and hydraulic resistances at the base (R_1, $R_2 \neq 0$), located on the water conveyance gallery (Fig. 5.8a) or on both sides of the turbines (Fig. 5.8b), automatically operating at constant power in a separate network. Both cases can be treated in a similar way. We shall present the case of a system of surge tanks located on both sides of the turbines [18, 19, 21].

The general equations of the oscillatory water motion in the system after a power manoeuvre from N_1 to N_2, with the usual notation are the following:

$$\begin{aligned}
&\frac{L_1}{g}\frac{dW_1}{dt} + Z_1 + P_1 + E_1 + R_1 = 0 \\
&\frac{L_2}{g}\frac{dW_2}{dt} - Z_2 + P_2 + E_2 - R_2 = 0 \\
&f_1 W_1 = \frac{dZ_1}{dt}F_1 + Q_T; \quad f_2 W_2 = -\frac{dZ_2}{dt}F_2 + Q_T \\
&\eta g(H + Z - K_T Q_T^2)Q_T = N_2
\end{aligned} \tag{6.143}$$

where

$$P_j = P'W_j\left|W_j\right|; \quad E_j = E_j'W_j\left|W_j\right|; \quad R_j = R_j'V_{sj}\left|V_{sj}\right|;$$
$$F_j = F_j(Z_j); \quad \eta = \eta(H,N); \quad j = 1,2 \tag{6.144}$$

In order to simplify notation we write

$$P_j + E_j = P_j'W_j\left|W_j\right| + E_j'W_j\left|W_j\right| = P^*W_j\left|W_j\right|; \quad j = 1,2 \tag{6.145}$$

We shall adopt a system of finite differences and mean values of the variables, similar to (6.137).

For the study of the general hydraulic stability of the plant operation in transient regimes, the system (6.143) in finite differences takes the form

$$\begin{aligned}
&\text{(a)} \quad \frac{L_1}{g}\frac{\mathrm{d}W_1}{\mathrm{d}t} + Z_{1\mathrm{m}} + P_1^*W_{1\mathrm{m}}\left|W_{1\mathrm{m}}\right| + R_1 = 0\\
&\text{(b)} \quad \frac{L_2}{g}\frac{\Delta W_2}{\Delta t} - Z_{2\mathrm{m}} + P_2^*W_{2\mathrm{m}}\left|W_{2\mathrm{m}}\right| - R_2 = 0\\
&\text{(c)} \quad f_1W_{1\mathrm{m}} = F_{1\mathrm{m}}\frac{\Delta Z_1}{\Delta t} + Q_{\mathrm{Tm}}\\
&\text{(d)} \quad f_2W_{2\mathrm{m}} = -F_{2\mathrm{m}}\frac{\Delta Z_2}{\Delta t} + Q_{\mathrm{Tm}}\\
&\text{(e)} \quad \eta g(H + Z_{1\mathrm{m}} - Z_{2\mathrm{m}} - K_TQ_{\mathrm{Tm}}^2)Q_{\mathrm{Tm}} = N_2
\end{aligned} \tag{6.146}$$

By eliminating successively the term Q_{Tm} between equations (c), (d), and then (d), (e) of the above system, and performing the computations, the system (6.146) becomes

$$\begin{aligned}
&\Delta Z_1 = -2\left[\frac{L_1}{g}\frac{\Delta W_1}{\Delta t} + Z_{1i} + P_1^*\left(W_{1i} + \frac{\Delta W_1}{2}\right)\left|W_{1i} + \frac{\Delta W_1}{2}\right| + R_1\right]\\
&\Delta Z_2 = 2\left[\frac{L_2}{g}\frac{\Delta W_2}{\Delta t} - Z_{2i} + P_2^*\left(W_{2i} + \frac{\Delta W_2}{2}\right)\left|W_{2i} + \frac{\Delta W_2}{2}\right| - R_2\right]\\
&\Delta W_1 = 2\left[\frac{f_{1\mathrm{m}}}{f_1}\frac{\Delta Z_1}{\Delta t} + \frac{F_{2\mathrm{m}}}{f_1}\frac{\Delta Z_2}{\Delta t} + \frac{f_2}{f_1}\left(W_{2i} + \frac{\Delta W_2}{2}\right) - W_{1i}\right]\\
&(\Delta W_2)^3 + \alpha(\Delta W_2)^2 + \beta(\Delta W_2) + \gamma = 0
\end{aligned} \tag{6.147}$$

where the notation below is used:

$$\alpha = 6\left(W_{2i} + \frac{\Delta Z_2}{\Delta t}\frac{F_{2\mathrm{m}}}{f_2}\right)$$

$$\beta = 12\left(W_{2i} + \frac{\Delta Z_2}{\Delta t}\frac{F_{2\mathrm{m}}}{f_2}\right)^2 - \frac{4}{K_\mathrm{T}f_2^2}\left(H + Z_{1i} - Z_{2i} + \frac{\Delta Z_1 - \Delta Z_2}{2}\right)$$

$$\gamma = 8\left[\left(W_{2i} + \frac{\Delta Z_2}{\Delta t}\frac{F_{2m}}{f_2}\right)^3 + \frac{N_2}{\eta g K_T f_2^3} - \frac{\left(W_{2i} + \frac{\Delta Z_2}{\Delta t}\frac{F_{2m}}{f_2}\right)\left(H + Z_{1i} - Z_{2i} + \frac{\Delta Z_1 - \Delta Z_2}{2}\right)}{K_T f_2^2}\right] \tag{6.148}$$

For the case where the influence of the plant operating within a power system is taken into account, the following regulating equation can be admitted:

$$\eta g Q_T(H + Z_1 - Z_2 - K_T Q_T^2) = N_2\left[1 + \frac{3}{2}\frac{Y_1 - Y_2}{H_o}(1 - K)\right] \tag{6.149}$$

and the system of motion equations can be similarly dealt with. With these specifications, the system (6.147) can be written in a simplified form

$$\begin{aligned} \Delta Z_1 &= \Phi_1(\Delta W_1, \Delta Z_1); \quad \Delta Z_2 = \Phi_2(\Delta W_2, \Delta Z_2) \\ \Delta W_1 &= \Psi_1(\Delta Z_1, \Delta Z_2, \Delta W_2); \quad \Delta W_2 = \Psi_2(\Delta Z_1, \Delta W_2) \end{aligned} \tag{6.150}$$

In order to solve this system, the following computation algorithm is used:

Stage I

$$\begin{aligned} &\Delta Z_1^p = \Delta_1'; \quad \Delta Z_2^p = \Delta_1'' \\ &\Delta W_2 = \Psi_2(\Delta_1', \Delta_1'') = \delta_1''; \quad \Delta W_1 = \Psi_1(\Delta_1', \Delta_1'', \delta_1'') = \delta_1' \\ &\Delta Z_1^c = \Phi_1(\delta_1', \Delta_1'); \quad \Delta Z_2^c = \Phi_2(\delta_1'', \Delta_1'') \\ &\left|\Delta Z_1^p - \Delta Z_1^c\right| = \left|\varepsilon_1'\right| < \varepsilon; \quad \left|\Delta Z_2^p - \Delta Z_2^c\right| = \left|\varepsilon_1''\right| < \varepsilon \end{aligned}$$

Stage II

$$\begin{aligned} &\Delta Z_1^p = 2\Delta_1' = \Delta_2'; \quad \Delta Z_2^p = 2\Delta_1'' = \Delta_2'' \\ &\Delta W_2 = \Psi_2(\Delta_2', \Delta_2'') = \delta_2''; \quad \Delta W_1 = \Psi_1(\Delta_2', \Delta_2'', \delta_2'') = \delta_2' \\ &\Delta Z_1^c = \Phi_1(\delta_2', \Delta_2'); \quad \Delta Z_2^c = \Phi_2(\delta_2'', \Delta_2'') \\ &\left|\Delta Z_1^p - \Delta Z_1^c\right| = \left|\varepsilon_2'\right| < \varepsilon; \quad \left|\Delta Z_2^p - \Delta Z_2^c\right| = \left|\varepsilon_2''\right| < \varepsilon \end{aligned}$$

Stage III

$$\begin{aligned} &\Delta Z_1^p = \Delta_1' - \left(\frac{\Delta_1' - \Delta_2'}{\varepsilon_1' - \varepsilon_2'}\right)\varepsilon_1' = \Delta_3' \\ &\Delta Z_2^p = \Delta_1'' - \left(\frac{\Delta_2'' - \Delta_1''}{\varepsilon_1'' - \varepsilon_2''}\right)\varepsilon_1'' = \Delta_3'' \\ &\Delta W_2 = \Psi_2(\Delta_3', \Delta_3'') = \delta_3''; \quad \Delta W_1 = \Psi_1(\Delta_3', \Delta_3'', \delta_3'') = \delta_3' \\ &\Delta Z_1^c = \Phi_1(\delta_3', \Delta_3'); \quad \Delta Z_2^c = \Phi_2(\delta_3'', \Delta_3'') \\ &\left|\Delta Z_1^p - \Delta Z_1^c\right| = \left|\varepsilon_3'\right| < \varepsilon; \quad \left|\Delta Z_2^p - \Delta Z_2^c\right| = \left|\varepsilon_3''\right| < \varepsilon \end{aligned}$$

In the iteration process, for the first values of ΔZ_1, ΔZ_2 one uses their values at the previous computation step while for the first step an arbitrary set of values, consistent with the phenomenon, is chosen. The iterative computation is performed according to the specifications made for the case of one surge tank. The computation method can easily be written in an algorithmic language and used on middle capacity computers.

6.4. Applications to hydroelectric power plants

6.4.1. The Somes–Mariselu hydroelectric power plant

A large hydraulic study for this plant has been performed during the design stage. The surge tank has been sized for maximum oscillations and hydraulic stability and has been checked by several computation methods [18, 20]. Some of the results are presented here in a concise form. The main characteristics (Fig. 6.10) of the plant are: $L = 8{,}130$ m; $d = 4.40$ m; $L_T = 740$ m; $d_T = 3.40$ m; $H = 470$ m; $N = 215$ MW. The surge tank of the plant is of cylindrical form ($D_1 = 6$ m), with upper chamber ($D_2 = 20$ m), gallery shaped lower chamber ($\ell = 150$ m, $d = 5$ m) and diaphragm at the base. The head loss in the tank diaphragm has been admitted of the form $R = 0.00416 Q_c^2$.

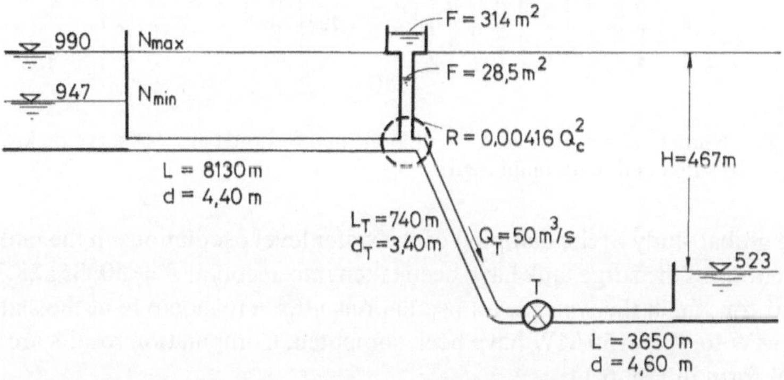

Figure 6.10a. Scheme of the Somes–Mariselu hydroelectric power plant.

Figure 6.10b. Somes–Mariselu hydroelectric power plant: machine room.

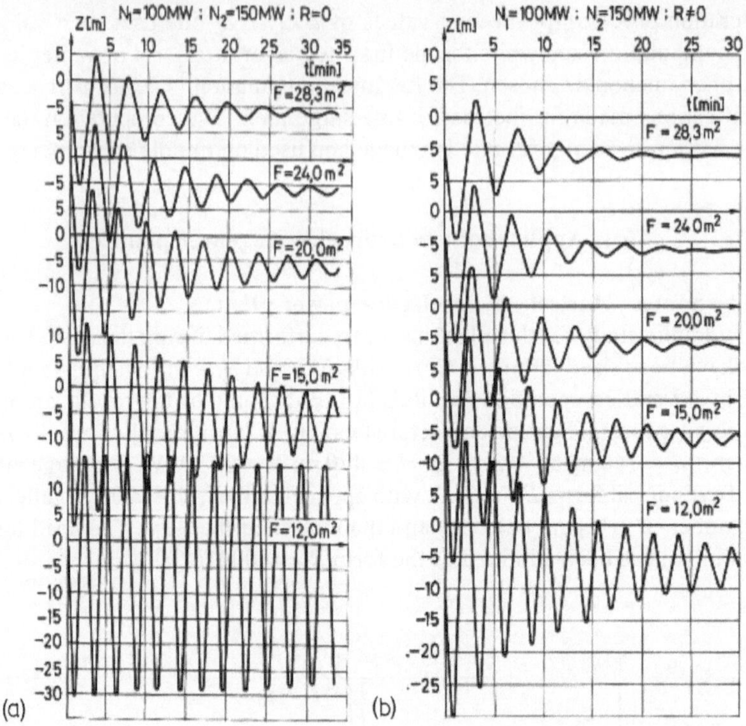

Figure 6.11. Numerical simulation of oscillations in the surge tank: (a) surge tank without diaphragm; (b) surge tank with diaphragm.

In the global study of the damping of the water level oscillations in the tank, several cross-sections of the surge tank have been taken into account: $F = 50$; 35; 28.3; 24; 20; 15; 12 m^2 for which the water level oscillations after a manoeuvre at the station from $N_1 = 10$ MW to $N_2 = 150$ MW have been computed. Computation results are shown in graphical form in Fig. 6.11.

For the design solution of the surge tank with the above dimensions, computations have been carried out for several operation manoeuvres, and the results are synthetically presented in Fig. 6.12 [18]. We mention the stability safety coefficient

$$n = \frac{F}{F_{\mathrm{Th}}} = \frac{28.3}{15.6} = 1.82$$

An analysis of the general hydraulic stability of the surge tank can be conducted also in a different way, by means of an analytical method [22]. One computes the coefficient $\alpha = f^2 R'/F^2 P'$, introduced by us, and gets $\alpha = 0.87$.

The extreme power manoeuvre: $N_1 = 0$, $N_2 = 215$ MW is considered and the cross-sections of the tank $F = 12$; 20; 28.3 m^2 are analysed. One plots the curves (Φ) and the curves family (ψ_c), by means of which an analysis of the general hydraulic stability of the tank (Fig. 6.13) can be performed. A(Y, V) and B(Y, V) denote the points corresponding to the total starting manoeuvre and the variant $\alpha = 0.61$ has also been taken into account. One should note the agreement between these results and the results obtained by numerical simulation: the cross-section $F = 12$ m^2 yields an unstable operation of the hydroelectric power plant, while the cross-section $F = 28.3$ m^2,

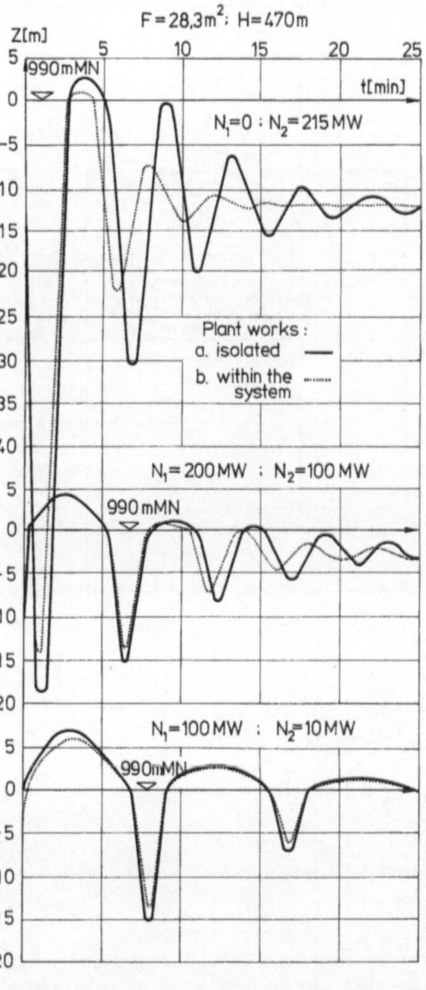

Figure 6.12. Computed oscillations in the surge tank, for several design variants.

adopted for the design of the hydroelectric power plant, ensures a stable operation at transient regimes.

We note that effective application of the method begins with the plotting of curves (Φ) for several values of F and α, eliminating for the beginning the general instability situations of the hydroelectric power plant, by which the corresponding (Φ) curves cut the Y axis at points of positive abscissas (Fig. 6.13).

6.4.2. Hydroelectric power plant with two surge tanks

Let us consider the case of a hydroelectric power plant protected by two surge tanks located on both sides of the turbines, with the following main characteristics: $L_1 = 5{,}000$ m; $d_1 = 7$ m; $L_T = 300$ m; $d_T = 5.6$ m; $L_2 = 500$ m; $d_2 = 7$ m; $H = 130$ m; $N = 180$ MW and analyse by numerical and analytical methods the general hydraulic stability of the system (Fig. 6.8b). The surge tanks are of constant cross-section, with diaphragms at the base. The head losses in the diaphragms have been admitted of the form $R_1 = R_2 = 0.001Q_c^2$.

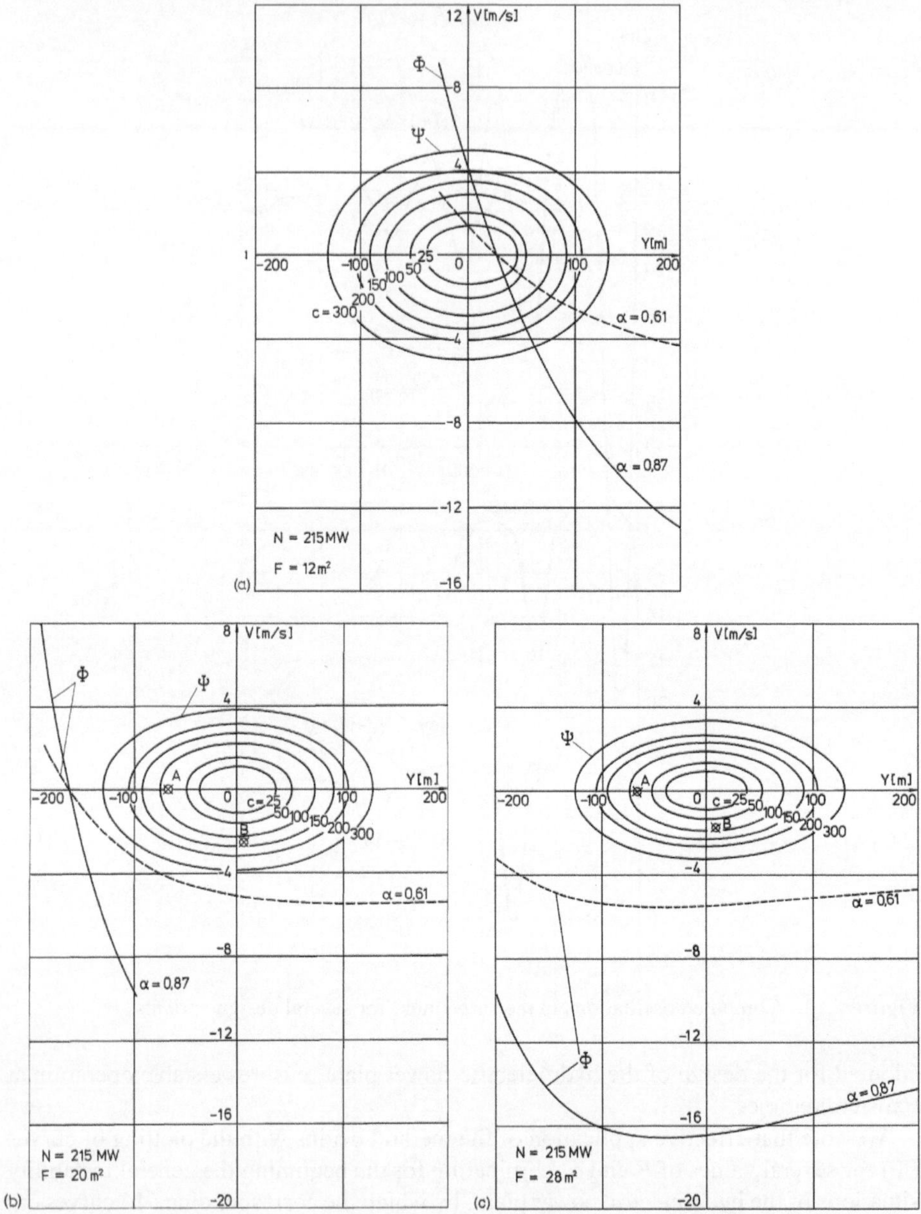

Figure 6.13. Study of the general hydraulic stability of the operation of the hydroelectric power plant by means of the curves Φ and ψ_c: (a) unstable working; (b, c) stable working.

Numerical simulation of the transient water motion in the system at a power manoeuvre in the station from $N_1 = 100$ MW to $N_2 = 150$ MW, led to the results presented in graphical form in Fig. 6.15 [19, 22]. The possibility of general hydraulic instability occurring at some cross-sections of the tanks and the suggestive computational reflection of this phenomenon should be mentioned. A more detailed investigation of the

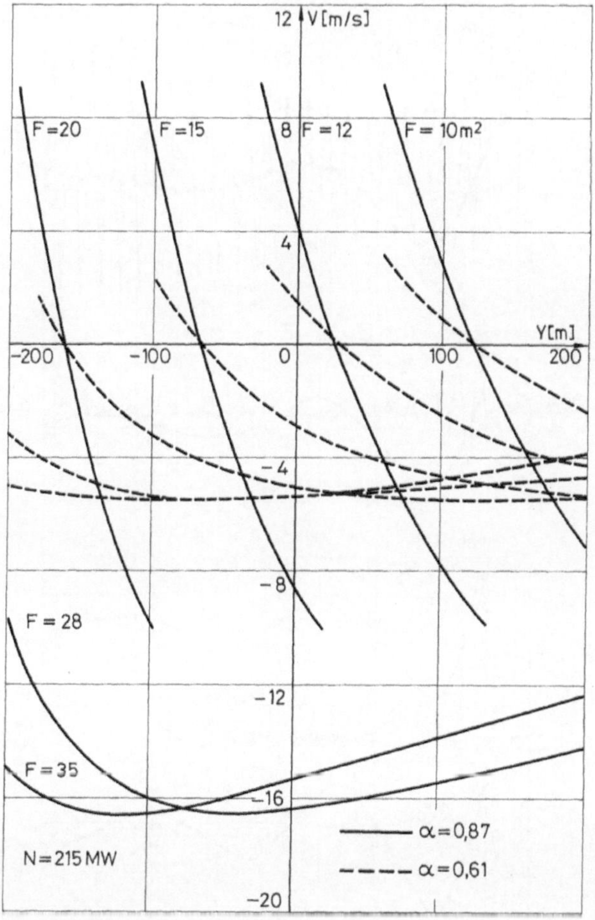

Figure 6.14. Representation of the curves Φ.

damping of water level oscillations in the two surge tanks has been performed for the cross-sections

$$F_1 = 350\,\text{m}^2, \quad F_2 = 130\,\text{m}^2$$

and for several manoeuvres at the station chosen from the most unfavourable in the operation of the hydroelectric power plant [18, 20].

The analysis of the general hydraulic stability of the system has been made under the assumption $\alpha_1 < 1$, $\alpha_2 < 1$ and the cases $\alpha_1 = \alpha_2 = 0.75$ and $\alpha_1 = \alpha_2 = 0.87$ [12] have been analysed.

Computations have been performed for the following values of the cross-sections: F_1; $F_2 = 200$; 300; ... 700; 800 m^2 (Fig. 6.16), and two regimes of the hydroelectric power plant operation, $N_2 = 100$ and 150 MW, have been taken into account. Computations results for F_1; $F_2 = 800$; 400 m^2 and $N_2 = 150$ MW are given in graphical form in Fig. 6.17. Since this example assumes equal cross-sections for the water conveyance gallery and the tailrace ($d_1 = d_2 = 7$ m), the resulting curves Φ_1, Φ_2 coincide with ψ_{c1} and ψ_{c2} respectively and analysis of the system stability could be conducted by means

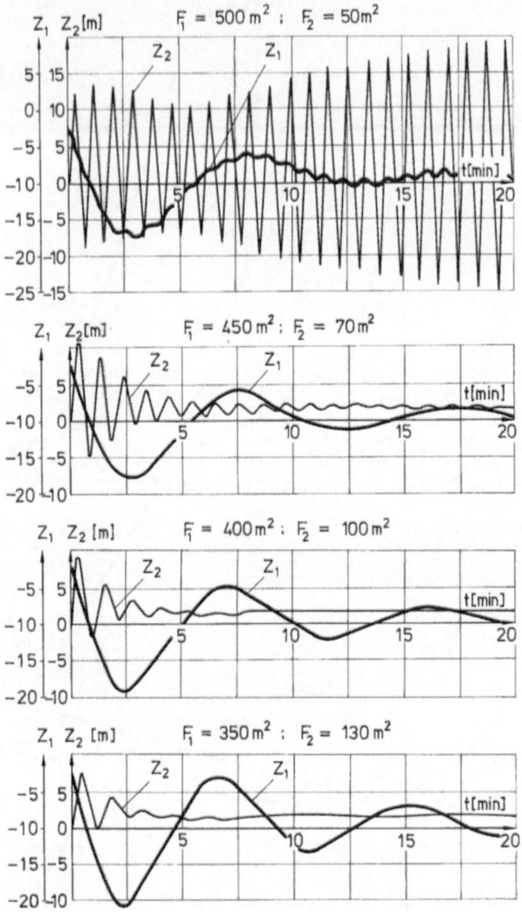

Figure 6.15. Oscillations computed in the system for different cross-sections of the surge tanks ($N_1 = 100$ MW, $N_2 = 150$ MW).

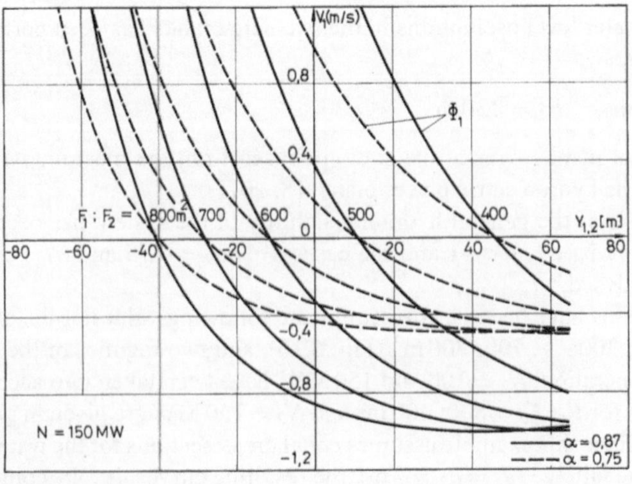

Figure 6.16. Graph of the $\Phi_1 = \Phi_2$ curves.

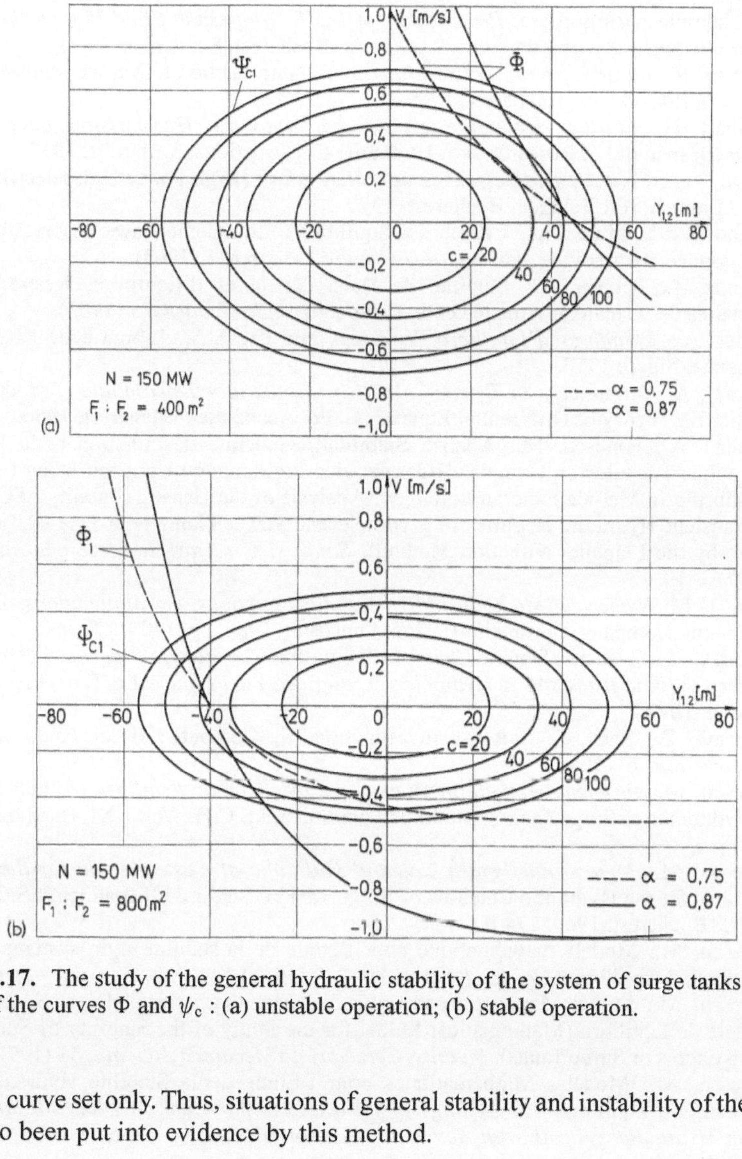

Figure 6.17. The study of the general hydraulic stability of the system of surge tanks by means of the curves Φ and ψ_c : (a) unstable operation; (b) stable operation.

of Φ, ψ_c curve set only. Thus, situations of general stability and instability of the system have also been put into evidence by this method.

References

1. Anderson, A.: A Novel Type of Surge Shaft Instability. *Proc. Inst. Civ. Engess*, Part 2, September, 1979.
2. Aronovici, G., Kartvelisvili, N., Liubimtev, K.: *Ghidravliceskii Udar i Uravnitelnie Reservoari.* Nauka, Moscow, 1968.
3. Arsenie, D., Halanay, A., Popescu, M.: Liapunov Techniques for the Study of Hydraulic Stability, *Proceedings of the International Conference on Unsteady Flow and Fluid Transients*, Durham, United Kingdom, 1992, pp. 291–294.
4. Arsenie, D.: Un Critèrium de Stabilité et Quelques Inégalités Nouvelles Obtenues par la Méthode de la Fonction de Liapunov dans le Cas de Moyennes Oscillations dans

des Cheminées d'Équilibre. *Proceedings of IAHR Symposium Fluid Motion Stability in Hydraulic Systems with Automatic Regulators*, Bucharest, September 26–29, 1976.

5. Beliman, R.: *Introducere in Analiza Matriceala*, (Introduction to Matrix Analysis) (from Russian). Ed. Tehnica, Bucharest, 1969.
6. Bernhart, H.: *Einfluss der Schlieszeit auf die Druckstos. Transmission durch Wasserschlosser.* Institute of Researches of Hydraulic Constructions, Karlsruhe, 1977.
7. Brown, J.G.: *Centrale Hidroelectrice de Putere Mare* (High Power hydroelectric plants) (from English). Ed. Tehnica, Bucharest, 1970.
8. Escande, L.: Stabilité d'une Chambre d'Équilibre à Étranglement avec asservissement de la Puissance Hydraulique. *Journal of Hydraulic Research*, 1 (1963).
9. Escande, L., Huron, R.: Stabilité de Deux Chambres d'Équilibre, Respectivement Solidaires des Canaux d'Amenée et de Fuite. *La Houille Blanche*, 5 (1953).
10. Gardel, A.: *Chambres d'Équilibre*, F. Rouge and Cie S.A., Librairie de l'Université, Lausanne, Suisse 1956.
11. Halanay, A.: *Introducere in Teoria Calitativa a Ecuatiilor Diferentiale* (Introduction to Qualitative Theory of Differential Equations). Ed. Academiei, Bucharest, 1956.
12. Halanay, A., Popescu, M.: Analiza Stabilitatii Generale de Functionare la Regimuri Hidraulice Tranzitorii a Uzinelor Hidroelectrice Prevazute cu Unul sau Doua Castele de Echilibru prin Metoda Functiei Liapunov (Analysis of the General Stability of Operation to Transient Hydraulic Regimes of Hydroelectric Power Plants with One or Two Surge Tanks by the Liapunov Function Method). *Studii si Cercetari de Mecanica Aplicata*, 1 (1979) 3–19.
13. Ixaru, G.L.: *Numerical Methods for Differential Equations and Applications.* D. Reidel Publishing Company, Dordrecht, Boston, Lancaster, 1984, pp. 170–177.
14. Jaeger, C.: Le System à Double Chambre d'Équilibre. *La Houille Blanche*, 4 (1957).
15. Jaeger, C.: *Fluid Transients in Hydroelectric Engineering Practice.* Ed. Blanckie, Glasgow, London, 1977.
16. Mosonyi, E., Nagy, I.: Stability Investigations by Computer. *Water Power and Dam Construction*, 7 (1964).
17. Popescu, M.: *Probleme Actuale in Domeniul Castelelor de Echilibru* (Actual Problems of Hydraulics of Surge Tanks). Studii de Hidraulica, I.S.C.H., Vol. XXI, Bucharest, 1969, 225 pp.
18. Popescu, M.: *Metode noi Pentru Calculul Hidraulic al Castelelor de Echilibru* (New Methods for the Hydraulic Calculus of Surge Tanks). Studii de Hidraulica, I.S.C.H., Vol. XXIII, Bucharest, 1969, 174 pp.
19. Popescu, M.: Modèle mathématique pour l'étude de la stabilité d'un système de deux chambres d'équilibre. *C.R. Acad. Sci. Paris*, 273 (1971) 129–132.
20. Popescu, M.: Model Matematic pentru Studiul Stabilitatii Castelelor si Sistemelor de Castele de Echilibru (Mathematical Model for the Study of the Stability of Surge Tanks and Systems of Surge Tanks). *Studii si Cercetari de Mecanica Aplicata*, 31 (1972) 27–49.
21. Popescu, M.: Modèles Mathématiques pour l'Étude de la Stabilité Hydraulique des Chambres d'Équilibre. *Proceedings of the IAHR Symposium on Fluid Motion Stability in Hydraulic Systems with Automatic Control*, Bucharest, September 26–29, 1976, pp. 216–227.
22. Popescu, M., Halanay, A.: Analyse de la Stabilité Hydraulique Générale de Fonctionnement aux Régimes Transitoires des Usines Hydroélectriques Munies d'un ou de Deux Chambres d'Équilibre. *XVIII Congres AIRH*, Cagliari, Italia, 1979 pp. 85–97.
23. Razvan, V.: *Teoria Stabilitatii* (Stability Theory). Ed. Stiintifica si Enciclopedica, Bucharest, 1987.
24. Sideriades, L.: Méthodes de Topologie Qualitative: Application à l'Étude des Cheminées d'Équilibre. *La Houille Blanche*, 4 (1962).
25. Stucky, A.: *Cours d'Aménagement des Chutes d'Eau. Chambres d'Équilibre.* Ecole Polytechnique de l'Université, Lausanne, 1958.
26. Vucetic J.: Nouveau Critère de Stabilité des Cheminées d'Équilibre en Régime Initial Non-Permanent. *Le Génie Civil*, 145 (1968).
27. Zicman, B.: Méthodes Nouvelles pour le Calcul des Cheminées d'Équilibre. *La Houille Blanche*, 5, (1953).

Chapter 7

Experimental Results in the Study of Mass Oscillations

The specialised literature records a great number of experimental studies about the mass oscillation phenomenon. Among those, we note the papers [1, 2, 6, 8], containing investigations carried out in laboratory installations and [3–5, 7], exhibiting measurements in hydraulic installations in operation.

7.1. Laboratory studies

7.1.1. Experimental results on the I.C.H. installation

In Chapter 2 we showed the laboratory installation designed and worked out at the Hydraulic Research Institute (I.C.H.) in Bucharest, on which all types of non-permanent motions in hydraulic systems under pressure have been studied.

For the study of the *mass oscillation* phenomenon, experiments have been carried out with the same range of permanent regime flows as for the study of the *waterhammer* phenomenon. The *mass oscillation* phenomenon has been experimentally simulated and measurements have been carried out on the fundamental hydraulic schemes I and II (Figs. 2.3 and 2.5), provided with protection installations of the surge tank and air chamber type [8, 9].

Hydraulic scheme I, with surge tank (I-A)

The scheme is shown in Figs. 2.3 and 2.4. The surge tank of the experimental installation ($H = 10.5$ m, $D = 200$ mm) is equipped at its upper and lower part with a graded surveyor's rod of 3 m length and a 2 atm gauge, respectively. The surveyor's rod allows direct reading of the instantaneous water level on a tank part made of transparent material.

Variations of the water level in the tank have been measured during experiments either by data recording on electronic devices, or by direct reading on the manometer, at 5 s time intervals. Sudden valve closing or opening manoeuvres have been performed, simulating the stopping or starting manoeuvres of a hydroelectric power plant protected by surge tank. Out of the great number of experiments carried out, Fig. 7.1 presents level measurements performed in the tank, for two regime flows. One ascertains the occurrence of harmonic damped oscillations at the sudden valve closing and of non-periodical oscillations at the valve opening manoeuvres. The harmonic oscillation period of the water level in the tank is 43 s.

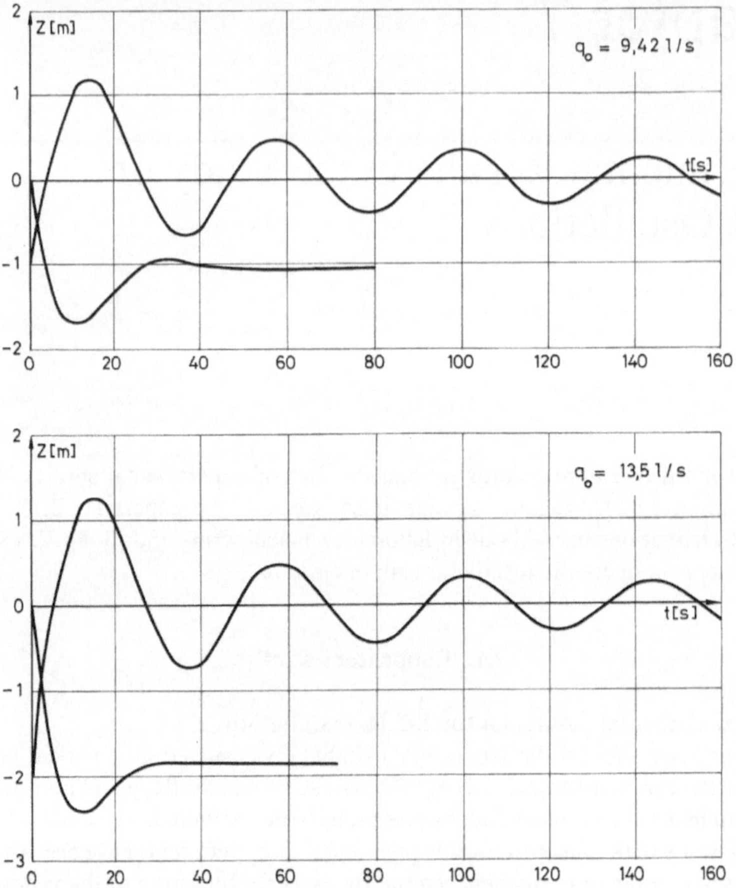

Figure 7.1. Scheme I-A: level measurements in the surge tank at sudden valve closing or opening manoeuvres, for two regime flows.

We note the slowly varying character of the water motion in the system (mass oscillation) caused by the surge tank in the installation (7.1) as compared to the rapidly varying water motion in the same system, in the absence of protection means (Fig. 2.7).

Hydraulic scheme I, with air chamber (I-B)

The scheme is shown in Figs. 2.3 and 2.4. The air chamber of the experimental installation ($L = 2$ m; $D = 300$ mm) is equipped with six vertically fitted small pipes with taps, ensuring the access of controlled quantities of the air in the chamber (between 10.5 and 63.5 dm^3), at regime pressure. The chamber is supplied with air from an electrically driven compressor. Experiments have been performed with simultaneous pressure recording at three points on the conduit, for several regime flows and volumes of the air cushion in the chamber. A part of these results are presented in Fig. 7.2, for the regime flow $q_o = 17.8$ l/s. One can experimentally notice that the air chamber effect on the non-permanent water motion in the system is similar to the effect encountered at the surge tank.

The analysis of the experimental results shows that the increase in the air cushion volume in the chamber leads to a decrease in the amplitudes and frequencies of water

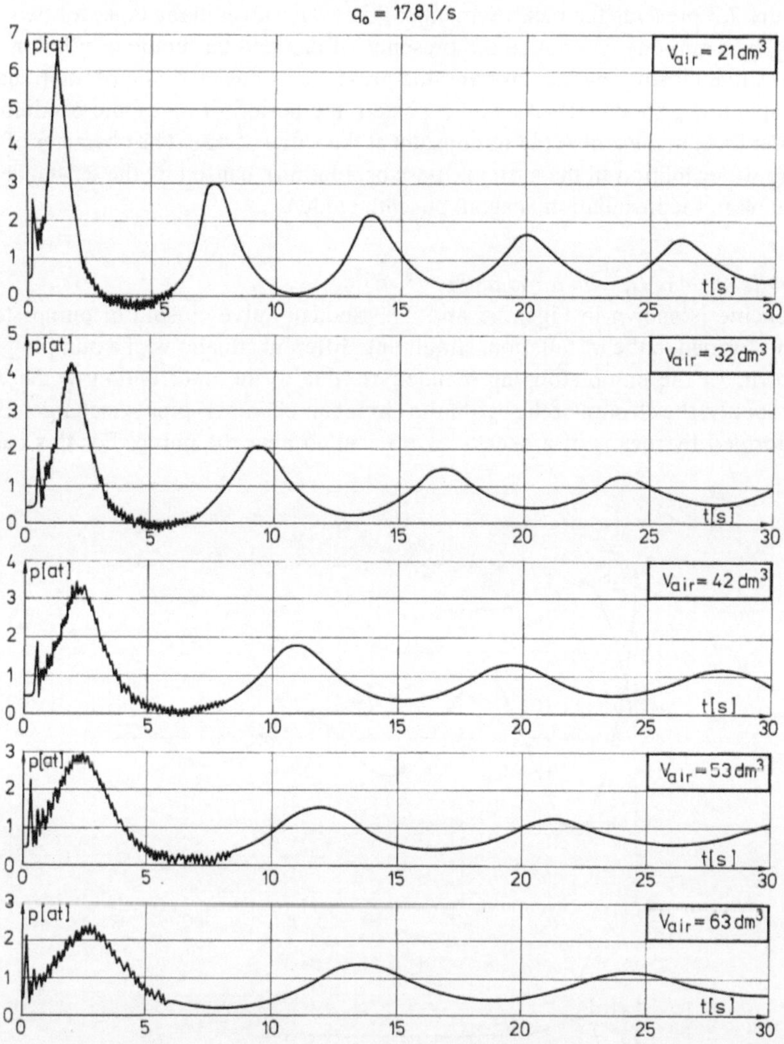

Figure 7.2. Scheme I-B: pressure measurements in cross-section of the pipe, at the sudden valve-closing manoeuvre, for different volumes of the air cushion in the air chamber.

pressure oscillations in the system. The pressure peak, appearing on diagrams immediately after the sudden closing of the valve, is due to the delay in the operation of the chamber, caused by the connection pipe to the chamber ($\ell_2 = 4.70\,\text{m}$).

Hydraulic scheme II, with surge tank (II-A)

The scheme is shown in Figs. 2.5 and 2.6. Experiments similar to the previously mentioned case of hydraulic scheme I with surge tank (I-A) have been made. In this case, the sudden valve closing and opening manoeuvres simulate the stop and start manoeuvres of a pumping station protected by a surge tank at its base. In the operation practice of pumping stations the sudden start manoeuvres are not used, the start of the station being normally controlled and performed slowly.

Figure 7.3 presents the measurements of the water level in the tank, for two regime flows. Here too, one can notice the presence of damped harmonic oscillations at the sudden valve closing manoeuvres and the presence of non-periodic oscillations at the valve opening manoeuvres. As was expected, the period of harmonic oscillations of the water level in the tank remains constant at the value of 43 s. The character of slowly varying water motion in the system (mass oscillation), caused by the existence of the surge tank in the installation is again put into evidence.

Hydraulic scheme II, with air chamber (II-B)
The scheme is shown in Figs. 2.5 and 2.6. Sudden valve closing or pump stopping were performed in the installation, simulating different situations of a pumping station shutdown. At the pump-stopping manoeuvre, due to the interruption of the electric power supply, the closure of the hydraulic circuit air chamber–pipe–discharge reservoir is performed by means of a check-valve installed near the pump. For this case too,

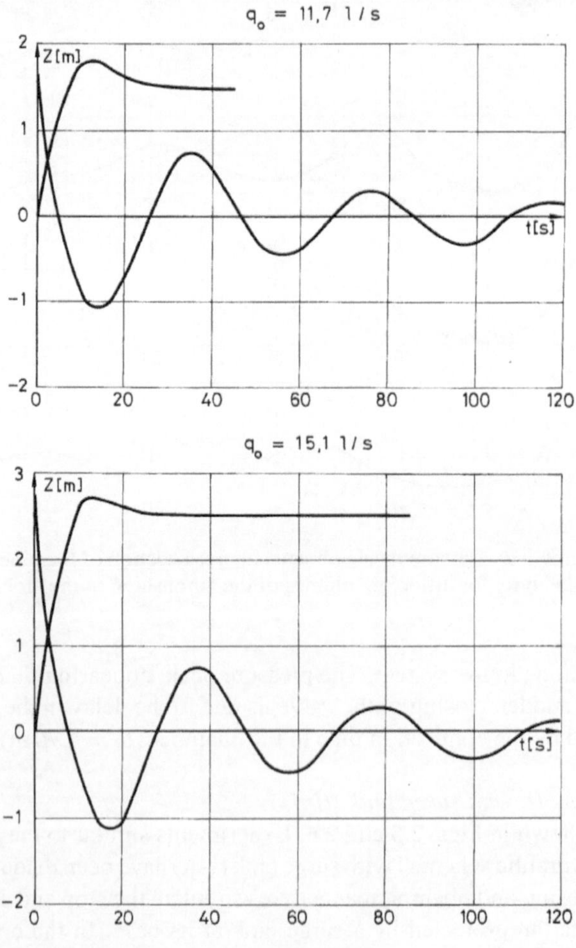

Figure 7.3. Scheme II-A: level measurements in the surge tank, at sudden valve closing or opening manoeuvres, for two regime flows.

experiments of simultaneous pressure recording has been performed at three points on the pipe, for several regime flows and volumes of the air cushions in the chamber.

The influence of the air chamber as a protection mean of a pumping station has been investigated in two variants: air chamber with simple connection pipe, and air chamber equipped with non-symmetrical hydraulic resistance device on the connection pipe.

Air chamber with simple connection pipe

In order to illustrate the influence of the air chamber on the non-permanent motion of water in the system, Fig. 7.4 presents part of the experimental results obtained for the two types of flow manoeuvres: (a) sudden valve closure, (b) stop of the pump.

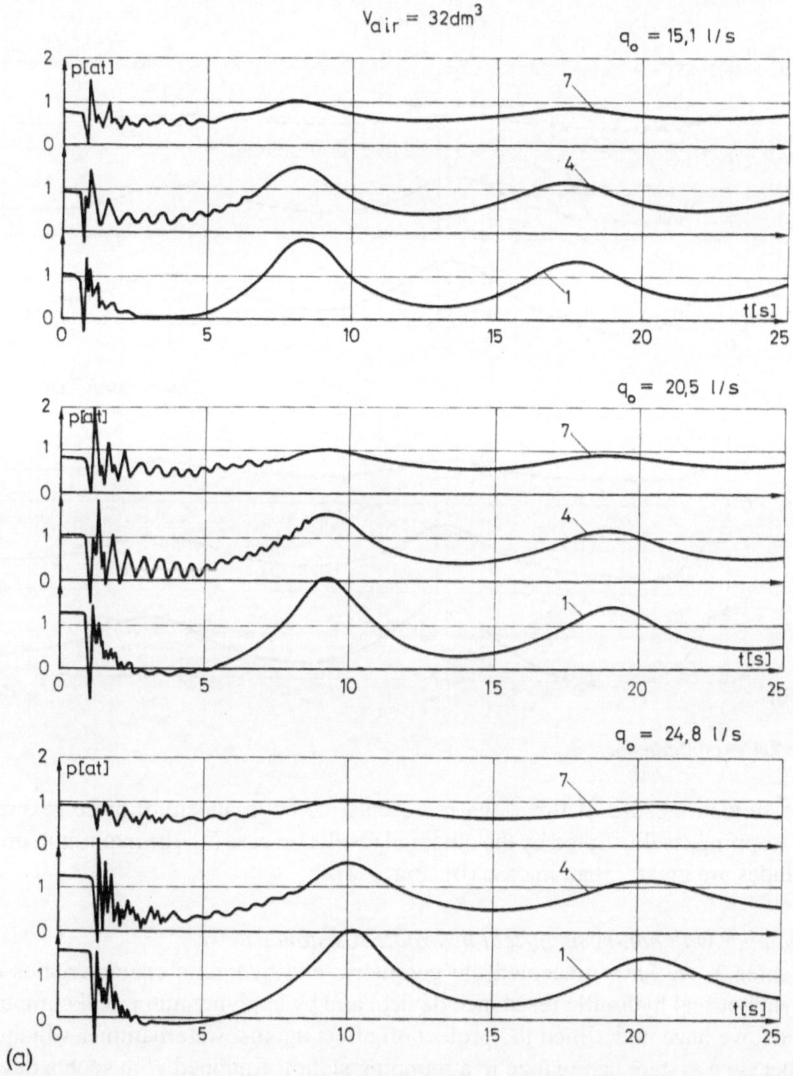

Figure 7.4. Scheme II-B: pressure measurements at three points of the pipe, for several regime flows (volume of the air cushion in the air chamber = 32 dm^3); (a) sudden valve closure; (b) stop of the pump.

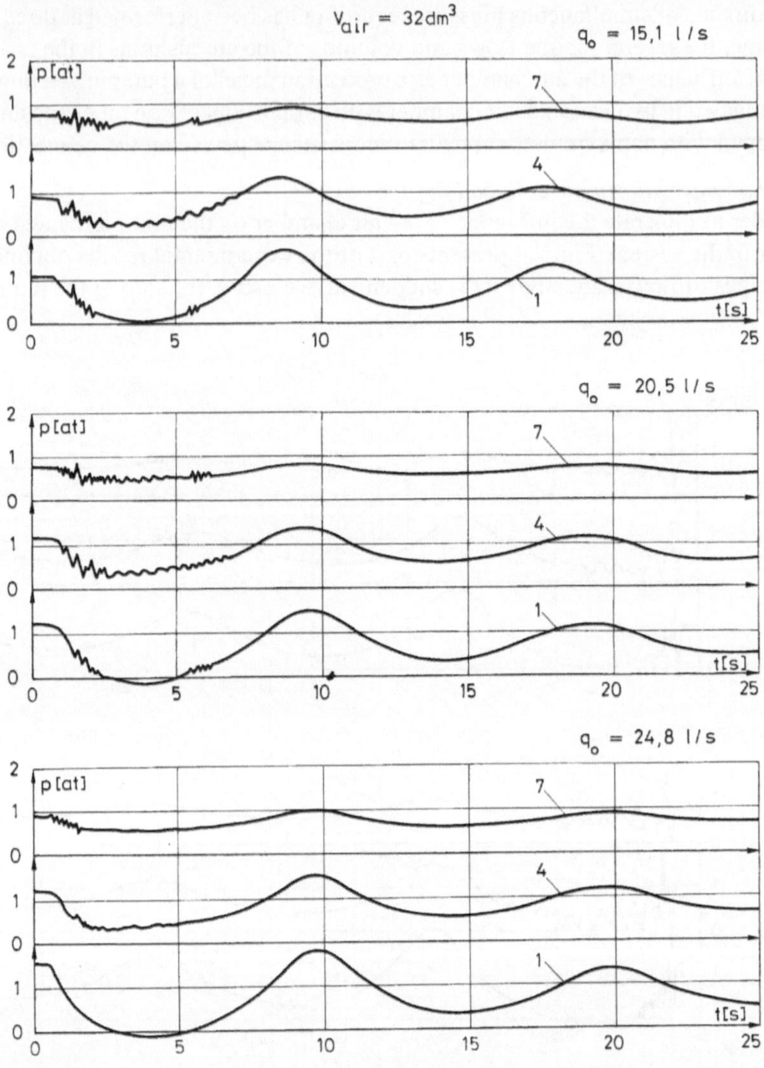

(b)

Figure 7.4. (*continued*).

The sudden closure of the valve represents a flow manoeuvre more severe than pump stopping, as illustrated by the obtained results: in case (a), the maximum pressure amplitudes are greater than in case (b) (Fig. 7.4).

Air chamber with non-symmetrical hydraulic resistance device

In Chapter 2, we have presented the geometric and hydraulic characteristics of the non-symmetrical hydraulic resistance device, and by applying numerical computation methods, we have underlined the protection effect against waterhammer which an air chamber type system can realize in a pumping station equipped with such a device.

In the present chapter, the *performance effect* of this protection system is also experimentally checked. The model of the laboratory device is shown in Fig. 7.5 [10]. In the experiments performed on hydraulic scheme II-B, endowed with air chamber

Figure 7.5. Model of the non-symmetrical hydraulic resistance device.

and non-symmetrical hydraulic resistance, a device with the characteristics $d = 0.4$ D, $\zeta_2/\zeta_1 = 5.25$, made of three conical pieces, was used.

Here too, experiments on a sudden valve-closing manoeuvre, as well as on pump stopping through interruption of electric power supply were performed. In order to exhibit the effect of non-symmetrical hydraulic resistance on the maximum pressures in the system at transient hydraulic regimes, the same experiments were performed as in the previous case of the air chamber with a simple connection pipe. The measurements results are shown in Fig. 7.6. A comparison between the experimental results in Figs. 7.4 and 7.6 illustrates the strong effect of the non-symmetrical hydraulic resistance on the maximum pressure amplitudes in the system, succeeding, in some cases, *to reduce the non-permanent regime maximum pressures below the values of the permanent regime maximum pressures*. This high performance effect was applied, for instance, to the two pumping stations of the Prut–Birlad derivation referred to in Chapter 3. The pressure peaks, of rapid variation, appearing in the diagrams in Figs. 7.4 and 7.6 for case (a) are due to the same cause presented when analysing the experimental results for scheme I-B.

The analysis of all the experimental results obtained with schemes I and II, in variants with or without protection means, confirms, also experimentally, that both the surge tank and the air chamber alter the quality of the non-permanent water motion in the system and that in the presence of these protection means *no waterhammer phenomenon occurs* either in the water conveyance galleries of hydroelectric power plants or in the discharge pipes of the pumping stations.

7.1.2. Experimental results obtained on a different installation

This installation has been designed and produced by the Polytechnical Institute of Timisoara, Romania, with the aim of performing some experimental investigations for paper [1]. Here are some technical and functional data on its main parts (Fig. 7.7):

A: Feeding pipe (Ø 150 mm) to the constant level tank equipped with check-valve.

B: Constant level water tank of metallic construction, with overflow and discharge pipe.

C: Discharge pipe (Ø 200 mm), ensuring the flow discharge from the tank overflow.

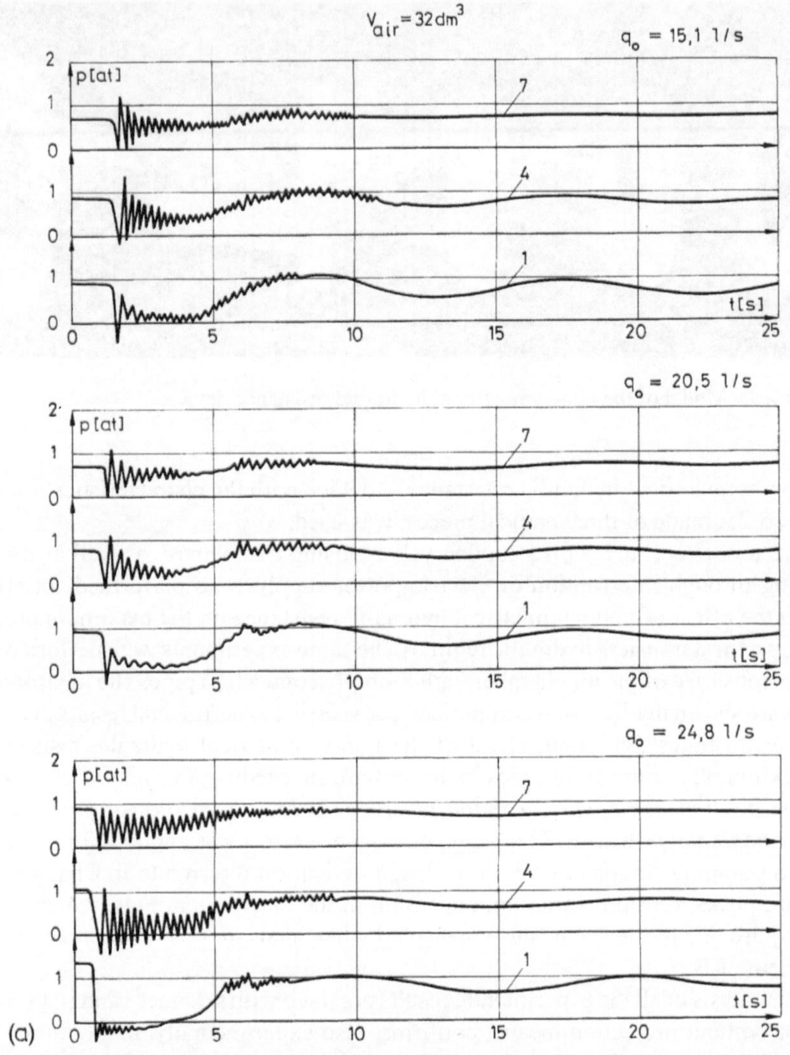

Figure 7.6. Scheme II-B: air chamber with asymmetric hydraulic resistance device: pressure measurements at three points of the pipe for different regime flow rates: (a) sudden valve closing; (b) stopping of the pump.

D: Metallic pipe of diameter Ø 108 mm and length $L = 180$ m.

E: Cylindrical surge tank, in three constructive variants, of diameters $D = 0.355$; 0.255; 0.155 m.

F: Rapid closing valve (Ø 100 mm), allowing regime flow regulation in the experimental installation.

G: Adjustment valve (Ø 100 mm) allowing sudden closure of the pipe.

H: End basin of constant level, allowing measurement of regime flows (q_o).

I: Water evacuation canal of the installation.

The installation has operated at permanent hydraulic regimes with flow rates $q_o = 10$–20 l/s, to which the corresponding speeds in the pipe are $v_o = 1$–2 m/s.

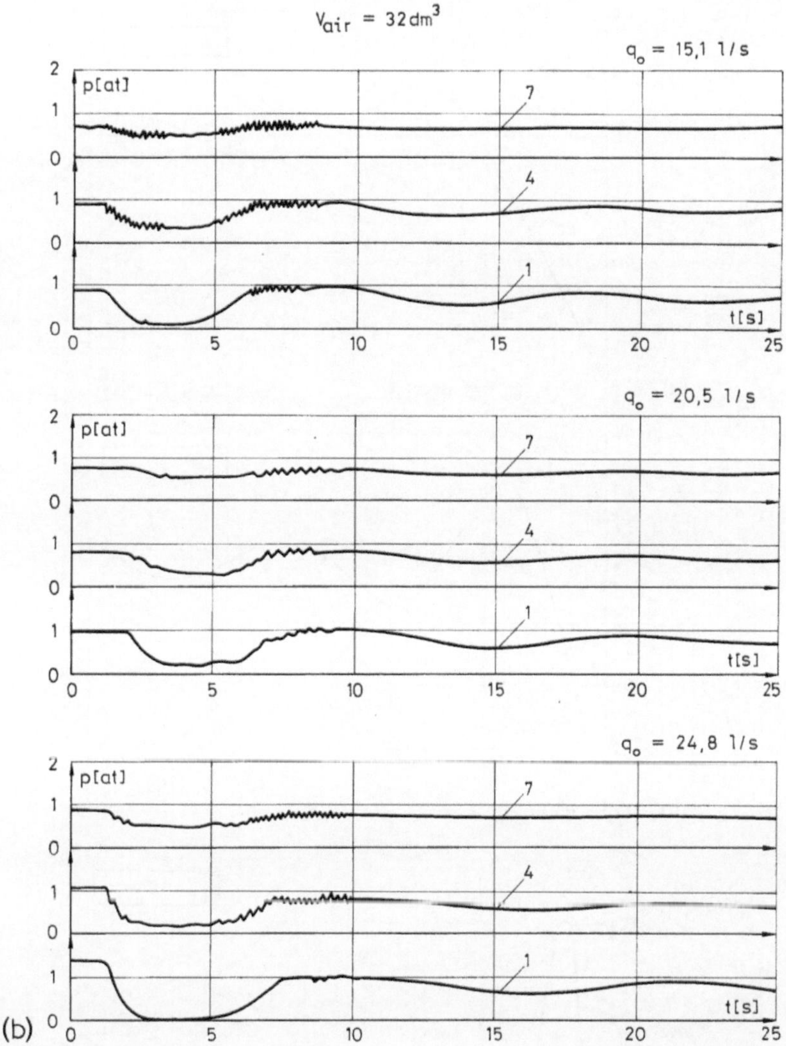

Figure 7.6. (*continued*).

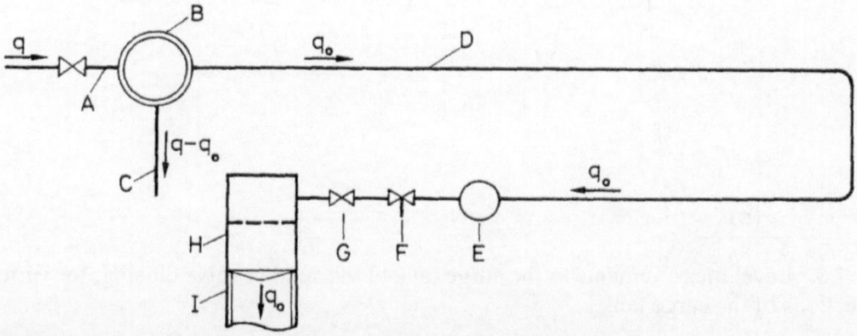

Figure 7.7. Scheme of the laboratory experimental installation.

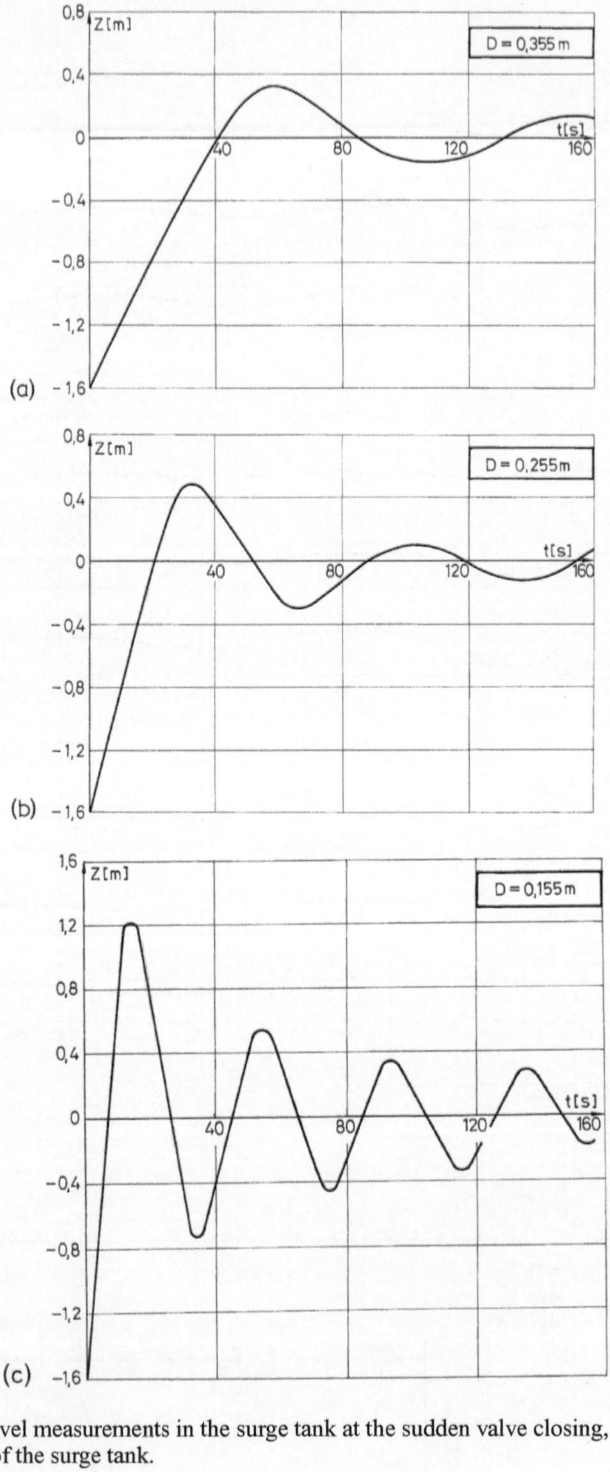

Figure 7.8. Level measurements in the surge tank at the sudden valve closing, for various cross-sections of the surge tank.

Level measurements at the sudden valve closure have been performed in the surge tank of the installation, for several regime flows. Variants of surge tanks with diameters $D = 0.355$; 0.255; 0.155 m have been successively investigated.

The obtained experimental results are exemplified by the graphs in Fig. 7.8, reproducing level variations in the surge tank of the installation, for the above

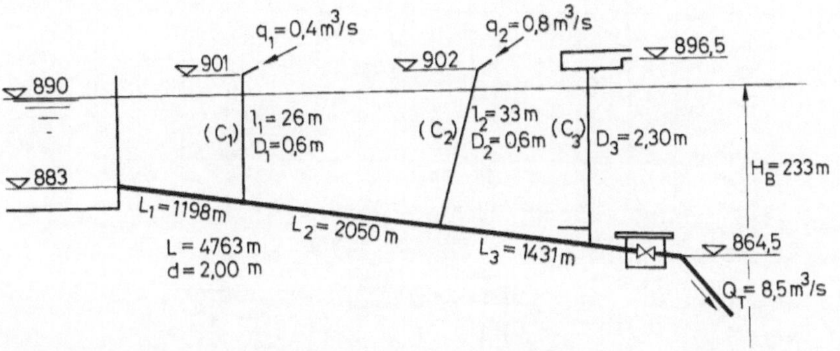

Figure 7.9. Scheme of the Moroieni hydroelectric power plant.

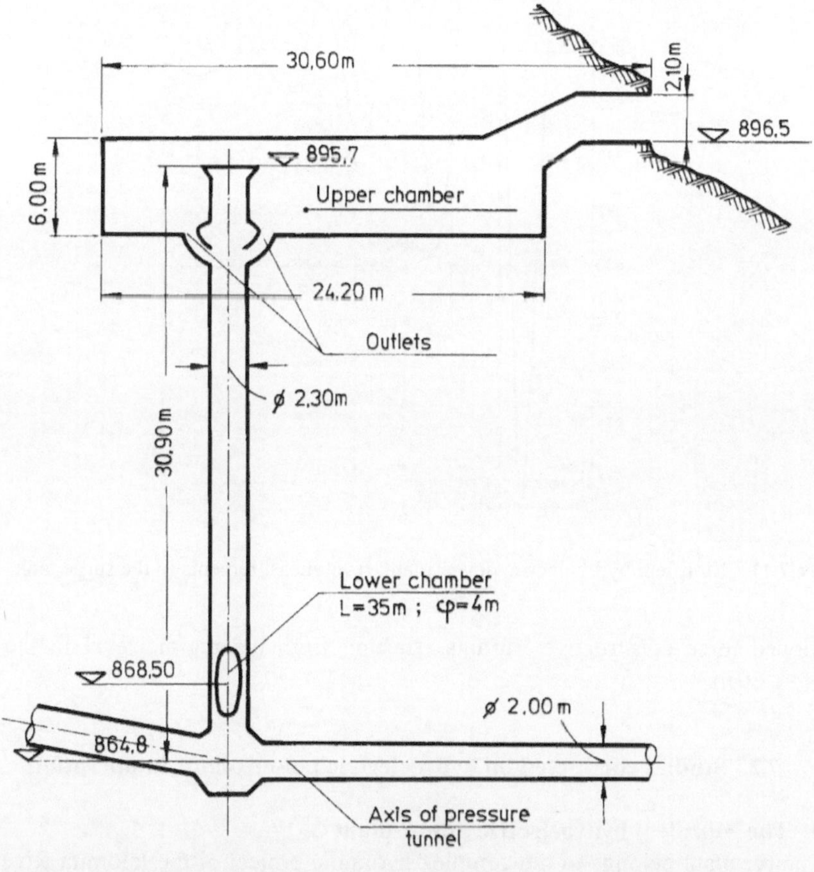

Figure 7.10. The main surge tank (C_3) of the Moroieni hydroelectric power plant.

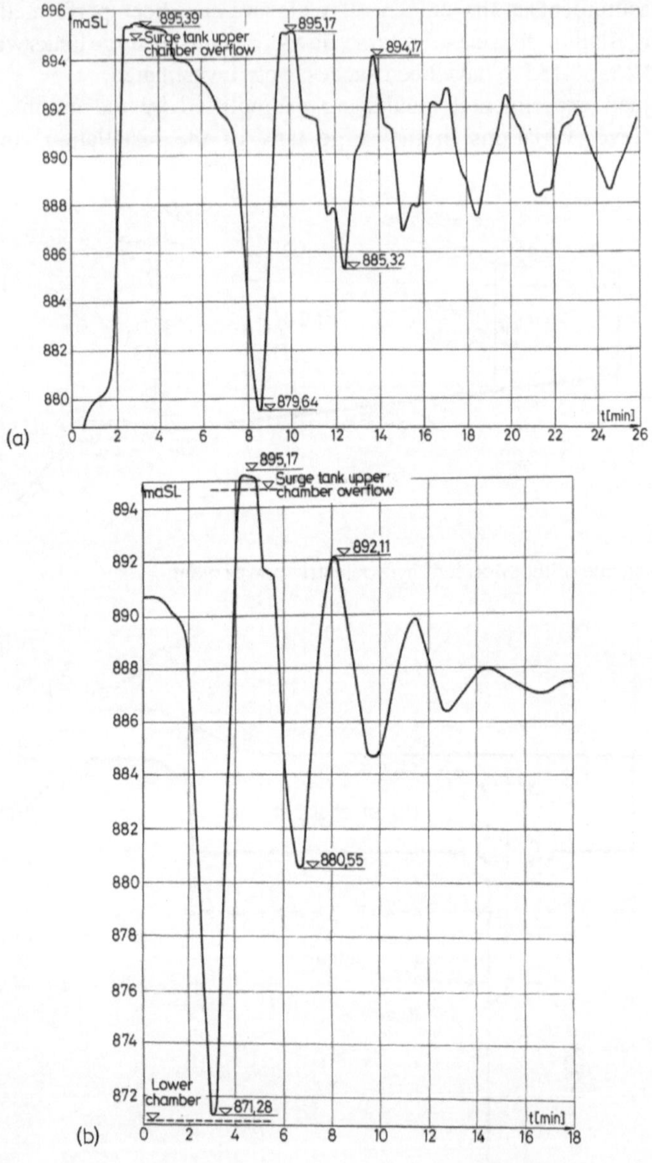

Figure 7.11. Moroieni hydroelectric power plant: level measurements in the surge tank.

mentioned three constructive variants, starting from the regime level in the tank $Z_o = -1.60$ m.

7.2. Studies conducted on hydroelectric power plants in operation

7.2.1. The Moroieni hydroelectric power plant

This power plant belongs to the complex hydraulic project of the Ialomita River, the first hydroelectric power plant of which, "Dobresti", was already in operation in 1930,

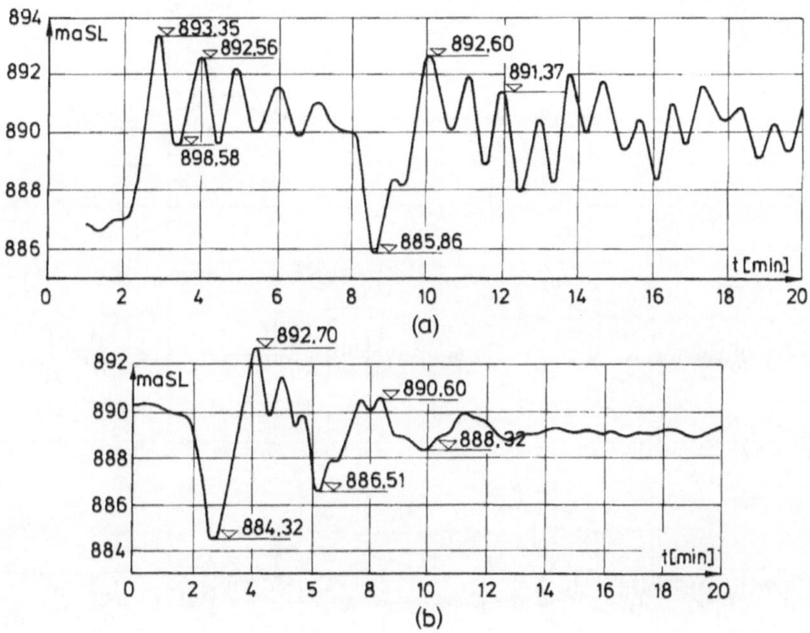

Figure 7.12. Moroieni hydroelectric power plant. Level measurements in the secondary surge tank (C_1) at: (a) the total stop manoeuvre of the plant; (b) the partial start manoeuvre at the plant.

with an installed power of 16 MW. The H.E.P. "Moroeni", designed by Prof. Cristea Mateescu, is provided with a water intake located immediately downstream of the H.E.P. "Dobresti" and uses the flow evacuated by this plant, to which the flow of secondary intakes is added directly connected to the water conveyance gallery, of the small rivers Rateiul and Raciul, with a total discharge of 8.5 m^3/s and a head $H = 233$ m [5].

The water conveyance to the plant consists of: a 4763 m long gallery, of diameter $D = 2$ m; a surge tank with chambers, having a 2.30 m diameter vertical well; the upper chamber is provided with a weir and the lower chamber consist of a 35 m long inclined gallery with a diameter of 4 m; the penstock is 584 m long and has a diameter varying at certain pipe lengths between 1.85 m and 1.50 m; finally, the station is equipped with two vertical shaft Pelton units and power generators of 8.25 MW power each (Figs. 7.9 and 7.10).

Measurements of the water level oscillations in the main surge tank (C_3) of the plant, as well as in the secondary surge tank (C_1) have been performed by a team of engineers of the Hydropower Research and Design Institute (ISPH) of Bucharest. Out of the measurements of extreme water oscillations in these tanks, we shall show several results concerning start/stop manoeuvres of the plant. For instance, Figs. 7.11 and 7.12 present the diagrams of the total shut-down manoeuvres of the plant working at 15 MW, in a 55 s time interval, as well as for the starting manoeuvres of the plant at 8 MW, in a 70 s time interval.

7.2.2. The Bicaz hydroelectric power plant

This is the first great hydroelectric power plant designed and built in Romania with an installed power $N = 210$ MW, a head $H = 146$ m and installed discharge $Q = 178$ m^3/s

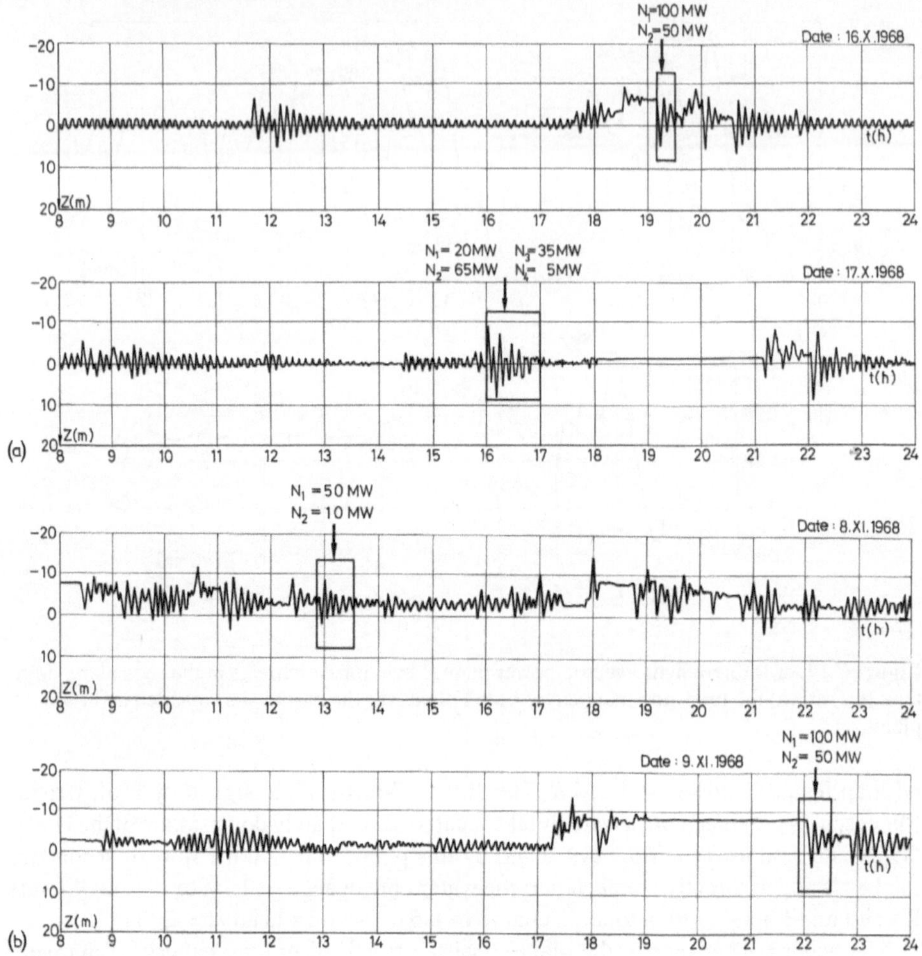

Figure 7.13. The Bicaz hydroelectric power plant: level measurements in the surge tank.

(Figs. 5.9 and 5.10). Among the main components of the hydroelectric power plant, we shall recall the following, with several of their technical characteristics: the gravity type concrete dam, of height $H = 127$ m, ensuring water storage of 1.23 billion m^3 water; the water conveyance gallery, of length $L = 4649$ m and diameter $d = 7$ m; the surge tank with upper chamber and hydraulic resistance (diameter of the well $D_1 = 26$ m, diameter of the upper chamber $D_2 = 33$ m); the penstock of metallic construction of length $L_T = 157$ m and, finally, the station equipped with four Francis units of 27.5 MW and two of 50 MW, in all 210 MW. Several measurements on the operation of this plant at transient hydraulic regimes have been undertaken. Level measurements in the main surge tank (C_3) of the hydroelectric power plant at: (a) the manoeuvre of total stopping of the plant, and (b) the manoeuvre of partial starting of the plant (Fig. 7.13).

Finally, reverting to the results of laboratory experimental studies presented in Chapters 2, 7 and 9 on the main types of non-permanent water motions in hydraulic systems under pressure occurring in engineering practice and performed on laboratory installations with special technical and functional characteristics, equipped with modern electronic measurement devices, we note that they bring necessary elements—of first importance—for the knowledge of certain physical aspects of the investigated phenomena. At the same time, they may provide reference material for further improvement of the present computation methods of the unsteady motion in hydraulic pressure systems [8, 9].

References

1. Arsenie, D.: *Contributii la Calculul Hidraulic al Castelelor de Echilibru* (Contributions to the Hydraulic Computation of the Surge Tanks). Ph.D. Dissertation, Institutul Politehnic Timisoara, 1974.
2. Escande, L.: *Experiences sur la Stabilité des Cheminées d'équilibre avec Etranglement.* Laboratoires d'Hydraulique et de Mécanique des Fluides, Travaux et Publications, Toulouse, 1960.
3. Franke, P.G.: *Oscillations in a surge tank—calculation and measurement.* IAHR Congress, London, 1963.
4. Granze, H., Schubert, J., Forrest, J.: Analysis of Field Measurements of Air Chamber Installations. *Proceedings of 2nd International Conference on Pressure Surges*, London, England, 1976.
5. Mateescu, C., Sipiceanu, M.: Uzina Hidroelectrica Moroieni, (Moroieni Hydroelectric Power Plant). *Energetica*, 12 (1966).
6. Naber, G.: Mass Oscillations in Surge Tanks. *Water Power and Dam Construction*, 6 (1964).
7. Pistilli, G.: *Condotte Elevatorie Munite di Camere d'Aria. Ricerca Sperimentale.* Universita di Napoli, No. 159, 1963.
8. Popescu, M., Arsenie, D.: *Metode de Calcul Hidraulic pentru Uzine Hidroelectrice si Statii de Pompare* (Methods of Hydraulic Computations for Hydroelectric Power Plants and Pumping Stations). Ed. Tehnica, Bucharest, 1987.
9. Popescu, M.: Rezultate Experimentale in Studiul Miscarii Nepermanente din Sistemele Hidraulice sub Presiune (Experimental Results in the Study of Unsteady Motions in Discharged Hydraulic Systems). *Studii si Cercetari de Mecanica Aplicata*, 4 (1982) 443–458, 5 (1982) 595–609.
10. Popescu, M.: *Modul de Resistenta Hidraulica si Baterie din Elemente Fixe Modulare* (Module of Hydraulic Resistance and Battery of Modular Fixed Elements). Invention, Patent O.S.I.M., Romania, Nr. 71286, 1979 (Gold Medal, 23rd International Exhibition of Invention, Geneva, Switzerland, 1995).

Chapter 8

Hydroelectric Power Plants and Pumping Stations Designed with Complex Hydraulic Schemes

8.1. Hydraulic computation of transient regimes for hydroelectric power plants designed with complex schemes

Hydraulic power plants designed with complex hydraulic schemes belong to the category of the most modern hydroelectric power plants and appeared as an engineering solution for integral and rational management of the hydroenergetic potential of rivers in a given geographical zone. Complex hydraulic schemes for hydroelectric power plants can be briefly defined as systems with several reservoirs, pressure tunnels, surge tanks, penstocks and hydroelectric stations.

This book approaches the hydraulic computation of transient regimes for some complex hydraulic schemes more frequently used in engineering practice. We shall adopt a simplified notation for them by letters A, B, C and D. Out of the results obtained in the field, we shall present several numerical and analytical computation methods and exemplify them with concrete cases [12, 25, 31, 36].

8.1.1. Numerical methods

In a previous book [12] we have minutely dealt with the application of these methods to complex hydraulic schemes and we have presented in an accessible form the numerical computation method for hydraulic scheme A. In the following, we shall present a similar computation method for hydraulic scheme B and we shall give computation examples for all hydraulic schemes considered [24].

We assume an automatic operation of the hydroelectric power plant at constant power, in a separate network. The general equations of the oscillatory water motion in the system, with usual notation, are the following (cf. Fig. 8.3 below):

$$Y = H_1 - H_2$$

$$\frac{L_1}{g}\frac{\mathrm{d}W_1}{\mathrm{d}t} + P_1 + Z_1 + R_1 + E_1 = 0$$

$$\frac{L'}{g}\frac{\mathrm{d}W'}{\mathrm{d}t} + P' + Z + R + E = Z_1 + R_1$$

$$\frac{L''}{g}\frac{\mathrm{d}W''}{\mathrm{d}t} + P'' + Z + R + E = Z_2 + R_2 - Y$$

$$\frac{L_2}{g}\frac{\mathrm{d}W_2}{\mathrm{d}t} + P_2 + Z_2 + R_2 + E_2 = 0 \tag{8.1}$$

$$f_1 W_1 = F_1 \frac{\mathrm{d}Z_1}{\mathrm{d}t} + f' W'$$

$$f_2 W_2 = F_2 \frac{\mathrm{d}Z_2}{\mathrm{d}t} + f'' W''$$

$$f' W' + f'' W'' = F \frac{\mathrm{d}Z}{\mathrm{d}t} + Q_\mathrm{T}$$

where $Q_\mathrm{T} = Q_\mathrm{t}(t)$ is the flow rate relation for computing the extreme oscillations in the tanks, and $\eta g(H_1 + Z + R - K_\mathrm{T} Q_\mathrm{T}^2) Q_\mathrm{T} = N_2$ is the adjustment relation for the study of hydraulic stability of the system, in which, for longitudinal and local head losses, we accepted relations of the form

$$P_j = P_j' W_j \left| W_j \right|, \quad R_j = R_j' V_{\mathrm{s}j} \left| V_j \right|, \quad j = 1, 2 \tag{8.2}$$

and for the surge tanks cross-sections, general relations of the form

$$F = F(Z), \quad F_j = F_j(Z_j), \quad j = 1, 2 \tag{8.3}$$

Kinetic terms have been expressed by the relations

$$E_j = E_j' W_j \left| W_j \right|, \quad j = 1, 2 \tag{8.4}$$

Efficiency curves of the turbines have been introduced in computations by general relations of the form

$$\eta = \eta(H, N) \tag{8.5}$$

In addition, the following relation is used:

$$P_j + E_j = P_j' W_j \left| W_j \right| + E_j' W_j \left| W_j \right| = P_j^* W_j \left| W_j \right|, \quad j = 1, 2 \tag{8.6}$$

In practical computations by numerical methods, tabulatory expression of some of these expressions is preferred. For instance, in the study of hydraulic stability of the operation of the system, in order to solve equation system (8.1) we write the finite difference equations and use the mean values of variables for the time interval Δt.

After performing the computations and the convenient grouping of terms, the system (8.1) becomes

$$\frac{L_1}{g}\frac{\Delta W_1}{\Delta t} + P_1^* \left(W_{1i} + \frac{\Delta W_1}{2} \right) \left| W_{1i} + \frac{\Delta W_1}{2} \right| + R_1' \left(\frac{F_{1\mathrm{m}}}{F_{1\mathrm{s}}} \frac{\Delta Z_1}{\Delta t} \right) \left| \frac{F_{1\mathrm{m}}}{F_{1\mathrm{s}}} \frac{\Delta Z_1}{\Delta t} \right|$$
$$+ Z_{1i} + \frac{2Z_1}{2} = 0$$

$$\frac{L'}{g}\frac{\Delta W'}{\Delta t}+P'^{*}\left(W_i'+\frac{\Delta W'}{2}\right)\left|W_i'+\frac{\Delta W'}{2}\right|+R'\left(\frac{F_m}{F_s}\frac{\Delta Z}{\Delta t}\right)\left|\frac{F_m}{F_s}\frac{\Delta Z}{\Delta t}\right|$$

$$+Z_i+\frac{\Delta Z}{2}=Z_{1i}+\frac{\Delta Z_1}{2}+R_1'\left(\frac{F_{1m}}{F_{1s}}\frac{\Delta Z_1}{\Delta t}\right)\left|\frac{F_{1m}}{F_{1s}}\frac{\Delta Z_1}{\Delta t}\right|$$

$$\frac{L''}{g}\frac{\Delta W''}{\Delta t}+P''^{*}\left(W_i''+\frac{\Delta W''}{2}\right)\left|W_i''+\frac{\Delta W''}{2}\right|+R'\left(\frac{F_m}{F_s}\frac{\Delta Z}{\Delta t}\right)\left|\frac{F_m}{F_s}\frac{\Delta Z}{\Delta t}\right|$$

$$+Z_i+\frac{\Delta Z}{2}=Z_{2i}+\frac{\Delta Z_2}{2}+R_2'\left(\frac{F_{2m}}{F_{2s}}\frac{\Delta Z_2}{\Delta t}\right)\left|\frac{F_{2m}}{F_{2s}}\frac{\Delta Z_2}{\Delta t}\right|-Y$$

$$\frac{L_2}{g}\frac{\Delta W_2}{\Delta t}+P_2^{*}\left(W_{2i}+\frac{\Delta W_2}{2}\right)\left|W_{2i}+\frac{\Delta W_2}{2}\right| \tag{8.7}$$

$$+R_2'\left(\frac{F_{2m}}{F_{2s}}\frac{\Delta Z_2}{\Delta t}\right)\left|\frac{F_{2m}}{F_{2s}}\frac{\Delta Z_2}{\Delta t}\right|+Z_{2i}+\frac{2Z_2}{2}=0$$

$$\Delta Z_1=\frac{\Delta t}{F_{1m}}\left[f_1\left(W_{1i}+\frac{\Delta W_1}{2}\right)-f'\left(W_i'+\frac{\Delta W'}{2}\right)\right]$$

$$\Delta Z_2=\frac{\Delta t}{F_{2m}}\left[f_2\left(W_{2i}+\frac{\Delta W_2}{2}\right)-f''\left(W_i''+\frac{\Delta W''}{2}\right)\right]$$

$$Q_T=f'\left(W'+\frac{\Delta W'}{2}\right)+f''\left(W_i''+\frac{\Delta W''}{2}\right)-F_m\frac{\Delta Z}{\Delta t}$$

$$\Delta Z=2\left[\frac{N_2}{\eta g Q_T}+K_T Q_T^2-R'\left(\frac{F_m}{F_s}\frac{\Delta Z}{\Delta t}\right)\left|\frac{F_m}{F_s}\frac{\Delta Z}{\Delta t}\right|-H_1-Z_i\right]$$

In the presentation of the computation algorithm of the proposed method, we shall use the following symbolic form of system (8.7):

$$\begin{aligned}
&\Delta W_1=\Phi_1(\Delta Z_1), && \Delta W'=\Phi_2(\Delta Z_1,\Delta Z),\\
&\Delta W''=\Phi_3(\Delta Z_2,\Delta Z), && \Delta W_2=\Phi_4(\Delta Z_2),\\
&\Delta Z_1=\Psi_1(\Delta W_1,\Delta W'), && \Delta Z_2=\Psi_2(\Delta W_2,\Delta W''),\\
&Q_T=\Phi_o(\Delta Z,\Delta W',\Delta W''), && \Delta Z=\Psi_o(Q_T)
\end{aligned} \tag{8.8}$$

The computation algorithm is conceived in steps, in an iterative development. Its simplified presentation runs as follows:

Step I

$I_A \quad \Delta Z^p=\Delta_{o1}$

$I_1 \quad \Delta Z_1^p=\Delta_{11},\quad \Delta Z_2^p=\Delta_{21},\quad \Delta W_1=\Phi_1(\Delta_{11}),\quad \Delta W'=\Phi_2(\Delta_{11},\Delta_{o1})$

$\Delta W''=\Phi_3(\Delta_{21},\Delta_{o1}),\quad \Delta W_2=\Phi_4(\Delta_{21})$

$\Delta Z_1^c=\Psi_1(\Delta W_1,\Delta W'),\quad \Delta Z_2^c=\Psi_2(\Delta W_2,\Delta W'')$

$\left|\Delta Z_j^p-\Delta Z_j^c\right|=|\varepsilon_{j1}|<\varepsilon,\quad j=1,2$

$$\text{I}_2 \quad \Delta Z_1^{\text{p}} = \Delta_{12} = 2\Delta_{11}, \quad \Delta Z_2^{\text{p}} = \Delta_{22} = 2\Delta_{21}$$

$$\Delta W_1 = \Phi_1(\Delta_{12}), \quad \Delta W' = \Phi_2(\Delta_{12}, \Delta_{\text{o}1})$$

$$\Delta W'' = \Phi_3(\Delta_{22}, \Delta_{\text{o}1}), \quad \Delta W_2 = \Phi_4(\Delta_{22})$$

$$\Delta Z_1^{\text{c}} = \Psi_1(\Delta W_1, \Delta W'), \quad \Delta Z_2^{\text{c}} = \Phi_2(\Delta W_2, \Delta W'')$$

$$\left|\Delta Z_j^{\text{p}} - \Delta Z_j^{\text{c}}\right| = |\varepsilon_{j2}| < \varepsilon, \quad j = 1, 2$$

$$\text{I}_3 \quad \Delta Z_j^{\text{p}} = \Delta_{j1} + \left(\frac{\Delta_{j2} - \Delta_{j1}}{\varepsilon_{j1} - \varepsilon_{j2}}\right)\varepsilon_{j1} = \Delta_{j3}$$

$$\Delta W_1 = \Phi_1(\Delta_{13}), \quad \Delta W' = \Phi_2(\Delta_{13}, \Delta_{\text{o}1})$$

$$\Delta W'' = \Phi_3(\Delta_{23}, \Delta_{\text{o}1}), \quad \Delta W_2 = \Phi_4(\Delta_{23})$$

$$\Delta Z_1^{\text{c}} = \Psi_1(\Delta W_1, \Delta W'), \quad \Delta Z_2^{\text{c}} = \Psi_2(\Delta W_2, \Delta W'')$$

$$\left|\Delta Z_j^{\text{p}} - \Delta Z_j^{\text{c}}\right| = |\varepsilon_{j2}| < \varepsilon, \quad j = 1, 2$$

$$\text{I}_\text{B} \quad Q_\text{T} = \Phi_\text{o}(\Delta_{\text{o}1}, \Delta W', \Delta W''), \quad \Delta Z_c = \Psi_\text{o}(Q_\text{T})$$

$$|\Delta Z^{\text{p}} - \Delta Z^{\text{c}}| = |\varepsilon_{\text{o}1}| < \varepsilon$$

Step II

$$\text{II}_\text{A} \quad \Delta Z^{\text{p}} = 2\Delta_{\text{o}1} = \Delta_{\text{o}2}$$

$\left.\begin{array}{l}\text{II}_1\\ \text{II}_2\\ \text{II}_3\end{array}\right\}$ Computations follow according to the algorithm in step I, with $\Delta Z^{\text{p}} = \Delta_{\text{o}2}$

$$\text{II}_\text{B} \quad Q_\text{T} = \Phi_\text{o}(\Delta_{\text{o}2}, \Delta W', \Delta W''), \quad \Delta Z^{\text{c}} = \Psi_\text{o}(Q_\text{T})$$

$$|\Delta Z^{\text{p}} - \Delta Z^{\text{c}}| = |\varepsilon_{\text{o}2}| < \varepsilon$$

Step III

$$\text{III}_\text{A} \quad \Delta Z^{\text{p}} = \Delta_{\text{o}1} + \left(\frac{\Delta_{\text{o}2} - \Delta_{\text{o}1}}{\varepsilon_{\text{o}1} - \varepsilon_{\text{o}2}}\right)\varepsilon_{\text{o}1} = \Delta_{\text{o}3}$$

$\left.\begin{array}{l}\text{III}_1\\ \text{III}_2\\ \text{III}_3\end{array}\right\}$ Computations follow according to the algorithm in step I, with $\Delta Z^{\text{p}} = \Delta_{\text{o}3}$

$$\text{III}_\text{B} \quad Q_\text{T} = \Phi_\text{o}(\Delta_{\text{o}3}, \Delta W', \Delta W''), \quad \Delta Z^{\text{c}} = \Psi_\text{o}(Q_\text{T})$$

$$|\Delta Z^{\text{p}} - \Delta Z^{\text{c}}| = |\varepsilon_{\text{o}3}| < \varepsilon$$

where the following notation has been used: ΔZ^{p} is the proposed value for ΔZ, ΔZ^{c} the computed value for ΔZ, Δ_{ij} is an abbreviated notation for ΔZ^{p}.

In the iterative process, one takes as first values of ΔZ, ΔZ_1 and ΔZ_2 their values at the previous computation step, and at the first computation step one can take an arbitrary set of acceptable values, compatible with the physical phenomenon.

If the accuracy conditions

$$|\Delta Z^{\text{p}} - \Delta Z^{\text{c}}| < \varepsilon, \quad |\Delta Z_j^{\text{p}} - \Delta Z_j^{\text{c}}| < \varepsilon, \quad j = 1, 2 \tag{8.9}$$

are simultaneously satisfied, the computation is considered to be finished for the respective time step and one proceeds to the following step. If not, the computation begins again with the IIIrd step, considering the carried out steps the IInd and the IIIrd steps as the Ist and the IInd ones. A similar iterative process takes place for each step in steps 1, 2, 3. The numerical solution of the problem has proved to be stable and rapidly convergent. In the case of the plant operation in a power system, one uses the corresponding adjusting equation, and the system of equation of motion is similarly solved.

A similar solution of the system of equations is undertaken when computing the extreme oscillations of water levels in the tanks, using instead of the adjustment relation, the flow-rate relation in system (8.1). The method can be easily written in an algorithmic language and used on a mean capacity computer.

Computation examples

In order to illustrate the practical application of numerical methods in the study of transient hydraulic regimes in the operation of hydroelectric power plants, a computation example is given for each of the analysed hydraulic schemes [25, 31].

Hydraulic scheme A. The case of a hydroelectric power plant with the following main characteristics (Fig. 8.1) has been chosen: $L = 5{,}000$ m; $d = 6.00$ m.

Power plant 1		*Power plant 2*	
$L_1 = 3{,}000$ m,	$d_1 = 5.00$ m,	$L_2 = 2{,}000$ m,	$d_2 = 5.00$ m
$L_{T1} = 300$ m,	$d_{T2} = 4.00$ m,	$L_{T2} = 150$ m,	$d_{T2} = 4.00$ m,
$H_1 = 400$ m,	$N_1 = 180$ MW,	$H_2 = 200$ m,	$N_2 = 100$ MW

The results of hydraulic computations illustrate the complex character of the oscillatory phenomenon in the system, as well as the domains of stable or unstable hydraulic operation of the plant (Fig. 8.2).

Hydraulic scheme B. In the computation example below, the case of a hydroelectric plant with the following main characteristics (Fig. 8.3) have been selected: $H_1 = H_2 = 275$ m ($Y = 0$); $N = 100$ MW; $L_1 = 5{,}600$ m; $L_2 = 9{,}200$ m; $L'' = 2{,}200$ m; $f = 12.5\ \mathrm{m}^2$; $L_T = 300$ m; $f_T = 9.75\ \mathrm{m}^2$.

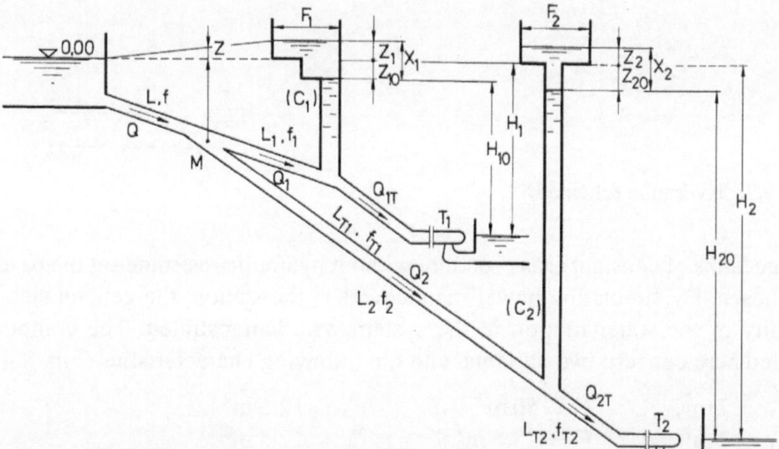

Figure 8.1. Hydraulic scheme A.

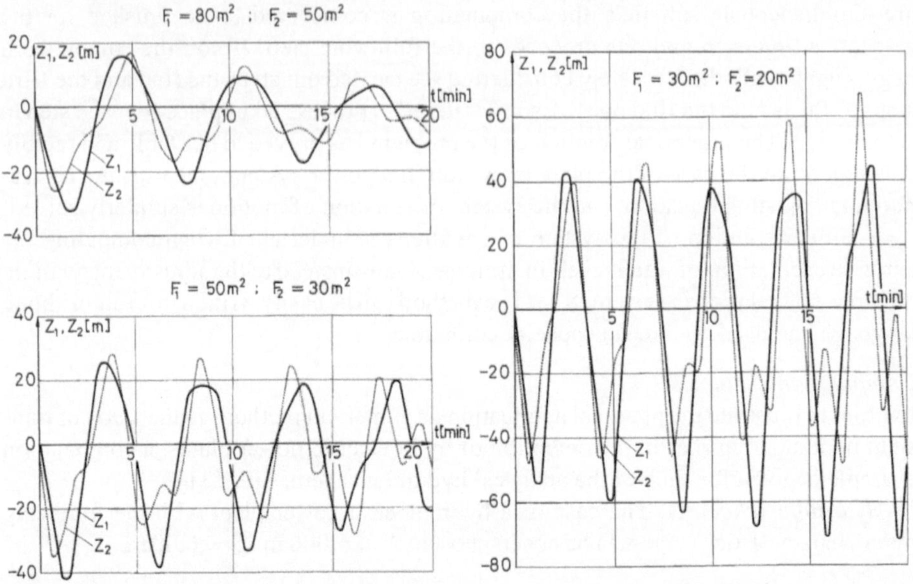

Figure 8.2. Curves of water level variation with time for various cross-sections of the water tanks ($R \neq 0$).

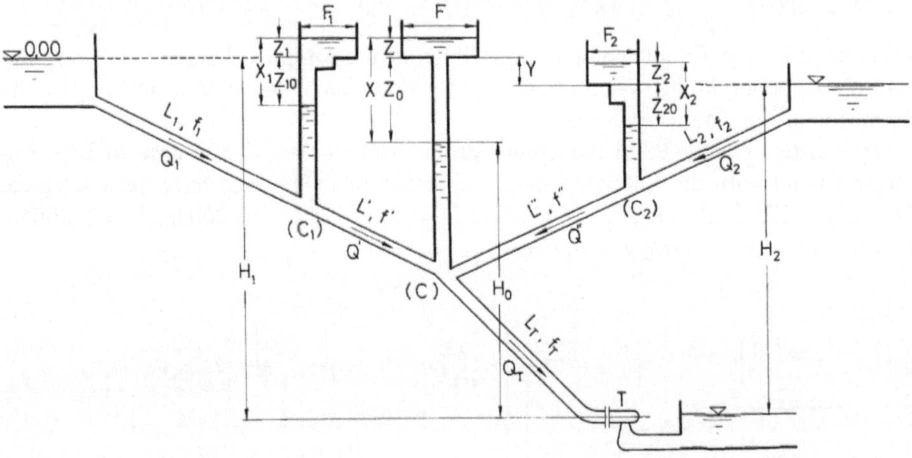

Figure 8.3. Hydraulic scheme B.

Surge tanks of constant cross-section, without hydraulic resistance at the bases have been chosen. By simulating power manoeuvres at the station, the general stability or instability of the water motion in the system was demonstrated. The computations presented here concern two variants with the following characteristics (Fig. 8.4):

(a) $F_1 = 7\,\text{m}^2$, $F = 50\,\text{m}^2$, $F_2 = 12.5\,\text{m}^2$
(b) $F_1 = 7\,\text{m}^2$, $F = 28.5\,\text{m}^2$, $F_2 = 12.5\,\text{m}^2$

Hydraulic scheme C. This computation example presents the case of a hydro-electric power plant with the following main characteristics (Fig. 8.5): $H = 500\,\text{m}$;

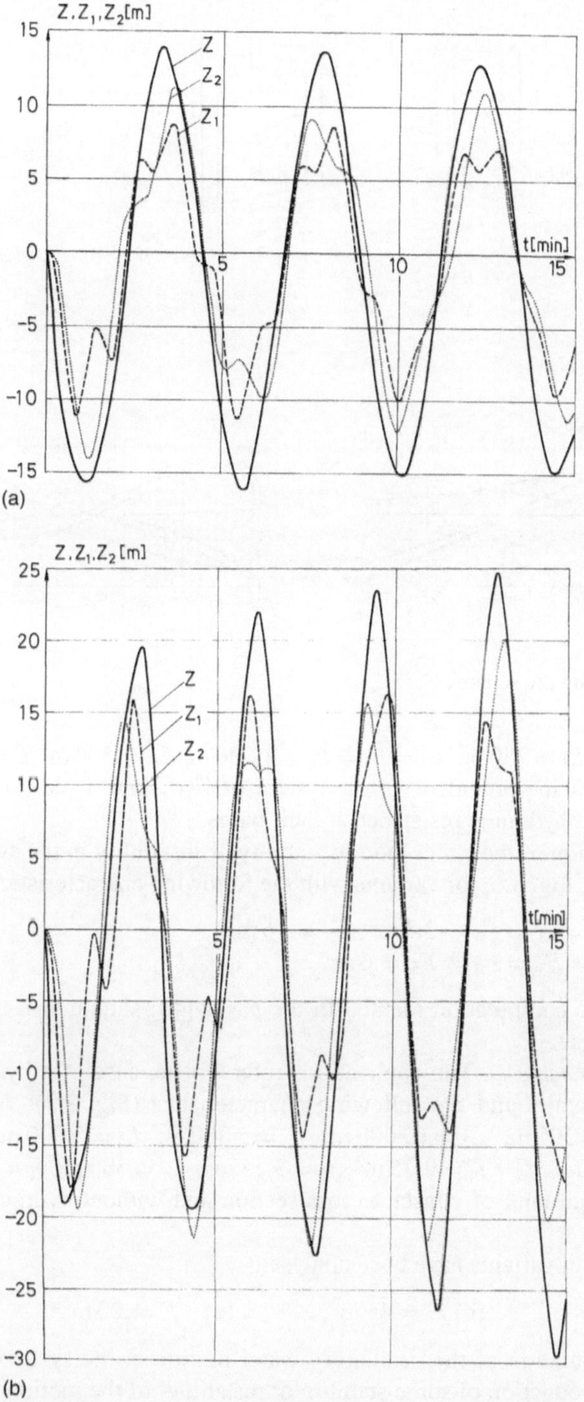

Figure 8.4. Hydraulic operation of the hydroelectric power plant at a starting manoeuvre with $N_2 = 50\,\mathrm{MW}$: (a) stable operation. (b) unstable operation.

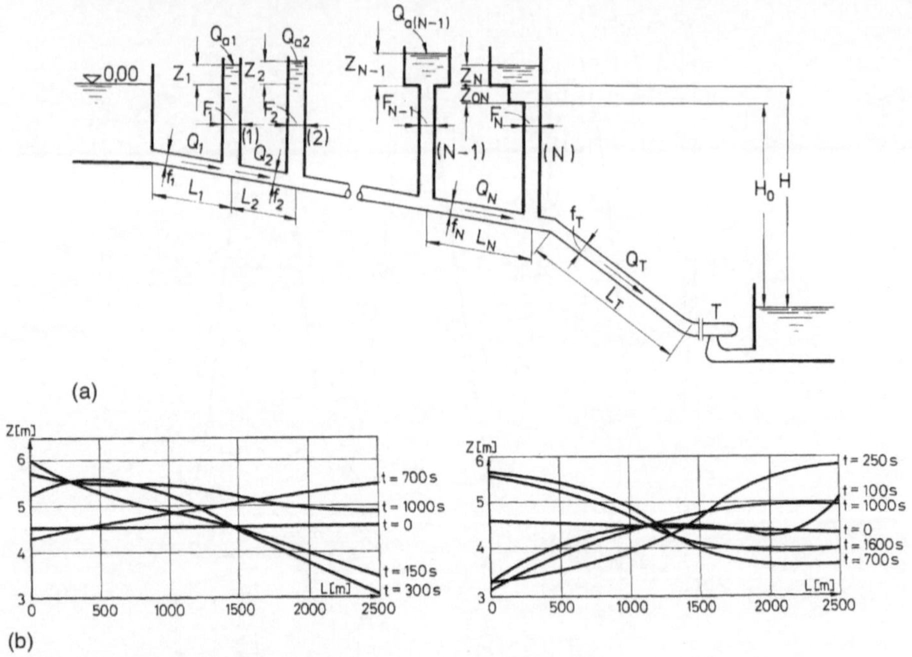

Figure 8.5. Hydraulic scheme C.

$N = 240\,\text{MW}$; $L_1 = 750\,\text{m}$; $L_2 = L_3 = L_4 = 1{,}500\,\text{m}$; $L_5 = 500\,\text{m}$; $f = 15.5\,\text{m}^2$; $L_T = 350\,\text{m}$; $f_T = 9.50\,\text{m}^2$ provided with a system of five surge tanks of constant cross-section, without hydraulic resistance at their bases.

The illustration of the water motion stability or instability in the system appears in the diagrams in Fig. 8.6, for variants with the following characteristics:

(a) $F_1 = F_2 = F_3 = F_4 = 3.5\,\text{m}^2, F_5 = 16\,\text{m}^2$
(b) $F_1 = F_2 = F_3 = F_4 = F_5 = 6\,\text{m}^2$

The numeric computation method in the paper [40] should also be noted for this hydraulic scheme.

Hydraulic scheme D. The applications refers to the case of a hydroelectric plant with five reservoirs and the following characteristics (Fig. 8.7): $H = 250\,\text{m}$; $N = 100\,\text{MW}$; $L_1 = 500\,\text{m}$; $L_2 = 10{,}000\,\text{m}$; $L_3 = 5{,}300\,\text{m}$; $L_4 = 5{,}000\,\text{m}$; $L_5 = 1{,}500\,\text{m}$; $f_1 = f_2 = 12.5\,\text{m}^2$; $f_3 = f_4 = 9.75\,\text{m}^2$; $f_5 = 3.15\,\text{m}^2$; $L_T = 300\,\text{m}$; $f_T = 20\,\text{m}^2$, we have chosen the surge tank of constant cross-section and without hydraulic resistance at the base.

The following variants have been analysed:

(a) $F = 80\,\text{m}^2$, (b) $F = 45\,\text{m}^2$, (c) $F = 20\,\text{m}^2$

Numerical simulation of the oscillatory water motion in the system allowed a very suggestive reproduction of some stability or instability of the motion (Fig. 8.8).

8.1.2. Analytical methods

When reviewing the present development of analytical computation methods of complex hydraulic schemes, one can see that only simplified case of complex hydraulic

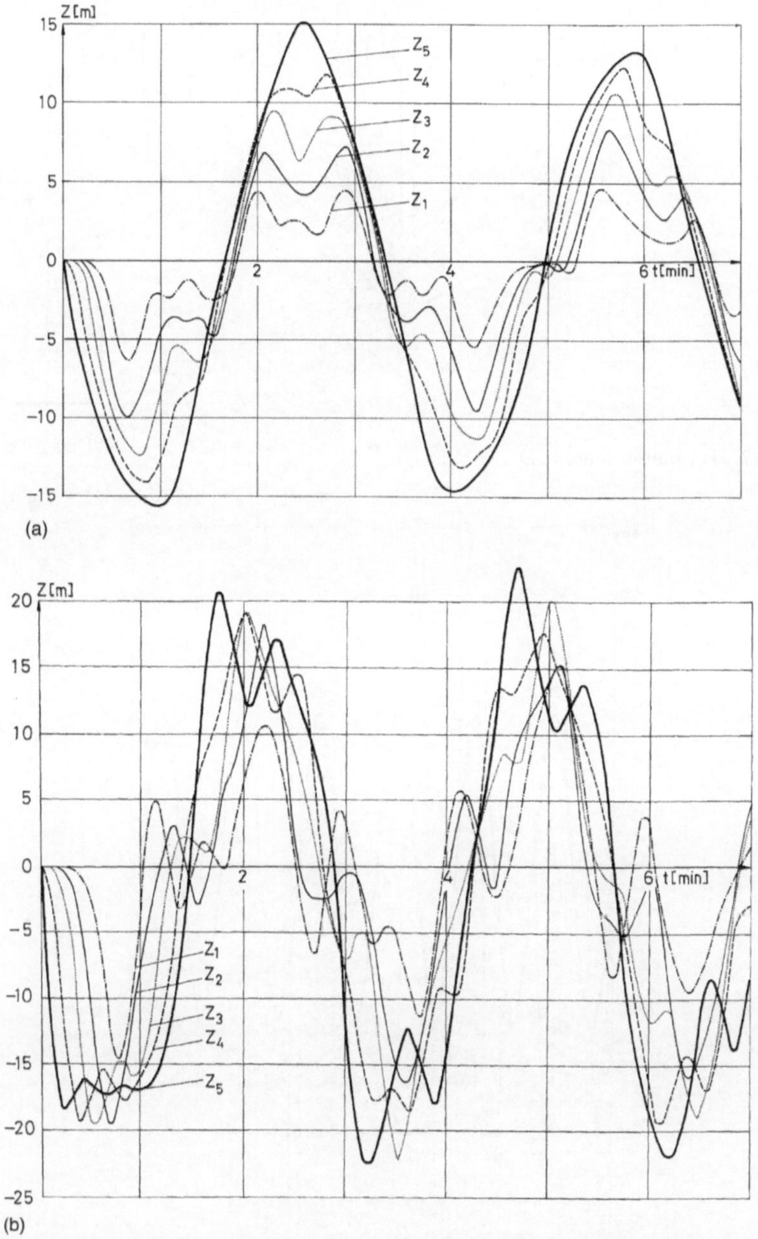

Figure 8.6. Hydraulic operation of the hydroelectric power plant at a starting manoeuvre with $N_2 = 50\,\text{MW}$: (a) stable operation, (b) unstable operation.

schemes, of the type of the one treated by Escande [8, 9] have been solved up to the present. The first noteworthy results concerning the analytic solution of complex hydraulic schemes of the type considered in this book have been communicated to the 2nd International Hydraulics Symposium IAHR, held in Bucharest in 1984 [24, 29].

Illustratively, we shall present an analytical method to solve *the problem of hydraulic stability in operation* of some complex hydraulic scheme which was treated numerically

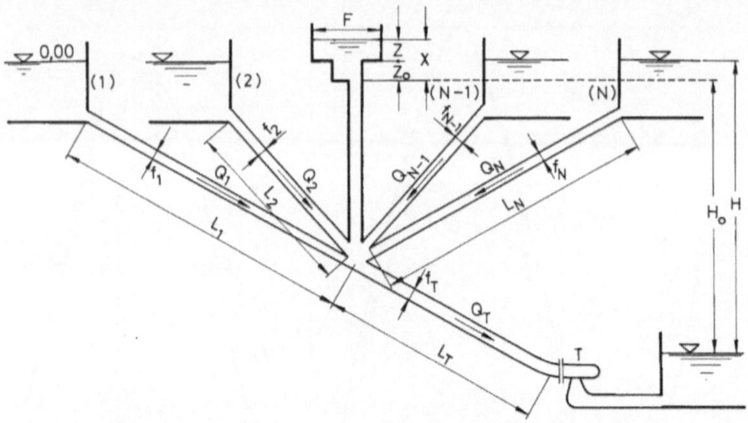

Figure 8.7. Hydraulic scheme D.

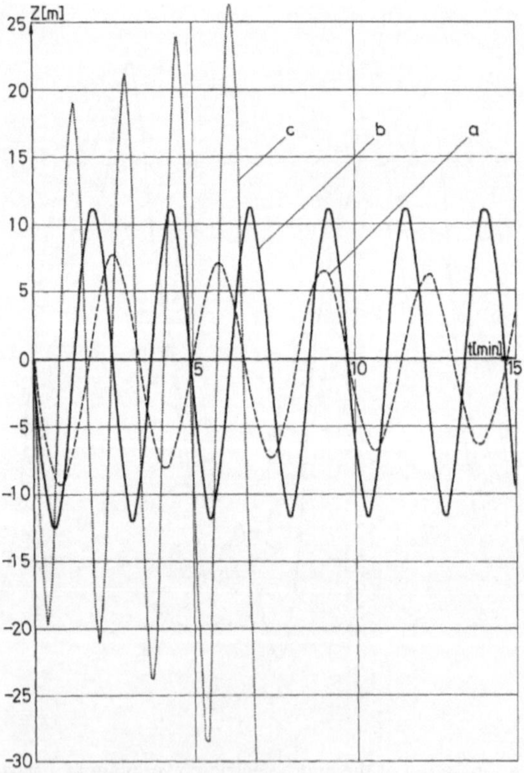

Figure 8.8. The hydraulic stable or unstable operation of the hydroelectric power plant at a starting manoeuvre of $N_2 = 50$ MW is illustrated.

above. Generally speaking, numerical methods give no preliminary information on the cross-section of surge tanks, nor do they yield any conclusion concerning the influence of various constructive elements of the hydraulic management on this critical cross-section. Thus, if we refer to hydraulic scheme B, there arises the natural need to obtain in design computations a preliminary information about the critical cross-section, defined as the value of the main surge tank cross-section, beyond which stability against small oscillations is guaranteed for the operation of the entire hydraulic system.

Such a critical cross-section will fulfil, in the case of complex schemes, the same function as Thoma's cross-section in the case of the basic hydraulic scheme with one surge tank. Recall that the stability problem is investigated under the conditions of automatic operation of the plant at constant power, the power manoeuvres being chosen from those which may generate instability in the system operation.

The mathematical model

Without affecting the generality of the method, in order to simplify some computations, we shall admit the following hypotheses, currently used in practice:

$$H_1 = H_2 = H(Y = 0), \quad f_1 = f', \quad f_2 = f'' \tag{8.10}$$

Under these conditions, the considered complex hydraulic scheme is described by the following system of differential equations (Fig. 8.3):

$$\begin{aligned} &\frac{L_1}{g}\frac{\mathrm{d}W_1}{\mathrm{d}t} + P_1 W_1^2 + Z_1 = 0, \quad \frac{L_2}{g}\frac{\mathrm{d}W_2}{\mathrm{d}t} + P_2 W_2^2 + Z_2 = 0 \\ &F_1 \frac{\mathrm{d}Z_1}{\mathrm{d}t} = f'(W_1 - W'), \quad F_2 \frac{\mathrm{d}Z_2}{\mathrm{d}t} = f''(W_2 - W'') \\ &\frac{L'}{g}\frac{\mathrm{d}W'}{\mathrm{d}t} + P' W'^2 + Z - Z_1 = 0, \quad \frac{L''}{g}\frac{\mathrm{d}W''}{\mathrm{d}t} + P'' W''^2 + Z - Z_2 = 0 \\ &F\frac{\mathrm{d}Z}{\mathrm{d}t} + Q_T = f'W' + f''W'' \end{aligned} \tag{8.11}$$

to which the adjustment relation at constant head

$$\eta g(H + Z)Q_T = N \tag{8.12}$$

is added. Here f', f'' denote the areas of the two tunnel cross-sections, F_1, F_2 the areas of the horizontal cross-sections of the two secondary surge tanks (capture wells), F the area of the main tank, L_1 the length of the tunnel from the reservoir to the secondary tank, W_1 the water velocity in this tunnel length, Z_1 the water level in the secondary surge tank, and L_2, W_2, Z_2 are similarly defined for the second tunnel and the second surge tank. Further on, L' is the tunnel length from the secondary surge tank (capture well) up to the main surge tank, W' the water speed in this tunnel length, Z is the water level in the main tank, L'' and W'' are similarly defined for the second tunnel, Q_T denotes the turbined flow.

When writing the model, hydraulic head losses in the tunnels have been written under the natural assumption that the velocities are positive. Local head losses at tank's bases have been neglected.

The loss coefficients P_1, P_2, P', P'' are computed by the usual formulae

$$P' = \frac{\lambda' L'}{2gd'}, \quad P_1 = \frac{\lambda' L_1}{2gd'}, \quad P'' = \frac{\lambda'' L''}{2gd''}, \quad P_2 = \frac{\lambda'' L_2}{2gd''} \tag{8.13}$$

whence

$$\frac{P'}{L'} = \frac{P_1}{L_1}, \quad \frac{P''}{L''} = \frac{P_2}{L_2}$$

For the adopted model, the steady regime is given by the relations

$$\begin{aligned} &Z_1^o = -P_1 W_1^{o2}, \quad Z_2^o = -P_2 W_2^{o2}, \quad W_1^o = W_o', \quad W_2^o = W_o'' \\ &Z_o - Z_1^o = -P' W_o'^2, \quad Z_o - Z_2^o = -P'' W_o''^2, \quad Q_o = f' W_o' + f'' W_o'' \\ &N^2 - \eta^2 g^2 (H + Z_o)^2 Q_o^2 = 0 \end{aligned} \tag{8.14}$$

and the steady regime water level in the tank Z_o is computed from equation

$$Z_o(H + Z_o)^2 + \frac{N^2}{\eta^2 g^2 \left(\dfrac{f'}{\sqrt{P' + P_1}} + \dfrac{f''}{\sqrt{P'' + P_2}} \right)^2} = 0 \tag{8.15}$$

admitting a solution in the interval $(-(H/3), 0)$ if

$$N^2 < \frac{4H^3}{27} \eta^2 g^2 \left(\frac{f'}{\sqrt{P_1 + P'}} + \frac{f''}{\sqrt{P'' + P_2}} \right)^2 \tag{8.16}$$

a condition always satisfied in practical situations.

We shall investigate the stability of the steady regime corresponding to this value of Z_o. The linearised system around this steady regime (hence describing small oscillations around this regime) can be written (using the same notation for the deviations as well)

$$\begin{aligned} &\frac{L_1}{g}\frac{dW_1}{dt} + 2P_1 W_1^o W_1 + Z_1 = 0, \quad \frac{L_2}{g}\frac{dW_{12}}{dt} + 2P_2 W_2^o W_2 + Z_2 = 0 \\ &F_1 \frac{dZ_1}{dt} = f'(W_1 - W'), \quad F_2 \frac{dZ_2}{dt} = f''(W_2 - W'') \\ &\frac{L'}{g}\frac{dW'}{dt} + 2P' W_1^o W' + Z - Z_1 = 0 \\ &\frac{L''}{g}\frac{dW''}{dt} + 2P'' W_2^o W'' + Z - Z_2 = 0 \\ &F \frac{dZ}{dt} = f' W' + f'' W'' + \frac{Q_o Z}{H + Z_o} \end{aligned} \tag{8.17}$$

This is a 7th order system of differential equations. It is well known that stability against small oscillations means that the characteristic equations have negative real parts.

Starting from the coefficients of the characteristic polynomial, one can write the necessary and sufficient conditions for stability, e.g. in the form given by the Hurwitz criterion. If it were true that such an analysis can easily be made in the given situations using a computer, a general analysis which puts into evidence the critical cross-section using the Hurwitz criterion is nevertheless hopeless. Such an analysis has been performed in reference [8] for the case when no capturing wells exist on the two pressure tunnels, hence the equations of the system are of order 3; even in this case, the analysis is very complicated and the result appears to be hardly interpretable.

In the present paper we propose to approach the problem by using the method of Liapunov's function. This approach has been previously considered with encouraging results by the authors, for the case of basic schemes with one and two surge tanks [29]. Although the obtained results are only sufficient, their analysis in some particular significant cases shows that they are not essentially different from the necessary ones.

Liapunov's function

Before describing the adopted Liapunov function we have to introduce some notation.

We denote by $\beta > 0$ the positive root of the second order equation

$$\left(X - \frac{2P_1 W_1^{\circ}}{L_1}\right)\left(X - \frac{2P_2 W_2^{\circ}}{L_2}\right) + \frac{H + Z_{\circ}}{Q_{\circ}}\left[\frac{f'}{L' + L_1}\left(X - \frac{2P_2 W_2^{\circ}}{L_2}\right) + \frac{f''}{L'' + L_2}\left(X - \frac{2P_1 W_1^{\circ}}{L_1}\right)\right] = 0 \tag{8.18}$$

The free term of the trinomial is

$$-\frac{2P_1 W_1^{\circ} P_2 W_2^{\circ}(H + 3Z_{\circ})}{L_1 L_2 |Z_{\circ}|} < 0 \tag{8.19}$$

hence there exists a unique positive root.

The sum of the trinomial's roots is

$$S = \frac{2P_1 W_1^{\circ}}{L_1} + \frac{2P_2 W_2^{\circ}}{L_2} - \frac{H + Z_{\circ}}{Q_{\circ}}\left(\frac{f'}{L' + L_1} + \frac{f''}{L'' + L_2}\right) \tag{8.20}$$

and, since the second root is negative it follows that

$$\beta > \frac{2P_1 W_1^{\circ}}{L_1} + \frac{2P_2 W_2^{\circ}}{L_2} - \frac{H + Z_{\circ}}{Q_{\circ}}\left(\frac{f'}{L' + L_1} + \frac{f''}{L'' + L_2}\right) \tag{8.21}$$

Finally, one easily sees that

$$\beta < \frac{2P_1 W_1^{\circ}}{L_1} + \frac{2P_2 W_2^{\circ}}{L_2} \tag{8.22}$$

and denoting

$$B = \frac{L'}{L' + L_1}\left(\beta - \frac{2P_2 W_2^{\circ}}{L_2}\right), \quad C = \frac{L''}{L'' + L_2}\left(\beta - \frac{2P_1 W_1^{\circ}}{L_1}\right) \tag{8.23}$$

it follows that B and C have different signs.

With this notation, the adopted Liapunov function is

$$\begin{aligned} &V(W_1, W_2, Z_1, Z_2, W', W'', Z) \\ &= \frac{F}{2}\left(\frac{\mathrm{d}Z}{\mathrm{d}t}\right)^2 + \frac{F_1}{2}\left(\frac{\mathrm{d}Z_1}{\mathrm{d}t}\right)^2 + \frac{F_2}{2}\left(\frac{\mathrm{d}Z_2}{\mathrm{d}t}\right)^2 + \frac{gf'}{2L'}(Z - Z_1)^2 + \frac{gf''}{2L''}(Z - Z_2)^2 \\ &\quad + \frac{gf'}{2L_1}Z_1^2 + \frac{gf''}{2L_2}Z_2^2 - \frac{gQ_{\circ}}{2(H + Z_{\circ})}\left(\frac{2P_1 W_1^{\circ}}{L_1} + \frac{2P_2 W_2^{\circ}}{L_2} - \beta\right)Z^2 \\ &\quad + \frac{\mu^2}{2}\left[Bf'\left(W' + \frac{L_1}{L'}W_1\right) + Cf''\left(W'' + \frac{L_2}{L''}W_2\right)\right]^2 \end{aligned} \tag{8.24}$$

where μ^2 will be selected so as to ensure the condition related to Liapunov's function derivative. We note that (dZ/dt), (dZ_1/dt), (dZ_2/dt) are linear forms in Z, W', W'', Z, Z_1, Z_2, W_1, W_2, hence V is a quadratic form in these variables. Since β satisfies the inequality (8.21), the part of this quadratic form explicitly containing Z, Z_1, Z_2 follows to be positively definite, hence it is positive and vanishes only for $Z = Z_1 = Z_2 = 0$, $dZ_1/dt = dZ_2/dt = 0$, hence $W_1 = W'$, $W_2 = W''$, $dZ/dt = 0$, further $f'W' + f''W'' = 0$ whence

$$Bf'\left(1 + \frac{L_1}{L'}\right) W' + Cf''\left(1 + \frac{L_2}{L''}\right) W'' = 0 \tag{8.25}$$

These relations lead to $W' = W'' = 0$ (B and C having different signs, the two linear forms cannot be proportional). The special case $B = C = 0$ corresponds to the perfectly symmetrical situation $P_1 W_1^{\text{o}}/L_1 = P_2 W_2^{\text{o}}/L_2$; this is a limiting case, to be analysed separately. Hence, V follows to be positively defined.

After some computations too long to be reproduced here, we get

$$\begin{aligned}
-\frac{1}{g}\frac{dV}{dt} = {} & a_{11}\left(\frac{dZ}{dt}\right)^2 + a_{22}\left(\frac{dZ_1}{dt}\right)^2 + a_{33}\left(\frac{dZ_3}{dt}\right)^2 + 2a_{12}\frac{dZ}{dt}\frac{dZ_1}{dt} + 2a_{13}\frac{dZ}{dt}\frac{dZ_2}{dt} \\
& + 2a_{14}\frac{dZ}{dt}\left[Bf'\left(W' + \frac{L_1}{L'}W_1\right) + Cf''\left(W'' + \frac{L_2}{L''}W_2\right)\right] \\
& + 2a_{24}\frac{dZ_1}{dt}\left[Bf'\left(W' + \frac{L_1}{L'}W_1\right) + Cf''\left(W'' + \frac{L_2}{L''}W_2\right)\right] \\
& + 2a_{34}\frac{dZ_2}{dt}\left[Bf'\left(W' + \frac{L_1}{L'}W_1\right) + Cf''\left(W'' + \frac{L_2}{L''}W_2\right)\right] \\
& + a_{44}\left[Bf'\left(W' + \frac{L_1}{L'}W_1\right) + Cf''\left(W'' + \frac{L_2}{L''}W_2\right)\right]^2
\end{aligned} \tag{8.26}$$

where

$$\begin{aligned}
a_{11} &= \left(\frac{2P_1W_1^{\text{o}}}{L_1} + \frac{2P_2W_2^{\text{o}}}{L_2} - \beta\right)F - \frac{Q_{\text{o}}}{g(H + Z_{\text{o}})} \\
a_{22} &= \frac{2P_1W_1^{\text{o}}}{L_1}F_1, \quad a_{33} = \frac{2P_2W_2^{\text{o}}}{L_2}F_2, \quad a_{12} = -\frac{BL_1F_1}{2L'} \\
a_{13} &= -\frac{CL_2F_2}{2L''}, \quad a_{23} = 0, \quad a_{14} = \frac{1}{2}(1 + \mu^2 DF) \\
a_{24} &= \frac{\mu^2 D}{2}\frac{L_1f_1}{L_1 + L'}, \quad a_{34} = \frac{\mu^2 D}{2}\frac{L_2f_2}{L_2 + L''} \quad a_{44} = \mu^2\beta \\
D &= -\left(\beta - \frac{2P_1W_1^{\text{o}}}{L_1}\right)\left(\beta - \frac{2P_2W_2^{\text{o}}}{L_2}\right) = \frac{H + Z_{\text{o}}}{Q_{\text{o}}}\left(\frac{Bf'}{L'} + \frac{Cf''}{L''}\right)
\end{aligned} \tag{8.27}$$

and

$$D > 0 \tag{8.28}$$

According to the general theory, stability is ensured if $-(1/g)(dV/dt)$ is semipositively defined [11]. As the adopted written form shows, we consider $-(1/g)(dV/dt)$

as a quadratic form in the variables dZ/dt, dZ_1/dt, dZ_2/dt and $Bf'[W' + (L_1/L')W_1] + Cf''[W'' + (L_2/L'')W_2]$.

Sylvester's conditions are

$$a_{11} > 0, \quad a_{11}a_{22} - a_{12}^2 > 0, \quad a_{11}a_{22}a_{33} - a_{12}^2 a_{33} - a_{13}^2 a_{22} > 0 \tag{8.29}$$

to which the condition of the determinant being positive is added. The conditions (8.29) reduce to

$$a_{11} > \frac{a_{12}^2}{a_{22}} + \frac{a_{13}^2}{a_{33}} \tag{8.30}$$

We write the matrix of the form

$$\begin{pmatrix} A & b \\ b^* & \mu^2 D\dfrac{\beta}{D} \end{pmatrix}$$

whose determinant is $(\det A)\,(\mu^2 D(\beta/D) - b^* A^{-1} b)$; since the first conditions imply $\det A > 0$ the last Sylvester condition writes

$$\mu^2 D\frac{\beta}{D} - b^* A^{-1} b > 0 \tag{8.31}$$

The vector b has the coordinates a_{14}, a_{24}, a_{34}, and A is the symmetric matrix with elements a_{ij}, $i = 1, 2, 3$; $j = 1, 2, 3$.

Denoting by A_{ij} the entries of the matrix A^{-1}, and writing the equality $Y = \mu^2 D$ we get the condition

$$\begin{aligned} &A_{11}(1 + YF)^2 + 2A_{12}Y(1 + YF)\frac{L_1F_1}{L_1 + L'} + 2A_{13}Y(1 + YF)\frac{L_2F_2}{L_2 + L''} \\ &+ 2A_{23}Y^2\frac{L_1F_1}{L_1 + L'}\frac{L_2F_2}{L_2 + L''} + A_{22}Y^2\left(\frac{L_1F_1}{L_1 + L'}\right)^2 \\ &+ A_{33}Y^2\left(\frac{L_2F_2}{L_2 + L''}\right)^2 - 4Y\frac{\beta}{D} < 0 \end{aligned} \tag{8.32}$$

This inequality is possible only if the trinomial in Y has positive roots and Y takes a positive value, contained between them. Since the choice of Y is equivalent to the choice of the parameter μ^2, we have only to care about the trinomial having a positive root. Computation shows that the product of the roots is positive, hence the possibility to select μ^2 so that $-(1/g)(dV/dt)$ be positively defined as a quadratic form in the specified variables requires that the trinomial to have real positive roots, which (again, after computations too long to be reproduced here), leads to the conditions

$$\left(\frac{D}{2} + \frac{a_{22}a_{33}}{F_1F_2}\right)F > \frac{\beta Q_o}{g(H + Z_o)} + \left(\frac{a_{12}^2}{a_{22}} + \frac{a_{13}^2}{a_{33}}\right) - \frac{D}{2}\left(\frac{a_{12}}{a_{22}}\frac{L_1F_1}{L_1 + L'} + \frac{a_{13}}{a_{33}}\frac{L_2F_2}{L_2 + L''}\right) \tag{8.33}$$

$$\frac{a_{22}a_{33}}{F_1F_2}F > \frac{\beta Q_o}{g(H + Z_o)} + \left(\frac{L_1B}{L'}\right)^2\frac{a_{22}}{4\beta} + \left(\frac{L_2C}{L''}\right)^2\frac{a_{33}}{4\beta}$$

to which condition (8.30) is added under the form

$$\left(D\frac{a_{22}a_{33}}{F_1F_2}\right)F > \frac{\beta Q_o}{g(H + Z_o)} + \beta\left(\frac{a_{12}^2}{a_{22}} + \frac{a_{13}^2}{a_{33}}\right) \tag{8.34}$$

Critical cross-section of the main surge tank
We denote

$$F_{\text{cro}} = \frac{\beta Q_o F_1 F_2}{g(H+Z_o)a_{22}a_{33}} = \frac{\beta Q_o L_1 L_2}{g(H+Z_o)2P_1 W_1^2 \cdot 2P_2 W_2^o} \tag{8.35}$$

The second-order equation satisfied by β now leads to a second-order equation for F_{cro}, which after some manipulations becomes

$$\left(\frac{H+Z_o}{Q_o}\frac{\lambda'' W_o''}{d''}X-1\right)\left(\frac{H+Z_o}{Q_o}\frac{\lambda' W_o'}{d'}X-1\right)+\frac{H+H_o}{2|Z_o|}$$
$$\times\left\{\frac{Q_1^o}{Q_o}\left(\frac{H+Z_o}{Q_o}\frac{\lambda' W_o'}{d'}X-1\right)+\frac{Q_2^o}{Q_o}\left(\frac{H+Z_o}{Q_o}\frac{\lambda'' W_o''}{d''}X-1\right)\right\}=0 \tag{8.36}$$

where $Q_1^o = f'W_o'$, $Q_2^o = f''W_o''$.
We adopt the notation

$$\beta_1 = \frac{\beta L_1^o}{2P_1 W_1^o}, \quad \beta_2 = \frac{\beta L_2^o}{2P_2 W_2^o}, \quad \alpha = \sqrt{\frac{\lambda'' d'(L_1+L')}{\lambda' d''(L_2+L'')}} \tag{8.37}$$

Taking into account the equation satisfied by β, we obtain for β_1 and β_2 the equations

$$(X-1)(X-\alpha)+\frac{H+H_o}{2|Z_o|}\left[\frac{Q_1^o}{Q_o}(X-\alpha)+\frac{L_1+L'}{L''+L_2}\frac{f''W_o^1}{Q_o}(X-1)\right]=0$$
$$(X-1)\left(X-\frac{1}{\alpha}\right)+\frac{H+Z_o}{2|Z_o|}\left[\frac{Q_2^o}{Q_o}\left(X-\frac{1}{\alpha}\right)+\frac{L''+L_2}{L'+L_1}\frac{f'W_o''}{Q_o}(X-1)\right]=0 \tag{8.38}$$

The positive root of the first equation is β_1 while the positive root of the second equation is β_2. We further introduce the notation

$$F_{\text{cr1}} = \left[\frac{\beta Q_o}{g(H+Z_o)}+\beta\left(\frac{a_{12}}{a_{22}}+\frac{a_{13}}{a_{33}}\right)\right]\frac{1}{D+\dfrac{a_{22}a_{33}}{F_1F_2}} \tag{8.39}$$

After some calculation, we get

$$F_{\text{cr1}} = \frac{1}{\beta_1+\beta_2-\beta_1\beta_2}\left[F_{\text{cro}}+\beta_1(\beta_2-1)^2\left(\frac{L_1}{L'+L_1}\right)^2\alpha\frac{F_1}{4}\right.$$
$$\left.+\beta_2(\beta_1-1)^2\left(\frac{L_2}{L''+L_2}\right)^2\frac{1}{\alpha}\frac{F_2}{4}\right] \tag{8.40}$$

Similarly, denoting

$$F_{\text{cr2}} = \left[\frac{\beta Q_o}{g(H+Z_o)}+\beta\left(\frac{a_{12}^2}{a_{22}}+\frac{a_{13}^2}{a_{33}}\right)-\frac{D}{2}\left(\frac{a_{12}}{a_{22}}\frac{L_1F_1}{L'+L_1}+\frac{a_{13}}{a_{33}}\frac{L_2F_2}{L''+L_2}\right)\right]$$
$$\times\frac{1}{\dfrac{D}{2}+\dfrac{a_{22}a_{33}}{F_1F_2}}$$

$$F_{cr3} = \frac{\beta Q_o F_1 F_2}{g(H+Z_o)a_{22}a_{33}} + \left(\frac{L_1}{L'+L_1}\right)^2 + \left(\beta - \frac{a_{33}}{F_2}\right)^2 \frac{F_1F_2}{4\beta a_{33}}$$
$$+ \left(\frac{L_2}{L''+L_2}\right)^2 \left(\beta - \frac{a_{22}}{F_1}\right)^2 \frac{F_1F_2}{4\beta a_{22}} \quad (8.41)$$

we deduce after calculation

$$F_{cr2} = \frac{2}{1+\beta_1+\beta_2-\beta_1\beta_2}\left[F_{cro} + (\beta_2-1)^2\left(\frac{L_1}{L'+L_1}\right)^2 \alpha\frac{F_1}{4} + (\beta_1-1)^2\left(\frac{L_2}{L''+L_2}\right)\frac{1}{\alpha}\frac{F_2}{4}\right]$$
$$F_{cr3} = F_{cro} + \frac{1}{\beta_2}(\beta_2-1)^2\left(\frac{L_1}{L'+L_1}\right)^2\frac{F_1}{4} + \frac{1}{\beta_1}(\beta_1-1)^2\left(\frac{L_2}{L''+L_2}\right)^2\frac{F_2}{4} \quad (8.42)$$

The analysis using Liapunov's function shows that if

$$F > \max\{F_{cr1}, F_{cr2}, F_{cr3}\} = F_{cr} \quad (8.43)$$

the hydraulic operation of the system is stable against small oscillations, which solves the considered problem. Note that the properties of β imply: $\beta_1+\beta_2 > 1+\beta_1\beta_2(\beta_1-1)(\beta_2-1) < 0$; if $\alpha < 1$, we also have $\alpha < \beta_1 < 1 < \beta_2 < 1/\alpha$.

The coefficient α is in some sense measuring the asymmetry degree of the considered hydraulic scheme: if $\alpha = 1$, the scheme is symmetrical and we have $\beta_1 = \beta_2 = 1$. If α is close to 1, it follows that β_1, β_2 are also close to 1, and the three values F_{cr1}, F_{cr2}, F_{cr3} are close to one another and to F_{cro}.

In this case, F_{cro} is essentially Thoma's cross-section, which in the adopted notation reads

$$F_{Th} = \frac{Q_o}{g(H+Z_o)}\frac{L}{2PW_o} = \left(\frac{fd}{\lambda(H+Z_o)}\right) \quad (8.44)$$

The equation for determining the cross-section F_{cro} can also be written as:

$$\left(\frac{H+Z_o}{d''}\frac{\lambda'' Q_2^o}{Q_o}\frac{F_{cro}}{f''} - 1\right)\left(\frac{H+Z_o}{d'}\frac{\lambda' Q_1^o}{Q_o}\frac{F_{cro}}{f'} - 1\right) + \frac{H+Z_o}{2|Z_o|}$$
$$\times\left[\frac{Q_1^o}{Q_o}\left(\frac{H+Z_o}{d'}\frac{\lambda' Q_1^o}{Q_o}\frac{F_{cro}}{f'} - 1\right) + \frac{Q_2^o}{Q_2}\left(\frac{H+Z_o}{d''}\frac{\lambda'' Q_2^o}{Q_o}\frac{F_{cr}}{f''} - 1\right)\right] = 0 \quad (8.45)$$

In this form, the equation contains only dimensionless quantities (ratios of quantities of the same nature).

We also note that, in limit cases of strong asymmetry, when one of the feed tunnels is not significant, the main surge-tank critical cross-section is closer to Thoma's cross-section: the smaller the cross-section of the secondary surge tank is, or the closer to the reservoir it is.

Computation examples

(a) In order to check the hydraulic stability conditions obtained analytically, we shall consider the real case of hydraulic scheme B previously analysed by numerical methods, fixing for the beginning the cross-section of the capture wells to $F_1 = 7$ m, $F_2 = 12.5$ m.

Table 8.1. $F_1 = F_2 = 12.5\,\mathrm{m}^2$.

X	F_{cr1}	F_{cr2}	F_{cr3}	F_{cro}
1000	24.63	23.94	23.23	23.01
2000	27.71	26.63	25.63	25.50
3000	29.57	28.14	27.07	26.91
4000	29.19	28.08	27.13	26.91
5000	30.48	29.48	28.67	28.37
6000	30.52	29.72	29.14	28.74
7000	30.40	29.90	29.48	28.99
8000	30.19	29.81	29.74	29.11
9000	29.93	29.75	29.93	29.18
10000	29.64	29.64	30.08	29.20

Computations are made according to the presented methodology; by using relations (8.40), (8.42) and (8.43), we get the following value of the critical cross-section for small oscillations of the main surge tank:

$$F = F_{cr} = 29.5\,\mathrm{m}^2 \tag{8.46}$$

Admitting a safety coefficient against instability of the order $n = 1.6$–1.8, we find once again a good approximation of the surge tank cross-section F, selected by numerical methods (Fig. 8.4).

(b) By using the procedure here described, we have performed computation experiments, presented in Tables 8.1 and 8.2. These experiments confirm the practical character of the obtained results, as well as the possibility to apply these results in the design of surge tanks. The following fundamental data have been assumed:

$$H = 275\,\mathrm{m}, \quad N = 50\,\mathrm{MW}, \quad f' = f'' = 12.5\,\mathrm{m}^2,$$

$$L_1 + L' = X, \quad L_2 + L'' = 10000\,\mathrm{m},$$

$$L_1 = 0.75X, \quad L_2 = 2500\,\mathrm{m}$$

Finally, we would like to mention an analytical method of analysing the stability against small oscillations for the hydraulic scheme D, published in our paper [29].

8.2. Hydroelectric power plants and pumping systems with combined water conveyance structures

To the same category of modern hydroelectric power plants and pumping systems, conceived as complex hydrotechnical schemes, also belong those with combined water conveyance structures with automatic regulators, for instance the hydraulic schemes I and II analysed in the present paper (Figs. 8.9 and 8.10).

Hydraulic scheme I (hydroelectric power plant) consists of a supply reservoir, dam gate with automatic regulation of the water level in the channel, free surface channel, air chamber, pressure tunnel, surge tank and turbine with constant power automatic regulation.

Hydraulic scheme II (pumping system) consists of the same elements as scheme I, with the difference that instead of the turbine with constant power regulator it has a

Table 8.2. $F_2 = 12.5\,\mathrm{m}^2$.

X	L_1	F_1	F_{cr1}	F_{cr2}	F_{cr3}	F_{cro}
1000	0.1 X	12.5	24.63	23.93	23.21	23.01
		20.0	24.63	23.93	23.21	23.01
		30.0	24.63	23.93	23.21	23.01
	0.5 X	12.5	24.63	23.93	23.22	23.01
		20.0	24.63	23.93	23.22	23.01
		30.0	24.63	23.93	23.23	23.01
	0.98 X	12.5	24.63	23.94	23.24	23.01
		20.0	24.63	23.94	23.26	23.01
		30.0	24.65	23.95	23.28	23.01
5000	0.1 X	12.5	30.43	27.42	28.38	28.37
		20.0	30.43	27.42	28.39	28.37
		30.0	30.53	27.43	28.39	28.37
	0.5 X	12.5	30.48	27.48	28.51	28.37
		20.0	30.51	27.52	28.59	28.37
		30.0	30.55	27.56	28.69	28.37
	0.98 X	12.5	30.63	27.65	28.88	28.37
		20.0	30.75	27.79	29.19	28.37
		30.0	30.91	27.97	29.59	28.37
10000	0.1 X	12.5	29.20	29.20	29.21	29.20
		20.0	29.21	29.21	29.22	29.20
		30.0	29.22	29.22	29.23	29.20
	0.5 X	12.5	29.39	29.39	29.59	29.20
		20.0	29.51	29.51	29.82	29.20
		30.0	29.67	29.67	30.13	29.20
	0.98 X	12.5	29.95	29.95	30.70	29.20
		20.0	30.40	30.40	31.60	29.20
		30.0	31.00	31.00	32.80	29.20

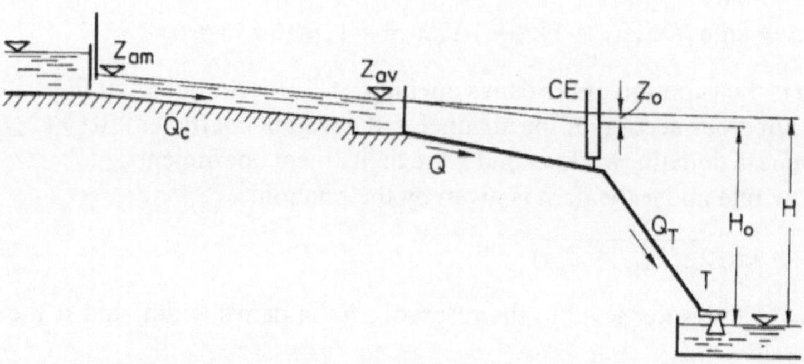

Figure 8.9. Hydraulic scheme I (hydraulic power plant).

turbopump with a cheek valve which closes itself at reserve flow and a discharge pipe in a constant level reservoir. The adjustment of the upstream gate is realized by means of an electrically driven device, according to the water level variation in two sections of the channel.

The investigation of the behaviour at transitory regimes of the hydraulic system corresponding to schemes I and II can be conducted by solving the equations of water

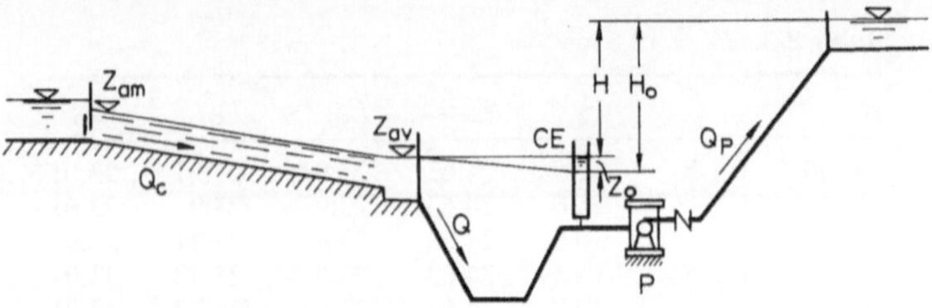

Figure 8.10. Hydraulic scheme II (pumping system).

motion either by analytical or numerical methods. Analytical methods generally imply the linearisation of the equations of motion as well as the adoption of simplifying computation assumptions the resulting solutions being therefore approximate [10]. Numerical methods allow in this case as well the solution of the equations of motion with any accuracy degree and, with the present use of electronic computers, can be applied on a large scale.

8.2.1. Equations of motion

In the present paper the problem of hydraulic calculations at transient regimes of systems conceived on hydraulic schemes of types I and II [19, 23, 27] is approached by means of numerical methods. We first establish the equations of motion:

Upstream dam gate

We admit a adjustment equation of the form

$$\Delta a = r\left[k\Delta Z_{\text{am}} + (1-k)\Delta Z_{\text{av}}\right] \tag{8.47}$$

or

$$\text{sign}\Delta a \neq \text{sign}\left[k\Delta Z_{\text{am}} + (1-k)\Delta Z_{\text{av}} \pm \varepsilon_z\right], \quad \Delta a = m\Delta t \tag{8.48}$$

where Δa is the variation of the dam's opening, ΔZ_{am} the variation of the water level in the upstream cross-section of the channel, k the weight coefficient ($0 \leq k \leq 1$), ε_z the insensitiveness domain, while m and r are adjustment coefficients.

The flow rate under the dam is given by the relation

$$Q_{\text{am}} = \mu ba\sqrt{2g(H_{\text{am}} - Z_{\text{am}})} \tag{8.49}$$

where H_{am} is the water level in the reservoir, b the dam's width, and μ the flow-rate coefficient.

The free surface channel

The water motion in the channel can be considered gradually varied, hence, described by the Saint Venant equations

$$B\frac{\partial Z}{\partial t} + \frac{\partial Q}{\partial x} = 0 \tag{8.50}$$

$$\frac{\partial Z}{\partial x} + \frac{1}{g\omega}\frac{\partial V}{\partial t} + \frac{V}{g}\frac{\partial V}{\partial x} + \frac{Q|Q|}{K^2} = 0 \tag{8.51}$$

where B is the width of the channel at the water surface, Z the water level in the channel, Q the flow, V the mean velocity in the cross-section, ω the flow cross-section area, K the flow modulus, x the distance along the channel, t the time, and g the acceleration due to gravity.

Air chamber

We shall write the continuity equation under the form

$$Q_c = \Omega \frac{dZ}{dt} + Q \tag{8.52}$$

where Q_c is the flow in the cross-section upstream the chamber, and Ω is the horizontal cross-section area of the chamber.

The tunnel–surge tank system

The general equations of the oscillatory motion in the system are

$$\frac{L}{g}\frac{dW}{dt} + Z_c + P + R + E = 0 \tag{8.53}$$

$$Wf = F\frac{dZ_c}{dt} + Q_T \tag{8.54}$$

where L is the tunnel length, W the water velocity in the tunnel, Z_c the water level in the surge tank, P the longitudinal head loss in the tunnel, R the local head loss in the surge tank strangulation, E the term corresponding to the kinetic energy of water in the cross-section in front of the tank, f the pressure tunnel cross-section area, F the area of the surge-tank horizontal cross-section, and Q_T the turbined flow.

The turbine (in the case of hydraulic scheme I). In order to compute the extreme water level oscillations in the system, we admit a general flow relation of the form

$$Q_T = Q_T(t) \tag{8.55}$$

In the study of hydraulic stability of the system, we assume the following adjustment equation, in the case of the plant operation, in a separate network

$$\eta g Q_T(H + Z_c + R - K_T Q_T^2) = N_2 \tag{8.56}$$

where H is the total head, N the turbines' power, η the turbines' output, and K_T is the longitudinal head loss coefficient in the penstock.

The pump (in the case of the hydraulic scheme II). We admit a general flow-rate relation of the form

$$Q_P = Q_P(t) \tag{8.57}$$

If there exists a flap on the discharge pipe at the pump, the oscillatory motion in the channel–tunnel system can be considered independent of the water motion in the discharge pipe. With these specifications, we are now in a position to explore the computation algorithm.

The computation of the combined motion in the water conveyance structures of the system is performed simultaneously, by various methods, for the water motion in both the channel, and for the pressure tunnel, the conditions in the common cross-section of the channel and the gallery being satisfied.

8.2.2. Numerical method

In order to compute the non-permanent water motion in the feed channel, a computation method which is an improved variant of the algorithm in reference [5] is suggested.

Expressing the velocity V as a function of the flow rate by the relation

$$V = \frac{Q}{\omega} \tag{8.58}$$

eqns. (8.50) and (8.51) take the form

$$B\frac{\partial Z}{\partial t} + \frac{\partial Q}{\partial x} = 0 \tag{8.59}$$

$$\frac{\partial Z}{\partial x} + \frac{1}{g\omega}\frac{\partial Q}{\partial t} + \frac{2Q}{g\omega^2}\frac{\partial Q}{\partial x} - \frac{Q^2}{g\omega^3}\frac{\partial \omega}{\partial x} + \frac{Q|Q|}{K^2} = 0 \tag{8.60}$$

The initial and boundary conditions associated with these equations consist in the assumption at an arbitrary moment of the flow rates and the water levels repartition in any cross-section of the channel.

$$Z_{t=t_o} = Z(x), \quad Q_{t=t_o} = Q(x) \tag{8.61}$$

as well of the flow conditions upstream and downstream the channel, as a function of the inflows into the channel. (In the case of rivers, the boundary conditions are usually expressed as: a known water discharge hydrograph at the flow in section $Q = Q(t)$; a known relationship between the discharge and the levels at the flow-out section $f(z, Q) = 0$ or other conditions.)

In order to write eqns. (8.59) and (8.60) in finite differences at a time step Δt, consider a channel length Δx limited by the computation sections i, $i+1$ numbered in a downstream sequence (Fig. 8.11). Adopting the following finite difference expression for the derivatives of functions $Z(x, t)$ and $Q(x, t)$ [1, 5]

$$\frac{\partial f}{\partial t} = \frac{1}{2\Delta t}[f(x, t\Delta t) + f(x + \Delta x, t + \Delta t) - f(x, t) - f(x + \Delta x, t)] \tag{8.62}$$

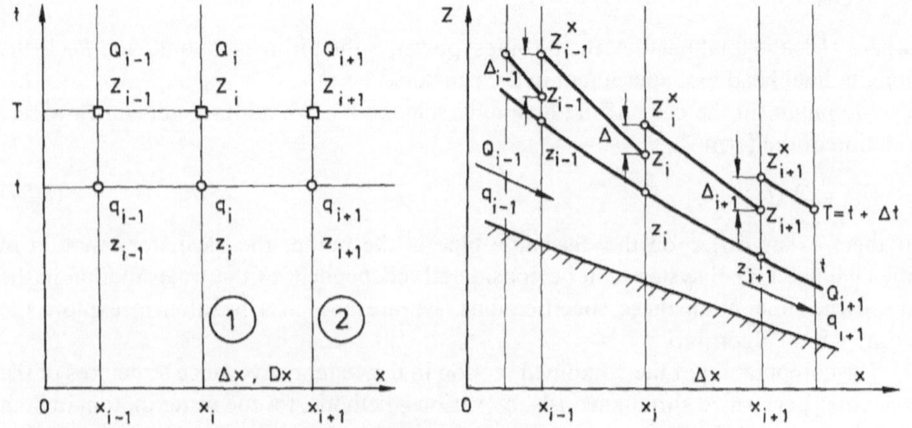

Figure 8.11. Representation of the computation method for a time step Δt.

$$\frac{\partial f}{\partial x} = \frac{1}{2\Delta x}[f(x+\Delta x, t+\Delta t) + f(x+\Delta x, t) - f(x, t+\Delta t) - f(x, t)] \tag{8.63}$$

the system (8.59), (8.60) becomes

$$\overline{\Delta Z} = -\left[\frac{1}{g\bar{\omega}}\frac{\Delta x}{\Delta t}\delta Q_{\mathrm{m}} + \frac{2\bar{Q}_{\mathrm{m}}}{g\bar{\omega}^2}\Delta\bar{Q} - \frac{\bar{Q}_{\mathrm{m}}^2}{g\bar{\omega}^3}\Delta\bar{\omega} + \Delta\bar{\omega}\frac{Q_{\mathrm{m}}|Q_{\mathrm{m}}|}{K^2}\right] \tag{8.64}$$

$$\Delta\bar{Q} = -\bar{B}\frac{\Delta x}{\Delta t}\delta Z_{\mathrm{m}} \tag{8.65}$$

where the following notation conventions have been applied:

- mean values at time $(t + \Delta t/2)$ are barred;
- mean values at the i section are denoted by the i index, and those in the mean section $(x + \Delta x/2)$ by the m index;
- the increment with respect to x is denoted by Δ, the one with respect to t, by δ.

The computation method relies on the idea that, within a computation step where the quantities z_i and q_i are known, quantities Z_i and Q_i at all computation sections at time $T + \Delta t$ are proposed. These quantities are to be successively corrected until the (Saint Venant) finite difference equations on all channel lengths, as well as the boundary conditions are satisfied. The quantities Q_i and Z_{i+1} being assumed as known, Q_{i+1} and Z_i can be computed on an arbitrary sector $\Delta x_i = x_{i+1} - x_i$, by solving through a direct iterative method the system of algebraic eqns. (8.64), (8.65).

Let $C_j(T)$ be a curve of the free water surface, defined by the values Z_i, $i = 1, 2, \ldots, n$, and adopted as first approximation. By applying the above algorithm from section to section in an upstream to downstream sequence, one can compute a new curve $C_j^*(T)$ of the free water surface, characterized by the values Z_i^*, $i = 1, 2, \ldots, n$.

The accuracy condition

$$|Z_i^* - Z_i| = |\Delta_i| < \varepsilon \tag{8.66}$$

is checked for each computation section, and if it is not satisfied, a new curve $C_{j+1}(T)$ of the free surface at time T is proposed, corresponding to the values $Z_i + \Delta_i^*$, where

$$\Delta_i^* = \frac{1}{4}[2\theta\Delta_i + (1-\theta)(\Delta_{i-1} + \Delta_{i+1})] \tag{8.67}$$

and θ is a coefficient below unit, between 0.5 and 1.

In the case of rivers, for the upstream section, Z_n^* is determined from the condition $f(Z_n^*, Q_n^*)$ and the check is made by a relationship of the form (8.67) as well.

For the first proposed curve $C_0(T)$ one can take the curve of the free water surface at moment t. The computation is considered to end for a time interval Δt when the condition (8.66) is satisfied for all the computation sections.

The proposed computation algorithm ensures a rapid convergence of the solution and a good numerical stability for adequately chosen Δx and Δt computation steps. One can take

$$\Delta t = (5\text{–}10)\frac{\Delta x}{|V| + \sqrt{g\dfrac{\omega}{B}}} \tag{8.68}$$

The computation method can easily be written in an algorithmic form and can be used on mean capacity electronic computers. For the oscillatory water motion in the tunnel–tank system one can use the previously presented computation algorithm for the case of a single surge tank.

The numerical solution of the problem is stable and rapidly convergent.

Computation example

We present some results of a hydraulic study on the computation of the extreme water level oscillations in the system as well as the stability of operation of a hydroelectric power plant designed as a hydraulic scheme of type I, with the following characteristics:

Upstream dam:	Adjustment of the gate is performed according to equation (8.48), with $k = 0; 0.5; 1$; $m = \alpha m_o$ and $\alpha = 1; 3$;
Feed channel:	Rectangular cross-section, with $B = 10$ m;
	$L_c = 2{,}500–5{,}000$ m; $i = 0.001$;
Air chamber:	$\Omega = 1{,}000\ \mathrm{m}^2$;
Pressure tunnel:	$D = 5$ m, $L = 5{,}000$ m;
Surge tank:	$F = 50–100\ \mathrm{m}^2$; $R = 0.001\ Q_c^2$;
Penstock:	$D_T = 4$ m; $L_T = 500$ m;
Hydroelectric power plant:	$Q_T = 60\ \mathrm{m}^3/\mathrm{s}$, $N = 200$ MW.

In the computation of extreme levels, manoeuvres of starting-up and total shutdown of the plant at the maximum flow rate $Q_T = 60\ \mathrm{m}^3/\mathrm{s}$ were considered. The results for a computation variant are shown in graphical form in Fig. 8.12. For the study of the hydraulic stability in operation of the plant, a power manoeuvre at the station with $N_1 = 120$ MW and $N_2 = 80$ MW has been considered, assuming $\Delta x = 500$ m.

Computations have been performed for several combinations of the values of quantities L_c, F, k and α, out of which we present the following (Fig. 8.13):

$$
\begin{array}{llll}
L_c = 5000\ \mathrm{m}, & F = 50\ \mathrm{m}^2, & k = 0, & \alpha = 1, \\
L_c = 5000\ \mathrm{m}, & F = 50\ \mathrm{m}^2, & k = 0.5, & \alpha = 3, \\
L_c = 2500\ \mathrm{m}, & F = 100\ \mathrm{m}^2, & k = 1, & \alpha = 1
\end{array}
$$

Numerical simulation of non-permanent water motion in the system for a large number of real cases in the operation of the plant allowed to put into evidence some very interesting cases of general instability of the water motion (Fig. 8.13a, b), as

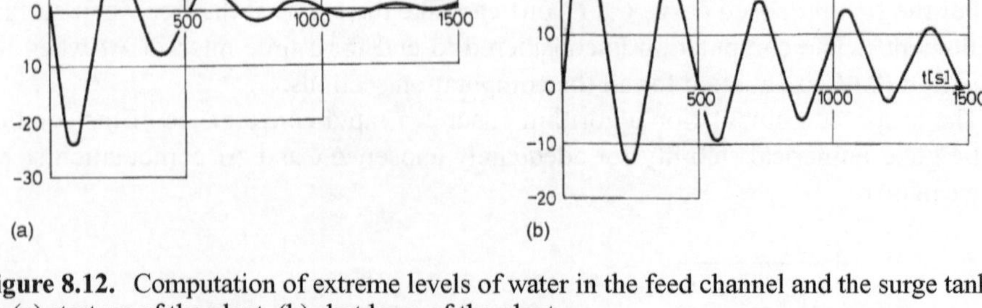

Figure 8.12. Computation of extreme levels of water in the feed channel and the surge tank, at: (a) start-up of the plant, (b) shutdown of the plant.

well as of the opportunities for their transformation into stable oscillatory motions (Fig. 8.13c). The notation h and Z_c has been used for the water depth in the channel and the water level in the surge tank respectively.

8.2.3. On the use of numerical models for unsteady regimes in open channels

The use of numerical models to simulate the unsteady flows in rivers and open channels has become a current practice for many, if not all, studies and projects of hydraulic engineering. Many methods and models have been developed and considerable progress has been attained in this field [2, 17]. However, some aspects concerning either the physical substantiation of the mathematical models of the numerical analysis techniques or the requirements for the model and input data calibration and/or validation, can still be developed and improved. In this respect, by presenting, in a concise form, two numerical models already in current use, we discuss also some remarks and ideas.

The usual way to develop numerical models for unsteady flow regimes in rivers and open channels is: starting from the differential equations of motion, which are transposed in *finite difference* equations, approximating the derivatives by differences and thus obtaining a set of algebraic equations to be solved numerically through a direct or an iterative procedure. In this way, one uses a double approximation process:

- first when writing the differential equations by neglecting "infinitely small" terms of higher order, that follow from applying the physical conservation laws (mass, momentum, energy) to an infinitesimal element of the fluid continuum;
- then, when converting the differential equations into difference equations, for the dimensions of a finite element, in the approximation of derivatives by finite differences, one neglects higher order differences (and this implies a certain refinement of the computational network).

In order to diminish the effect of these approximations on the numerical solution (accuracy, convergence, stability), in the development of difference equations one uses so called "centred" or "symmetrical" approximation formulae or schemes; in the implicit schemes one introduces weighted factors with different values for initial and final differences.

As an alternative to this approach, we can consider directly a *finite element* and admit for instance, the same hypothesis of linear variation of the physical parameters over the computational step (finite element) and apply the physical laws (conservation laws) to the condition of such a finite element and to the hypothesis under consideration. In this way we can obtain directly the difference equations with a sound justification of the weighted factors (that can be different for various terms in the equations) in accordance with the conditions and hypothesis considered when writing the equations (applying the physical laws) for a finite element.

As compared to the first computation method, presented in the previous paragraph, in the second method, also concisely described here below, we shall use finite difference equations written directly by applying the conservation laws (mass, momentum, energy) to a finite element, considering the hypothesis of linear variation (of the dependent variables and channel characteristics) between computational points (over a computational step—finite element).

With the usual notation, and with the subscript 1 for the upstream section and 2 for downstream (Fig. 8.11), using small letters for the function values at time t and capitals for $T = t + \Delta t$, double subscript for averaged values over a Δx step and a bar for values averaged over Δt, the finite difference equations can be written as

$$Q_2 - Q_1 + q_2 - q_1 = \frac{Dx}{Dt}\bar{B}_{12}(z_1 + z_2 - Z_1 - Z_2) \tag{8.69}$$

$$\frac{Dx}{Dt}\left(\frac{Q_1 + Q_2}{2} - \frac{q_1 + q_2}{2}\right) + \bar{Q}_{12}(V_2 - V_1) = g\omega_{12}\left(\frac{Z_1 - Z_2}{2} + \frac{z_1 - z_2}{2} - J_{12}Dx\right) \tag{8.70}$$

the first equation being the (mass) continuity equation and the second one the dynamic (momentum) equation, with weighted factor $\theta = 1/2$ in the continuity equation and $\theta = 1/2$ or $\theta = 1$ in the second equation.

Under the conditions of the hypotheses considered, the momentum equation expresses correctly the momentum conservation law, as applied to the finite element, if the computational steps satisfy the relation

$$\bar{V}_{12}Dt < Dx \tag{8.71}$$

If this condition is not in accordance with the scale of the phenomenon (in the sense that the time variations are too slow and the increments over the time step Δt as required by (8.71) have the order of magnitude of the computational accuracy), then it means that the physical phenomenon would be better described by a numerical model using the energy conservation law, and then (8.70) can be written as

$$Q_{12}(V_2 - V_1) = g\Omega_{12}(Z_1 - Z_2 - J_{12}Dx) \tag{8.72}$$

and the restriction (8.71) is not required for Δt. When the partial differential equations are used as primitive forms in developing numerical models, these situations are treated by neglecting some terms (inertial terms) in the dynamic equation, while maintaining the same approximation scheme for the finite differences.

Equations (8.69) and (8.70), by a convenient grouping of the terms, can be written in the form

$$-\beta Q_1 + Z_1 + \beta Q_2 + Z_2 = C_{12} \tag{8.73}$$

$$S_1 Q_1 - Z_1 + S_2 Q_2 + Z_2 = E_{12} \tag{8.74}$$

with the unknown values Z_1, Q_1, Z_2 and Q_2 explicitly expressed and where the notation is

$$\beta = \frac{Dt}{\bar{B}_{12}Dx} \tag{8.75}$$

$$S = \frac{|Q|}{2K^2Dx} + \frac{Dx}{2g\omega_{12}Dt} \pm \frac{Q}{2g\omega_{12}\Omega} \tag{8.76}$$

and C_{12}, E_{12} grouping the rest of the terms that are functions of known values z_1, q_1, z_2 and q_2 only.

In many cases, for the scope of routing natural floods in river channels, when eqns. (8.72) can be used instead of (8.70), the above expressions can be simplified accordingly; however, for other unsteady regimes as those induced in river channels or canals by power plant operation, or various regulation systems (on irrigation canals, for instance) or in the case of a flood wave produced by a dam failure, the use of the dynamic equation in the form (8.70) is required.

As these finite difference eqns. (8.73), (8.74) are non-linear, an iterative procedure should be used to solve them. Let an approximate tentative solution by

$$Z_1 = Z_1^{\text{o}} - DZ_1 \tag{8.77}$$

where Z_1^{o} is the correct, wanted solution, ΔZ_1 the required correction for Z_1 to get Z_1^{o} and similarly for Q_1, Z_2 and Q_2. Substituting expressions of the form (8.77) in eqns. (8.73) and (8.74) one gets

$$-\beta DQ_1 + DZ_1 + \beta DQ_2 + DZ_2 = DDZC \tag{8.78}$$

$$S_1 DQ_1 - DZ_1 + S_2 DQ_2 + DZ_2 = -DDZM \tag{8.79}$$

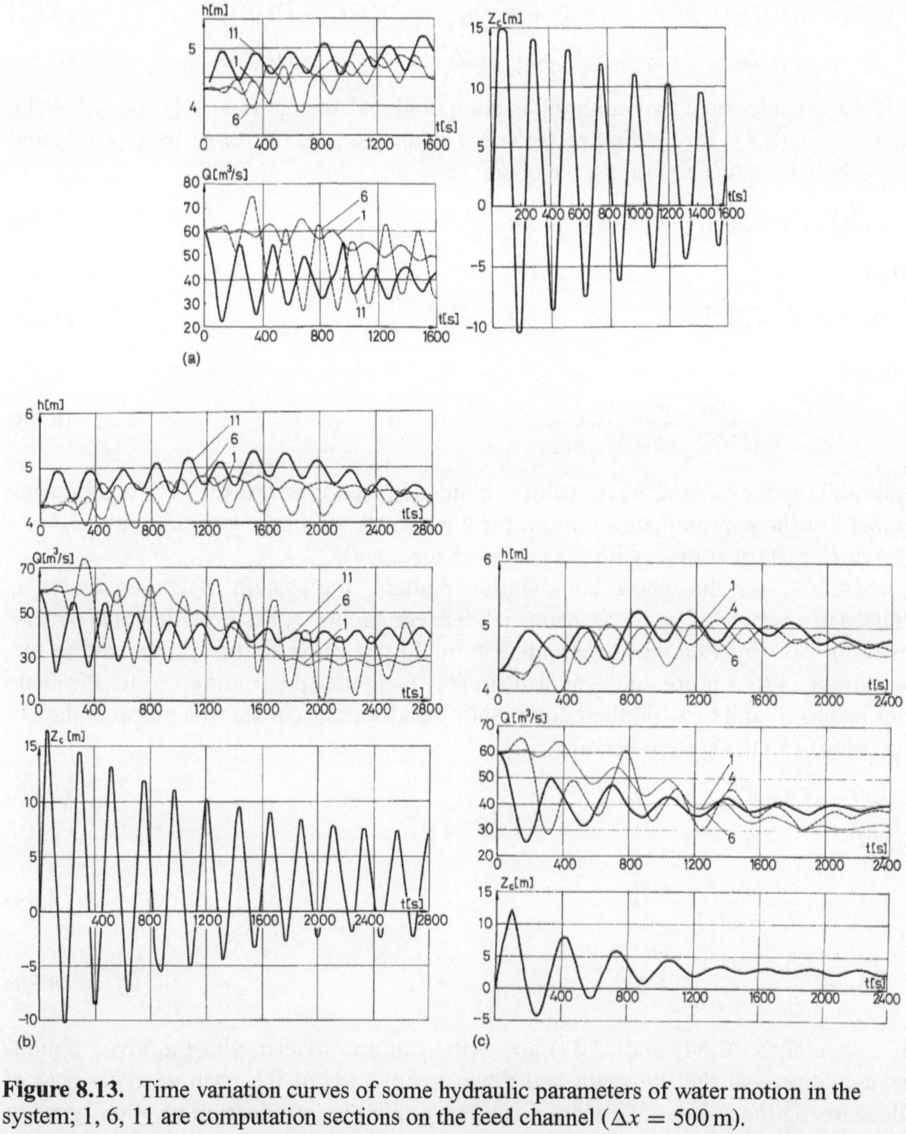

Figure 8.13. Time variation curves of some hydraulic parameters of water motion in the system; 1, 6, 11 are computation sections on the feed channel ($\Delta x = 500\,\text{m}$).

where the following notation has been introduced:

$$\begin{aligned} DZC &= \beta(Q_1 - Q_2 + q_1 - q_2) & DZM &= S_1Q_1 + S_2Q_2 \\ DZC_1 &= Z_1 + Z_2 - z_1 - z_2 & DZM_1 &= Z_1 - Z_2 \\ DDZC &= DZC - DZC_1 & DDZM &= DZM - DZM_1 \end{aligned} \tag{8.79'}$$

Thus, eqns. (8.73) and (8.74) can be written, in the final stage of the iterative process under the form

$$DZC - DZC_1 = 0 \tag{8.80}$$

$$DZM - DZM_1 = 0 \tag{8.81}$$

Adding and subtracting eqns. (8.78) and (8.79) we obtain:

$$(-S_1 - \beta)DQ_1 + 2DZ_1 + (-S_2 + \beta)DQ_2 = DDZC + DDZM \tag{8.82}$$

$$(-S_1 - \beta)DQ_1 + (S_2 + \beta)DQ_2 + 2DZ_2 = DDZC - DDZM \tag{8.83}$$

If we consider now two contiguous computational sectors $(i-1, i)$ and $(i, i+1)$, and use eqn. (8.83) for the first sector $(i-1, i)$ and eqn. (8.82) for sector $(i, i+1)$ and then subtract the two equations, we obtain

$$A_iDQ_{i-1} + B_iDQ_i + C_iDQ_{i+1} = D_i \tag{8.84}$$

where

$$D_i = Z_i^{i-1} - Z_i^{i+1} \tag{8.85}$$

and

$$Z_i^{i-1} = \frac{z_{i-1} + z_i}{2 + (DZC - DZM)_{i-1,i}} \tag{8.86}$$

(and similarly for Z_i^{i+1}) being the values obtained in section i, from the two neighbouring sectors, for the approximate solution; for the correct solution we should have $Z_i^{i+1} = Z_i^{i-1}$ or $D_i = 0$ (of course, with the accepted accuracy).

Therefore, for the whole computation domain, the system of $2N$ equations is reduced to a system of N simultaneously linear equations, with a three-diagonal matrix, easier to solve by a double-sweep algorithm. To this purpose the system can be further arranged to a more convenient form (for computer programming), to eliminate coefficients A_i and to avoid their computation and storage (in the first phase of the DS algorithm) and the system becomes [2]

$$DQ_i + C1_iDQ_{i+1} = C2_i \tag{8.87}$$

where

$$C1_i = \frac{C_i}{B_i - A_iC1_{i-1}} \tag{8.88}$$

$$C2_i = \frac{D_i - A_iC2_{i-1}}{B_i - A_iC1_{i-1}} \tag{8.89}$$

Of course eqns. (8.84) and (8.87) are written in accordance with the given boundary conditions, at the upstream and downstream ends of the channel. This way of schematising the system of equations allows easily the introduction of other types of intermediate boundary conditions, within the computational domain, as those given by inflows (from tributaries), outflows (deviations), weirs or gated dams with given or imposed level conditions, regulation gates etc., conditions that can be introduced in the computational algorithm without dividing the domain in several distinct stretches.

The iterative procedure goes on until the right term in eqn. (8.77), $D_i = Z_i^{i-1} - Z_i^i$, becomes zero, or close to zero within the proposed computational accuracy. The experience shows that for models developed in this way, with coefficients in eqn. (8.84) carefully expressed and adequate computational steps, the iterative procedure is rapidly convergent, and only few iterations are enough to attain a good accuracy, for most practical applications.

This computation method has also been used in many numerical models to simulate unsteady regimes in river channels or canals with flow and/or level regulating systems and also for dam failure flood wave propagation. The results obtained by this method and herein illustrated are for the case of a flood wave propagation on a natural river channel with a relative high slope; the flood wave is a rapidly raising wave consequent to a dam failure on a tributary. In Figs. 8.14 and 8.15 are shown the variation in time of the discharges in the sections upstream and downstream of the confluence and the level variation along the main channel at various time intervals.

With all the developments and improvements achieved so far in the numerical models for unsteady motion in open channels given by the literature in the field, many aspects regarding the stability and the convergence of the numerical solution are still an important concern for the analysts as well as for the users of these numerical models. Although the implicit difference schemes, that are mostly adopted for such models, are theoretically unconditionally stable, in practical applications an important problem still remains to be solved, i.e. the correct choice of computational steps Δx and Δt.

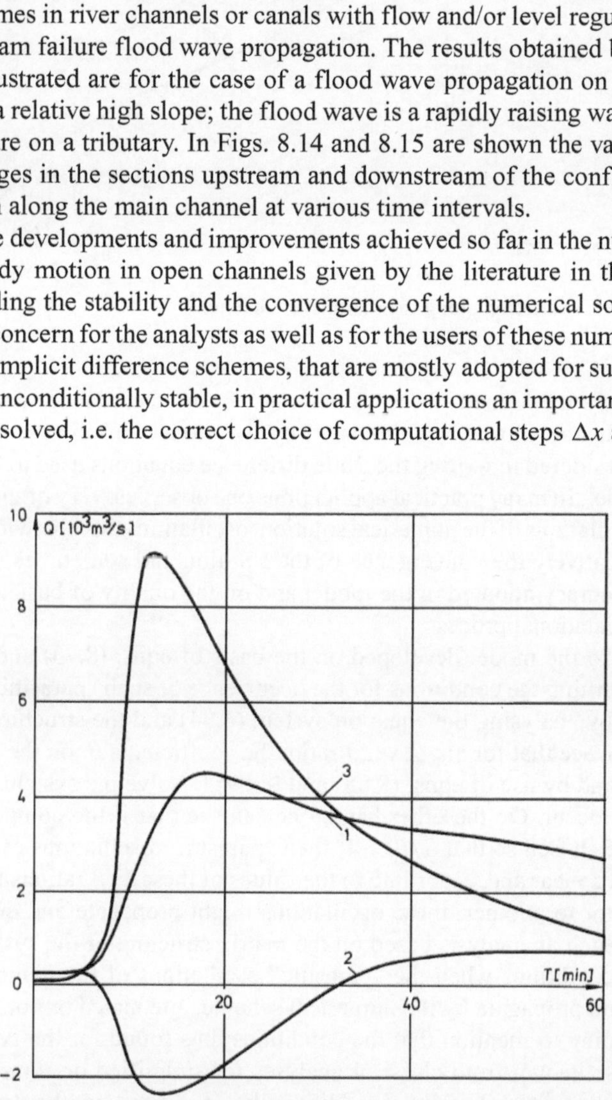

Figure 8.14. Time variation of discharges in variations sections, for rapidly raising flood wave propagation.

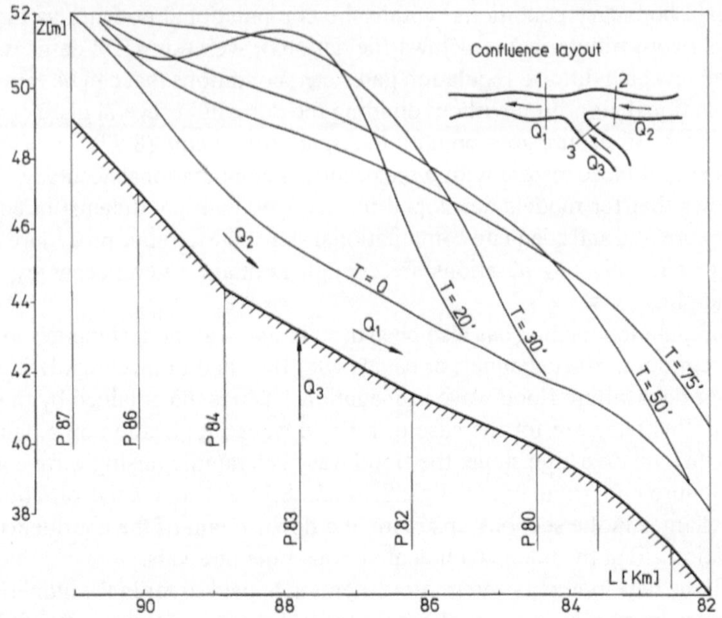

Figure 8.15. Level variation along the main river channel.

As previously mentioned, these values must be in accordance with the scale of the physical phenomenon to be simulated by the model and in agreement with the hypotheses considered in writing the finite difference equations used to build upon the numerical model. In many practical applications one observes, very often, the so-called "parasitic" oscillations of the numerical solution, oscillations that are twice undesirable: they affect negatively the convergence of the solution and sometimes (depending on the level of accuracy adopted in the model and on the quality of basic data) they can spoil the computational process.

If we refer to the model developed on the basis of eqns. (8.73) and (8.74), some remarks, concerning the conditions for the occurrence of such "parasitic" oscillations, can be drawn by analysing the equation system (8.84) and the structure of its coefficients. One can see that for any $\Delta x/\Delta t$ ratio, the coefficients B_i on the main diagonal are dominant and by use of eqns. (8.88) and (8.89) to solve the system, no numerical instability can occur. On the other hand, one can see that if the computational steps Δx and Δt are chosen so that $S/\beta \gg 1$, then "parasitic" oscillations of the numerical solution might appear and, according to the values of these S/β rations from one computational sector to another, these oscillations might propagate and even amplify to some extent. Such an analysis based on the matrix structure of the system (8.84) can point out the conditions when the "parasitic" oscillations of the numerical solution, might occur and propagate by the numerical scheme, attenuated or not.

It is interesting to mention that the conditions thus found for the $\Delta x/\Delta t$ ratio are similar to those known from classical analyses for simplified or complete equations (with a large inertia term) and that, additionally, conditions can be derived from the effect of the friction term, for the $\Delta Z_{12}/H$ ratio, i.e. between the level variation and the mean depth over a computation section. Let us also note that relation (8.67) represents an original form of what we call a dissipative interference often used in such numerical models for preventing and/or attenuating the "parasitic" oscillations. With the method presented in the previous paragraph, this interference is applied in the iterative process itself (and not to the values already computed at a time step t_i), which ensures it a special stabilizing effect.

To conclude, we can say that, in practical applications of the method presented herein, with special care given to the magnitude of computational steps Δx and Δt, in the sense of the above remarks, good results have been obtained both in the accuracy of the solutions and with respect to the computational time and effort required.

8.3. Hydraulic computation of transient regimes for pumped–storage hydroelectric power plants

A modern technical solution for energy storage in periods when the produced power of electric plants is larger than the power consumed by the system, so that it could be used in periods of energy consumption peaks, is presented by Pumped–Storage Hydroelectric power plants (HPS). These plants are characterised by the fact that the electric energy is produced by the turbining of a water volume which is partially or totally pumped back in the high-reservoir.

Thus, HPSs perform an electrical energy transfer in time, from the periods when this energy is available to periods when there exists a high energy demand. This energy is accumulated in the form of potential hydraulic energy during the interval between the two phases of the process: pumping and turbining. The main drawback of the electrical energy—of not being stored—is thus eliminated.

Although the electrical energy thus obtained is smaller that the consumed one, the advantage consists in the price of this energy which could be 3–4 times smaller than the one of the consumed energy. We also emphasize that HPS may play the role of regulating stations in a hydroelectric power plant system, i.e. in the frequency–power regulation, reactive energy regulation, and network voltage control.

In most cases, the HPS is made by means of a hydraulic feed channel connecting the two water chambers, the station building with the turbining and pumping devices being located somewhere along it. In some cases, the chamber of a normal hydroelectric plant is used as upstream or downstream reservoir for the HPS. When water is pumped from large rivers, the downstream reservoir is unnecessary.

According to Prof. Dorin Pavel, the pumped–storage hydroelectric power plants may be classified into closed-circuit and open-circuit plants, each of these categories being in turn divided into mixed or pure pumped storage plants, whether there exists or not an additional discharge in the upstream section (q) generated by gravity (Figs. 8.16 and 8.17) [21]. Literature in the field presents a large number of pumped storage hydroelectric systems either of closed, or open circuit type [13, 16, 37]. The first pumped–storage hydroelectric power plants have been realised in Europe, during the last century. Lately the construction of HPS has been extended on a wide scale. About 200 HPS were in operation or under construction in the 70s in 29 countries, out of which more than a half were located in Germany, Japan, USA and Switzerland [3]. In Romania for instance, up to the present, several HPS with open circuit have been constructed such as: Lotru-Ciunget, Sebes-Sugag and Gilceag, Dragan-Remeti, etc.

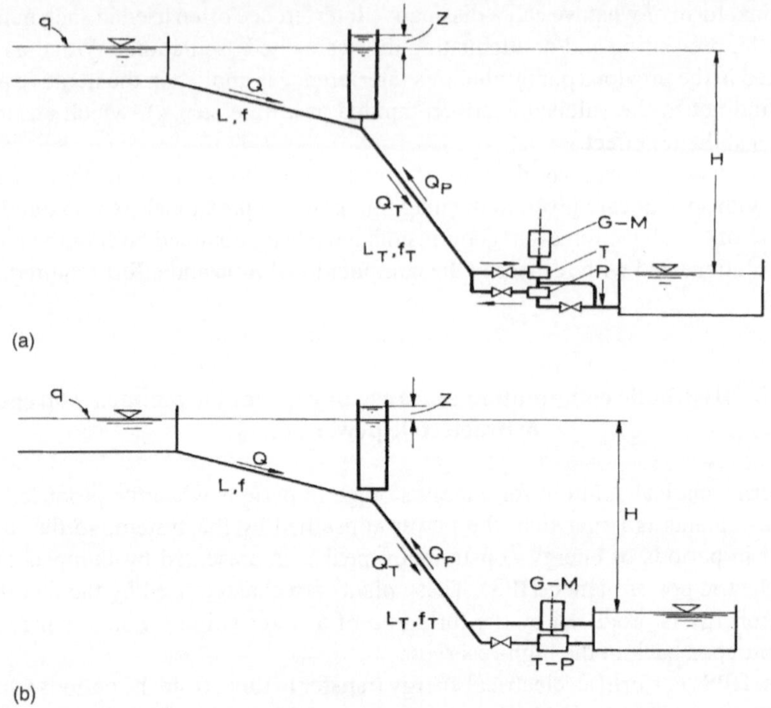

Figure 8.16. Scheme of a closed circuit HPS, in two variants.

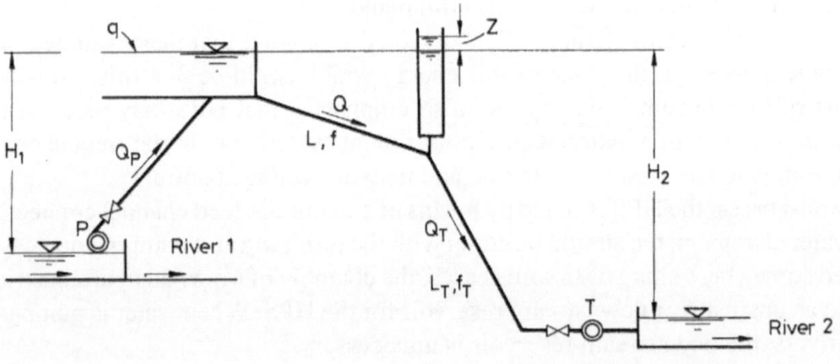

Figure 8.17. Scheme of a HPS accumulation through pumping with open circuit.

The possibilities for this type of plants in Romania have been analysed in reference [37]. There are also several studies containing practical suggestions for the construction of closed circuit HPS in Romania [21, 26].

At the international scale, the problem of constructing closed-circuit HPS was discussed and sustained in [13], by also taking into account the present increase in the number of nuclear and electric plants in the production of electric energy and the electric energy consumption graphs specific to each country, depending on their development stage. An analysis of the world energy consumption situation over the last decades shows that there exists a constant demand of hydrocarbon and that, in many countries,

Table 8.3. Technical parameters of several significant reversible pump–turbine units.

HPS description	Turbining power (MW)	Regime head (m)	Pumping head (m)	Regime discharge (m^3/s)	Rotation rate (rot/min)
Racoon Mountain, USA	400	305	305	314	300
Ludington, USA	343	107.7	113.6	125.4	112.5
Bear, Swamp, USA	320	228.6	240.8	108	225
Kabara, Japan	318	317.5	339.3	72	257
Okutataragi, Japan	310	406	423	80.6	300
Blenheim Gilboa, USA	300	338.3	357.3	100	257

the hydropower potential of rivers has been exhausted and consequently, a significant number of nuclear stations have been built. This gave an input to the use of HPS as top power stations. The convenience of a HPS implementation in a power-generating system as well as the extent of such projects must be considered by taking into account both the technical and economic parameters which define the respective HPS and numerous other factors such as: the configuration of the head loss curves, the characteristics of the existing power stations and, as far as the construction is concerned, the possibilities to build additional thermal and hydroelectric power plants of different types to cover the continually higher need for energy.

In the HPS operation regime one can identify two characteristic regimes (semi-cycles) that together form the turbining–pumping cycle. Their technical and operational features and, consequently, the role and functions they fulfil within the hydropower system are determined by the HPS hydro generating units. In the following, we shall present several theoretical and practical aspects regarding the ternary groups and reversible pump–turbine units which equip most of the HPS (Fig. 8.16).

Ternary groups were used for many HPS built before 1970. With these units, the turbine and pump can each be dimensioned at their optimum so as to obtain the highest efficiency in both turbining and pumping regimes and hence, in the turbining–pumping cycle (Fig. 8.16a). The turbine is fitted with the electric generator as a fixed assembly so as to start operation directly from the off position and to meet the power needs of the system as soon as possible. In most cases, the pump is assembled so as to ensure a possible disconnection. Reversible pump–turbine units are more and more used in HPS due to the recent progress obtained in their design (Fig. 8.16b). Generally, the same rotation rate applies in both operating regimes, but the reversible hydraulic unit is specially dimensioned to obtain a compromise between the optimum regimes of both pump and turbine. For this reason, the global efficiency of reversible hydraulic units is lower than in the case of ternary groups units, the pump and turbine of which are each dimensioned for optimum regimes.

During the last decade, spectacular progress has been made in the design and construction of reversible pumping–turbine units, resulting in the increase of net power, a better efficiency in both operating regimes and higher heads. Illustratively, Table 8.3 gives the technical parameters of several significant reversible pump–turbine units designed for high power and high heads [6]. As seen in the above Table, a net power of 350–400 MW is currently obtained with reversible pump–turbine units, for head values of 400–500 m.

The efficiency of pump–turbine units in both operating regimes is very close to that of ternary group units. When comparing the two unit types, it is important to take into account the difference in power. As the reversible pump–turbine units are more powerful, from this point of view they are preferable to ternary group units [4, 13, 16]. In the present paper the problem of hydraulic computation of transient regimes for closed circuit HPS is approached by taking into account the design and operation particularities of this type of plants [26].

8.3.1. Hydraulic computation methods

(a) Analytical method

As shown in Chapter 5, in the study of non-permanent water motion in the pressure tunnel–tank system, for the maximum hydraulic jump computation one can use the equation of motion of the form

$$\frac{\mathrm{d}^2\zeta}{\mathrm{d}\tau^2} + \frac{1}{2}\left|\frac{\mathrm{d}\zeta}{\mathrm{d}\tau}\right|\frac{\mathrm{d}\zeta}{\mathrm{d}\tau} + \zeta = 0 \tag{8.90}$$

with initial conditions

$$\begin{aligned} &\tau = \tau_o \Rightarrow \zeta_{\tau=\tau_o} = \frac{Z_o}{Z_*} = \zeta_o \\ &\left.\frac{\mathrm{d}\zeta}{\mathrm{d}\tau}\right|_{\tau=\tau_o} = \frac{T_*}{Z_*}\left.\frac{\mathrm{d}Z}{\mathrm{d}t}\right|_{t=t_o} = \sqrt{-2(1+\eta)\zeta_o} \end{aligned} \tag{8.91}$$

and for the minimum hydraulic jump computation, the equation of motion written in the form

$$\frac{\mathrm{d}^2\zeta}{\mathrm{d}\tau^2} + \frac{1}{2}\left(\frac{\mathrm{d}\zeta}{\mathrm{d}\tau}\right)^2 + \alpha\frac{\mathrm{d}\zeta}{\mathrm{d}\tau} + \zeta + \frac{\alpha^2}{2} = 0 \tag{8.92}$$

with initial conditions

$$\tau = \tau_o \Rightarrow \zeta = 0, \quad \frac{\mathrm{d}\zeta}{\mathrm{d}\tau} = -\alpha \cong \sqrt{-2\zeta_f} \tag{8.93}$$

where the following non-dimensional quantities

$$\begin{aligned} &\zeta = \frac{Z}{Z_*}, \quad \tau = \frac{t}{T_*}, \quad \eta = \frac{h_d}{h_a} = \frac{\zeta_d d}{\lambda_\ell L} \\ &Z_* = \frac{fd}{(\eta \pm 1)\lambda_\ell F}, \quad T_* = \sqrt{\frac{LF}{gf}} \end{aligned} \tag{8.94}$$

have been adopted together with the notation: d is the diameter of the pressure tunnel, f the cross-section area of the tunnel, F the surge tank horizontal cross-section area, λ_ℓ the hydraulic resistance coefficient on the tunnels, n the pressure–tunnel roughness coefficient, h_a the longitudinal head loss in the tunnel, and h_d the local head loss in the diaphragm. In relation (8.94), the sign plus is taken when water velocities in the tunnel and in the surge tank have the same sign and the sign minus when these velocities have opposite signs.

In the case of HPS, the computation of extreme levels in the surge tank is performed for the following flow rate manoeuvres at the station:

- start-up of the turbines;
- shutdown of the turbines;
- shutdown of the pumps,

corresponding to the maximum turbined, respectively pumped flows in the plant. The start-up manoeuvre at the pumps is not significant because during operation it is performed gradually in time. Adopting the solutions of eqns. (8.90) and (8.92) to the manoeuvres specific to HPS, we obtain the following results for extreme hydraulic jumps in the surge tank (Fig. 8.16).

The start-up manoeuvre of the turbines. The minimum jump is computed from

$$\zeta_{\min} = \zeta_t - \frac{\sqrt{-2\zeta_f}}{1 + 0.35\sqrt{-\zeta_f} - \zeta_f} \tag{8.95}$$

where $\zeta_f = -(h_a/Z_*)$ (corresponding to the final permanent level).

The shutdown manoeuvre of the turbines. The maximum jump is computed from

$$\zeta_{\max} + \ln(1 - \zeta_{\max}) = \zeta_o - \ln(1 + \eta\zeta_o) \tag{8.96}$$

where $\zeta_o = -(h_a/Z_*)$ (corresponding to the initial permanent level).

The shutdown manoeuvre of the pumps. The minimum hydraulic jump is computed from

$$|\zeta_{\min}| + \ln(1 - |\zeta_{\min}|) = -\zeta_o - \ln(1 - \eta\zeta_o) \tag{8.97}$$

where $\zeta_o = -(h_a/Z_*)$ (corresponding to the initial permanent regime).

(b) Numerical method

One uses the equations of motion in the form

$$\frac{L}{g}\frac{dW}{dt} - Z + P + R = 0$$

$$Wf = F\frac{dZ}{dt} + Q_{T,P} \tag{8.98}$$

$$Q_{T,P} = \Phi(t)$$

with longitudinal and local head losses expressed by

$$P = P'W|W|, \quad R = R'V_s|V_s| \tag{8.99}$$

where the usual notation has been used. In this method, the system of eqns. (8.98) is written in finite differences and is solved iteratively for successive time steps by a specific computation algorithm (see Chapter 5).

For HPS with closed circuit, in which turbines and pumps are axially located (Fig. 8.16b), the start-up manoeuvre of the pumps is performed in operation by a previous start-up of the turbines, after which the corresponding valve manoeuvres are performed to ensure the start of the pumps. This flow-rate manoeuvre produces in a relatively brief time interval a change of the following sense in the penstock and may sometimes be the most unfavourable manoeuvre for the maximum jump in the surge tank [26]. Since it occurs at some specific types of discharge manoeuvres at the station, the most unfavourable levels in the surge tank can be detected only by means of numerical methods. In this case, hydraulic computations serve both for the adequate dimensioning of the surge tank, and for finding out the optimum discharge manoeuvres in operation.

The computation of waterhammer in the penstock is performed for the mentioned discharge manoeuvres at the station, the extreme pressures values being retained for a correct dimensioning of the penstock. For a more accurate computation of waterhammer in the penstocks of these types of plants one can use computation methods accounting for the characteristic curves of the devices, in all the four operation quadrants [6].

8.3.2. Computation examples and comparisons

Hydraulic computations have been performed using both computational methods in the case of a pumped–storage hydroelectric power plant with closed circuit (Fig. 8.16) having the following main characteristics:

$L = 5{,}000\,\text{m},\ d = 7\,\text{m}$	feed gallery
$L_T = 200\,\text{m},\ d_T = 5\,\text{m}$	penstock
$D_1 = 12\,\text{m},\ D_2 = 14\,\text{m},\ D_3 = 16\,\text{m}$	surge tank with three variants of diameters
$R = 0.001Q_c^2$	
$Q_T = Q_P = 125\,\text{m}^3/\text{s}$	
$H_{max} = 250\,\text{m},\ H_{min} = 170\,\text{m}$	

Computation results by the analytical method

Start-up of the turbines. The hypothesis of the empty lake and $n_{max} = 014$ has been admitted.

$$(Z_{min})_1 = -36.99\,\text{m}$$
$$(Z_{min})_2 = -31.04\,\text{m}$$
$$(Z_{min})_3 = -26.61\,\text{m}$$

Shutdown of the turbines. The hypothesis of the full lake and $n_{min} = 0.012$ has been admitted.

$$(Z_{max})_1 = +32.60\,\text{m}$$
$$(Z_{max})_2 = +26.92\,\text{m}$$
$$(Z_{max})_3 = +22.50\,\text{m}$$

Shutdown of the pumps. The hypothesis of the empty lake and $n_{min} = 0.012$ has been admitted.

$$(Z_{min})_1 = -32.60\,\text{m}$$
$$(Z_{min})_2 = -26.92\,\text{m}$$
$$(Z_{min})_3 = -22.50\,\text{m}$$

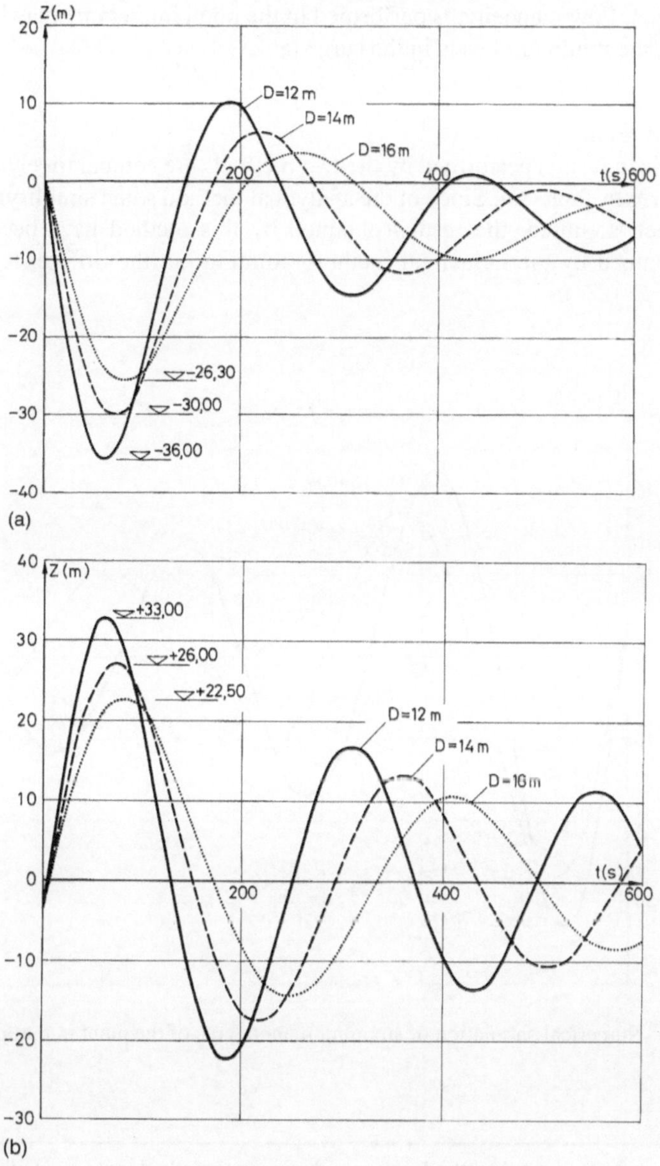

Figure 8.18. Numerical simulation of the starting and stopping manoeuvres of the plant in turbining regime: (a) starting the turbines; (b) stopping the turbines.

Computation results by the numerical method

Computations have been performed in the same hypotheses and results are graphically presented in Figs. 8.18 and 8.19. The analysis of maximum and minimum values of the hydraulic jumps in the surge tank shows that in the case of the given plant, the flow manoeuvres performed in the pumping regime no further restrictions are imposed as compared with similar manoeuvres in the turbined regime. In situations when the start-up manoeuvres of the turbines are submitted to some restrictions (for instance, with up to $0.8Q_{max}$), flow manoeuvres performed in the pumping regime may be viewed as determining the minimum levels in the surge tank.

Comparative results

The comparison results performed by the two methods are comparatively presented in a concise form in Table 8.4. Since in the analytical method some simplifying hypotheses have been admitted, the results obtained by this method have been compared to those obtained by the numerical method, pointing out the differences in percent ($\Delta = (Z^n - Z^a)/Z^a$).

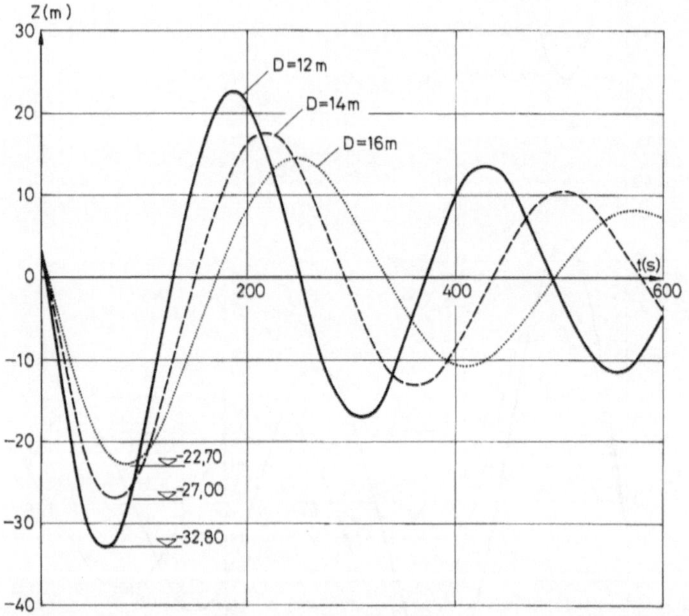

Figure 8.19. Numerical simulation of stopping manoeuvres of the plant in pumping regime.

Table 8.4.

Manoeuvre at central station	Analytical method (m)			Numerical method (m)			Differences (%)		
	Z_1	Z_2	Z_3	Z_1	Z_2	Z_3	Δ_1	Δ_2	Δ_3
Turbines starting (minimum levels)	−36.99	−31.04	−26.61	−36.00	−30.00	−26.30	2.7	3.5	1.2
Turbines stop (maximum levels)	+32.6	+26.92	+22.50	+33.00	+26.00	+22.50	1.2	3.5	0
Turbines stop (minimum levels)	−32.6	−26.92	−22.50	−32.80	−27.00	−22.70	0.6	0.3	0.9

A good agreement of the results can be noticed, the relative differences situated below 4% being of the same order of magnitude as the effect of the neglected factors in establishing the mathematical model.

8.4. Applications to several hydroelectric power plants and pumping stations in Romania

8.4.1. The Tismana hydroelectric power plant

The hydroelectric Cerna–Motru–Tismana complex system was conceived for putting into account the entire hydraulic and energy potential of the rivers within a large geographical area for industrial use and power supply. The general scheme of the system contains the Motru, Clocotis and Tismana hydroelectric power plants. The hydraulic scheme of the Tismana hydroelectric power plant is a modern constructive solution with the shallow wells that take in the Pociuia, Tismana, Tismanita and Bistricioara rivers directly connected to the two intakes (Fig. 8.20). It belongs to the hydraulic scheme group of the type B [20, 33, 35, 38, 43–45].

The hydraulic study has been performed numerically. In the operation scheme of the plant one can note the two intakes Motru ($L = 8{,}580$ m, $D = 3.60$ m) and Bistrita

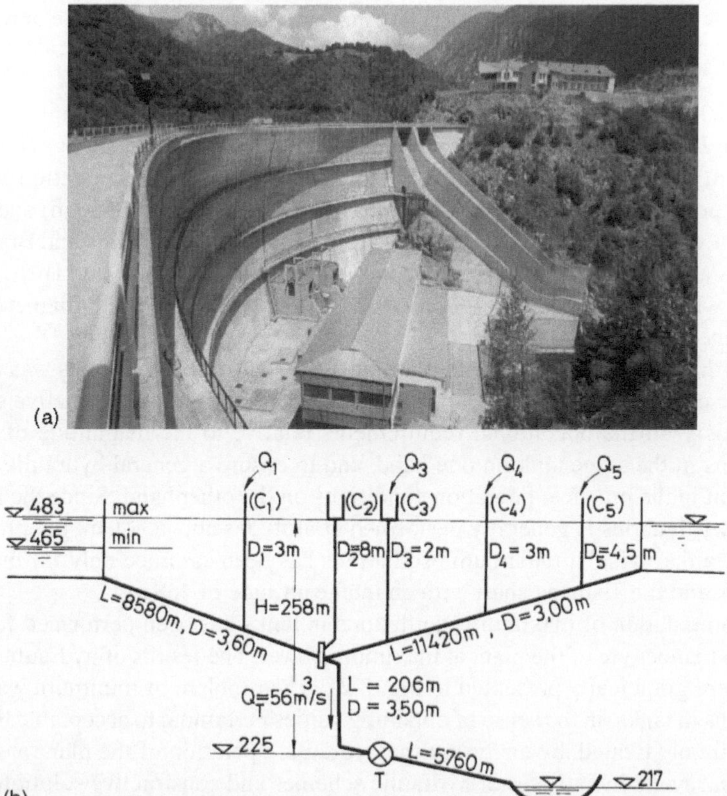

Figure 8.20. (a) Cerna–Motru–Tismana hydrotechnic complex: Central Cerna. (b) Scheme of the Tismana hydroelectric power plant.

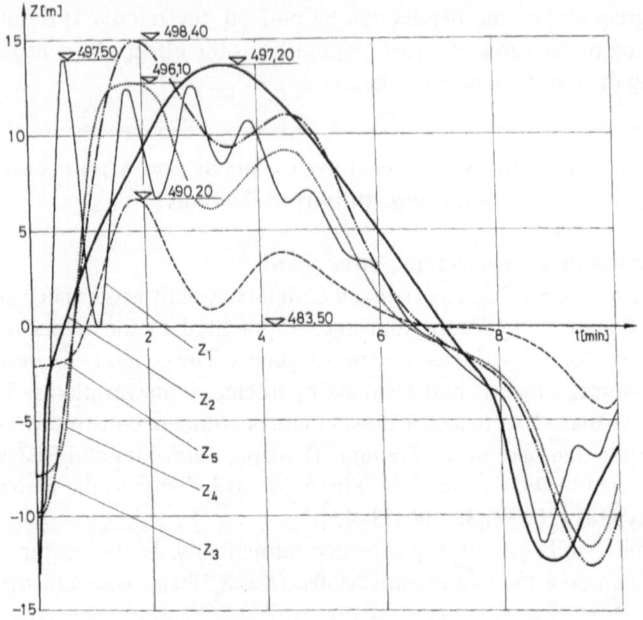

Figure 8.21. Maximum levels in tanks at the simultaneous stopping manoeuvre of two turbines.

($L = 11{,}420$ m, $D = 3.00$ m), the central surge tank with cylindrical upper chamber ($h = 21$ m, $D = 20$ m) and lower chamber consisting of an inclined gallery ($L = 130$ m, $D = 5.00$ m), the well ($H = 206$ m, $D = 3.50$ m), the underground station with two Francis type 50 MW turbines each, the free surface tailrace ($L = 4{,}150$ m) and the off-take ($L = 1{,}610$ m) (Fig. 8.20). The intakes from the Pocruia, Tismanita and Bistricioara tributaries are 4.50 m in diameter, directly connected to the Motru and Bistrita intake galleries, while the Tismana shaft consists of an upper chamber with diameter of 8 m and a connection pipe of length $L = 190$ m and diameters $D' = 2$ m and $D'' = 3$ m.

The hydraulic study of the entire plant operating at transient regimes was intended to find the correct dimensions for the surge tanks as well as the constructive solutions which satisfy all the operational requirements relative to the magnitude of extreme oscillations in the surge tank on one hand, and to ensure a general hydraulic stability of the plant in the practical operation conditions on the other hand. Since the intake of secondary tributaries is generally performed at altitudes above 500 m, the problem of the surge tanks sizing at maximum oscillations has been assumed only for the central surge tank and the Tismana shaft with an intake altitude of 468 m.

The computation of maximum oscillations in tanks has been performed for a total shutdown manoeuvre of the plant at maximum power. The results of hydraulic computations were graphically presented in Fig. 8.21. The problem of minimum water level oscillations in tanks, in the sense of obtaining values pertaining to acceptable limits has been separately treated, by analysing the hydraulic operation of the plant at transient regimes in several hypotheses of hydraulic schemes and constructive solutions for the surge tanks in the system (shafts and central surge tank).

Thus, the hydraulic operation of the entire system has been investigated for transient regimes at the start-up manoeuvre of the plant, with one or two turbines

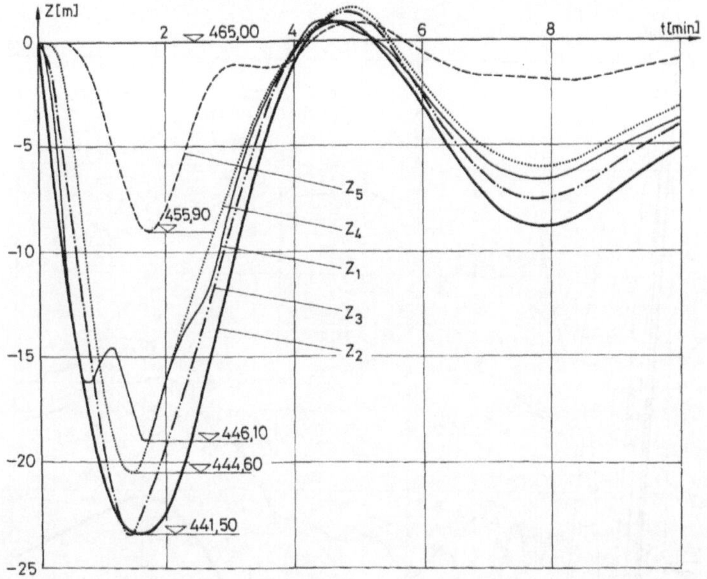

Figure 8.22. Minimum levels in surge tanks at the starting manoeuvre of the turbine.

(Figs. 8.22 and 8.23) in the hypothesis of hydraulic resistances at the tank base, of the following values:

$$R_1 = 0.0115Q_c^2, \quad R_2 = 0.0064Q_c^2, \quad R_3 = R_4 = R_5 = 0.05Q_c^2$$

From the analysis of computation results (Fig. 8.23) one ascertains that in this constructive solution adopted for the surge tanks, the minimum oscillations of water levels in some surge tanks fall below the admissible minimum levels (the case of surge tanks C_1 and C_4), and the attempts to correct the constructive scheme by adopting surge tanks of larger dimensions have not significantly improved the results [33, 43, 44]. In order to prevent the occurrence of such situations in the plant operation, for the Pocruia and Tismanita shafts (surge tanks C_1 and C_4) the constructive solutions of tank types equipped either with lower chambers or with increased hydraulic resistances have been analysed.

Out of the investigated variants, we present the results of hydraulic computations for the following two improvements of the initial solution, which led to a good operation of the entire hydraulic system:

(a) surge tank C_1 and C_4 equipped with lower chambers of volume 400 and 120 m^3, respectively (Fig. 8.24);
(b) surge tank C_1 and C_4 equipped with hydraulic resistances of values (Fig. 8.25)

$$R_1^* = 0.184Q_c^2 \quad \text{and} \quad R_4^* = 0.148Q_c^2$$

In the design of the plant the constructive solution (b) has been adopted as more economic.

On the thus corrected global hydraulic scheme of the plant one has made a check of the general hydraulic stability of the entire system in operation, at transient regimes, in the hypothesis of automatic adjustment at constant power. Part of the performed hydraulic computations is graphically presented in Fig. 8.26. The results of hydraulic

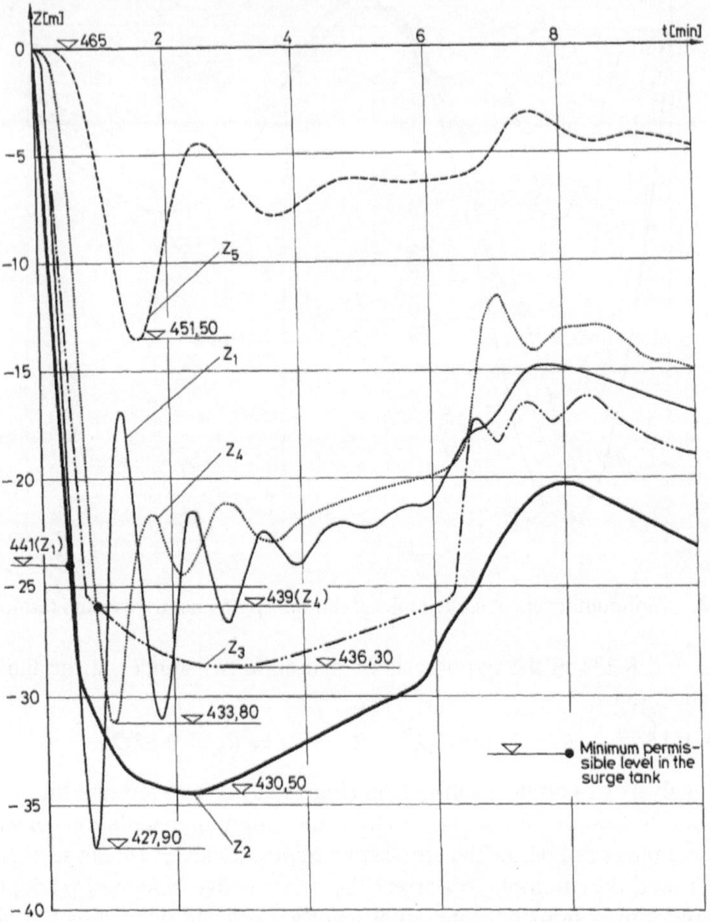

Figure 8.23. Minimum levels at simultaneous starting manoeuvre of two turbines.

computations have pointed out the complex nature of the global transient process of the water motion in the system, in fact impossible to investigate with the usual computation methods.

8.4.2. The Lotru hydroelectric power plant

The hydroelectric power plant Lotru is the largest on the interior rivers in Romania, with a power of 500 MW and head of over 800 m. The power station is in operation at the designed parameters since 1974. Based on the principle of full and rational use of the hydraulic potential of all rivers in the given area, the general hydraulic scheme of the complex system Lotru is remarkable by the large number and the diversity of construction works for intake and supply of water (gravitational and with pumping) in the main reservoir (Fig. 8.27) [34]. From the average inflow of 17 m^3/s, about 75% is obtained from secondary intakes.

Technical data concerning the main parts of the Lotru hydraulic scheme are as follows: (Figs. 8.27 and 8.28).

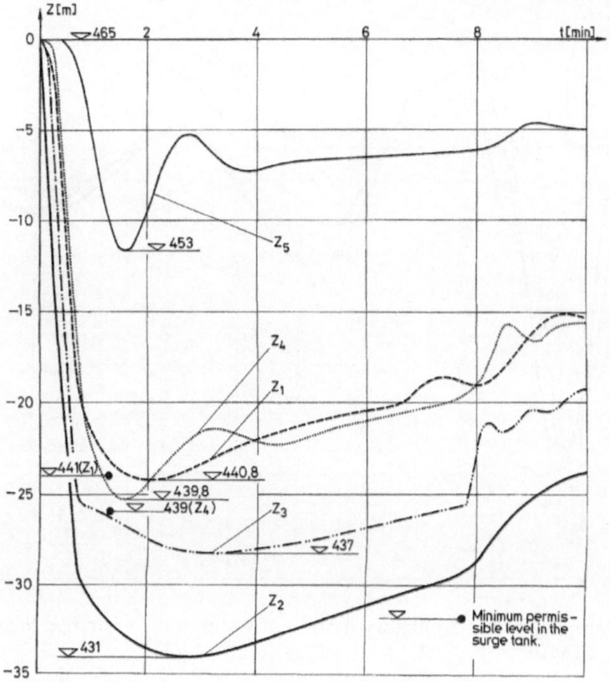

Figure 8.24. Minimum levels in surge tanks. Capture wells C_1 and C_4 are endowed with lower chambers of volume 400 m^3 and 120 m^3, respectively.

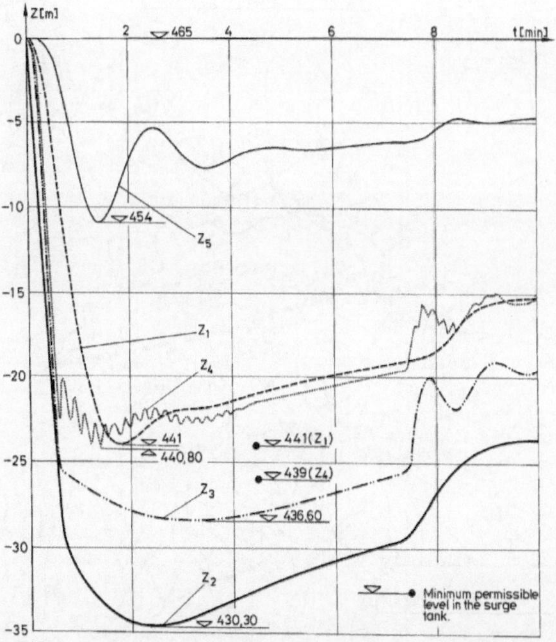

Figure 8.25. Minimum levels in surge tanks. The capture wells C_1 and C_4 are endowed at their base enlarged hydraulic resistance.

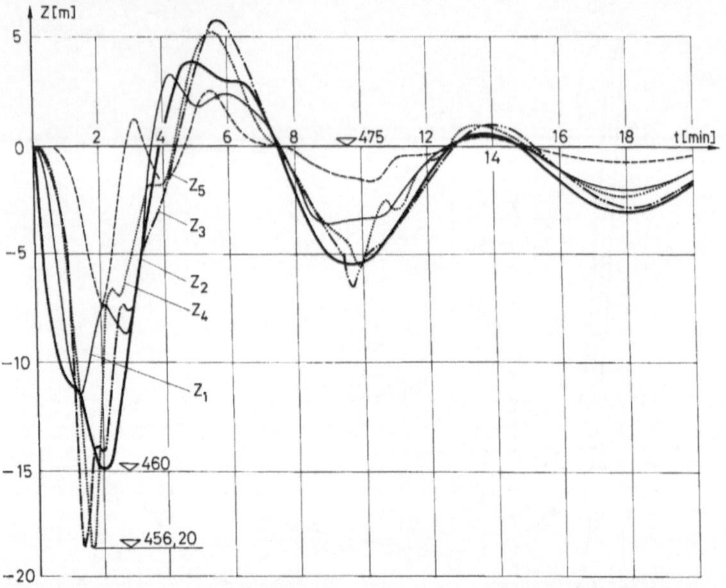

Figure 8.26. Damping of the oscillatory motion of water in the system at the starting of the plant with $N_2 = 40$ MW.

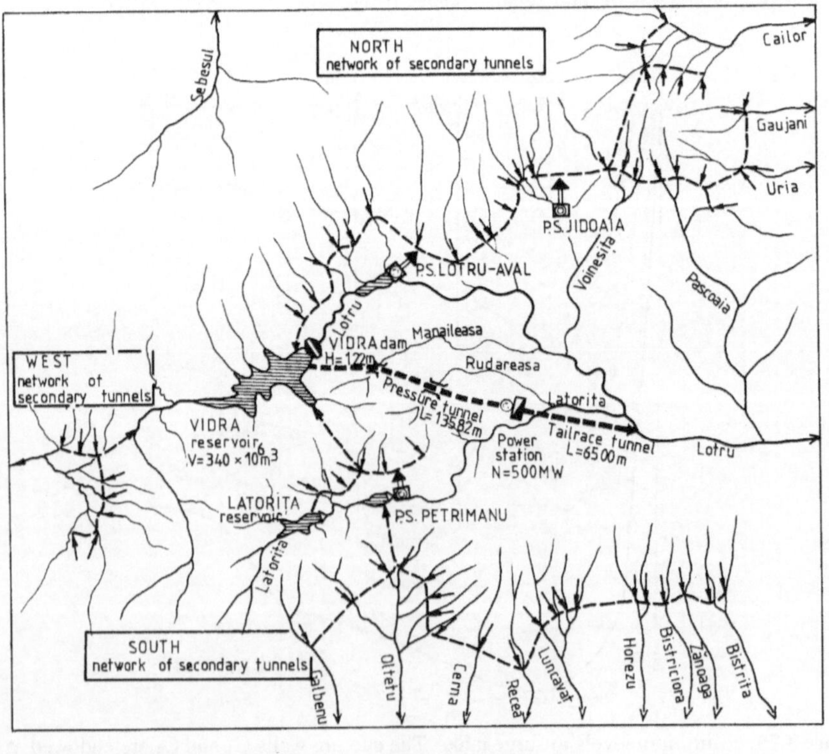

Figure 8.27. Scheme of the complex hydroelectric plant of Lotru.

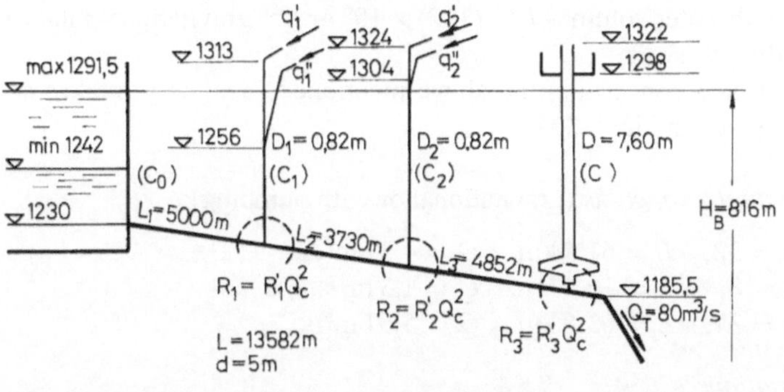

Figure 8.28. Scheme of the Lotru hydroelectric power plant.

Figure 8.29. Underground central station of the hydroelectric power plant Lotru.

Vidra dam, a rock fill dam with claystone core, height $H = 122$ m, realizing a reservoir with total volume $V = 340 \times 10^6$ m^3.
Main pressure tunnel of reinforced concrete, length $L = 13{,}582$ m, diameter $d = 5$ m.
Surge tank of differential type, designed in a special scheme.
Penstock of reinforced concrete and steel sheet, length $L_T = 1{,}320$ m, diameter $d_T = 4$ m.
Underground station equipped with three Pelton turbines with vertical axis each of 170 MW (Fig. 8.29).
Tailrace tunnel with free level flow, length $L_A = 6{,}500$ m, diameter $d_A = 5.35$ m.
Secondary intake shafts for rivers Manaileasa (C_1) and Rudareasa (C_2) directly connected to the main pressure tunnel $q_{i3} = 0.7$ m^3/s, $q_i = 0.5$ m^3/s, $q'_2 = 1$ m^3/s, $q''_2 = 0.6$ m^3/s.

The intake system for the rivers in the area and for the water supply in the reservoir consists of: 71 secondary intakes of different constructive types with intake flows $q = 1 + 10$ m^3/s; three networks of secondary pressure tunnels with a total length of 135 km and a cross-section area of $3.5 + 10$ m^3; three pumping stations with pumping heights $H_G = 180$–320 m and four dams with heights $H = 25$–50 m, creating secondary reservoirs with water volumes $V = (1–2) \times 10^6\,m^3$ for gravitational intake or pumping into the Vidra reservoir.

Technical data concerning the above mentioned secondary intakes and pumping stations are given in Fig. 8.27.

Secondary intake networks (gravitational or with pumping):

North: $n = 32$, $L = 61.7\,km$, $Q = 4.57\,m^3/s$,
West: $n = 8$, $L = 10.7\,km$, $Q = 1.3\,m^3/s$,
South: $n = 31$, $L = 63.8\,km$, $Q = 5.61\,m^3/s$

Pumping stations:

P.S. Jidoaia: $Q_P = 10\,m^3/s$, $H_G = 185\,m$, $N_P = 20\,MW$,
P.S. Lotru: $Q_P = 2\,m^3/s$, $H_G = 318\,m$, $N_P = 9\,MW$,
P.S. Petrimanu: $Q_P = 15\,m^3/s$, $H_G = 185\,m$, $N_P = 30\,MW$

The inclusion of these pumping stations in the general scheme of the complex hydropower system Lotru proved to be particularly efficient from technical and economic point of view, because of relatively small pumping height ($H_G = 180–320\,m$) with respect to the total head of the station $H_B \approx 800\,m$. The hydraulic scheme of the plant has been conceived as a modern solution, concentrating by secondary water conveyance structures over 60% of the stock, with the inlet pipes of the Nopteasa, Manaileasa and Rudareasa small rivers, directly connected to the main water conveyance structure. It belongs to the group of hydraulic schemes of the type C (Fig. 8.28).

We shall present the main results of a study performed by numerical methods, and aimed at analysing the hydraulic operation of the plant at transient regimes in several computation hypotheses, function of the constructive solutions for the surge tank and the inlet wells, so as to ensure a non-restrictive hydraulic operation [28, 30, 36]. The hydraulic study of the plant operation at transient regimes was intended at the same time to check the present condition of the surge tank and of the inlet wells, and some computation hypothesis which seldom appear in operation but imply very severe operation conditions. The numerical simulation of the total start manoeuvre of the plant at minimum level in the reservoir, in the hypothesis that hydraulic resistances and water levels ensured by the existing constructive solutions are neglected, has put into evidence the appearance of minimum levels which could cause air penetrating the gallery and forming pockets that could affect the operation of the station in normal regime (Fig. 8.30a).

Successive introduction in computations of a hypothesis close to the adopted solution showed that existing volumes and hydraulic resistances ensure that the respective parameter values fall between the computational limits indicated for a total safety (Fig. 8.30b). In this case, the minimum pressures in the feed gallery remain positive for the entire interval of the transient phenomenon in the system, thus confirming also on a numeric computation basis that the system was suitably dimensioned and corresponds to the operation demands for any case that may appear during operation. The numerical simulation of the total shutdown manoeuvre at the plant is illustrated in Fig. 8.31, containing also the variation of water levels in the intake shafts (Z_1, Z_2), and separately, the variation of levels in the shaft and in the upper chamber of the surge tank (Z_p, Z_c). In order to perform computations, the functions specific to the upper weir and the connecting ports between the well and the upper chamber were taken into account.

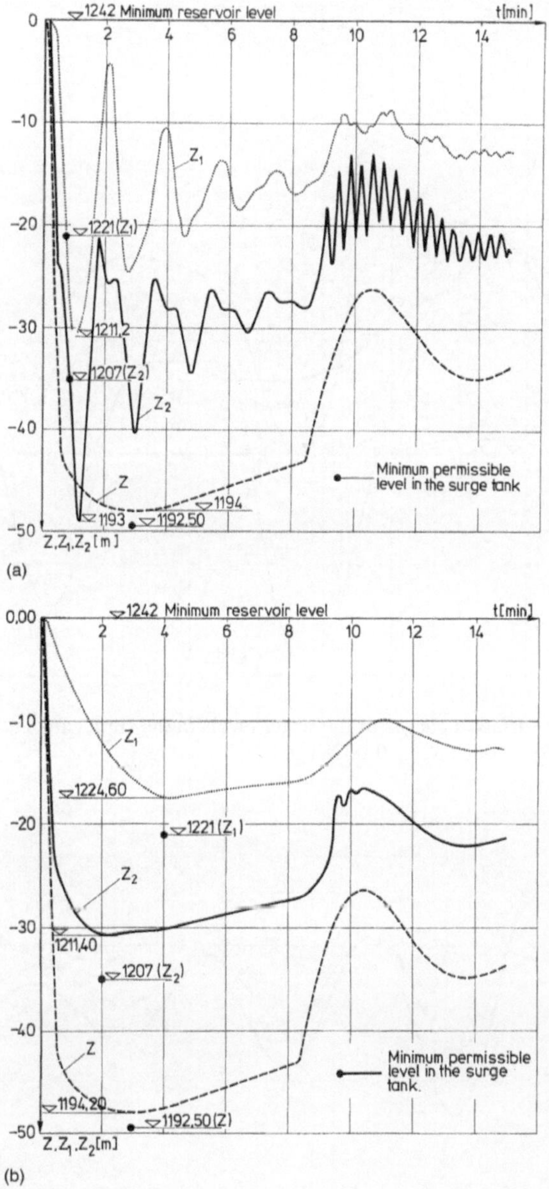

Figure 8.30. Time variation curves of the water levels in the surge tanks at the total starting of the plant: (a) $R_1 = R_2 = 0$; $R_3 = 0.00281Q_c^2$; $V_{\text{inf}\,1} = V_{\text{inf}\,2} = 0$; (b) $R_1 = R_2 = 6Q_c^2$; $R_3 = 0.00281Q_c^2$; $V_{\text{inf}\,1} \neq V_{\text{inf}\,2} \neq 0$.

To check the general operation stability of the plant at transient regimes, several manoeuvres for different heads of the station have been analysed in the hypothesis of automatic regulation on constant power; Fig. 8.32 illustrates graphically the damping of the oscillatory motion in the system at a power variation in the station, with $N_1 = 170\,\text{MW}$ and $N_2 = 340\,\text{MW}$ [33, 34].

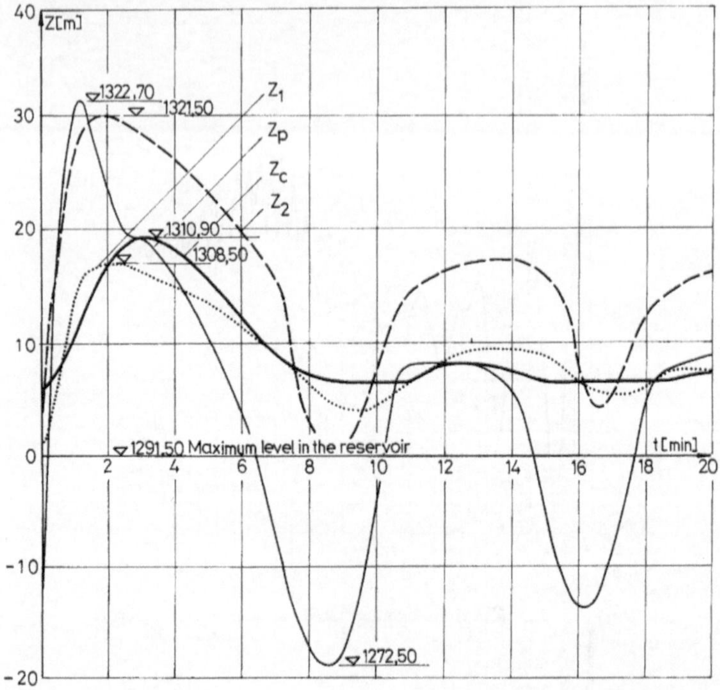

Figure 8.31. Time variation curves of the water levels in the surge tanks at the total stopping of the plant: $R_1 = R_2 = 6Q_c^2$; $R_3 = 0.00281Q_c^2$.

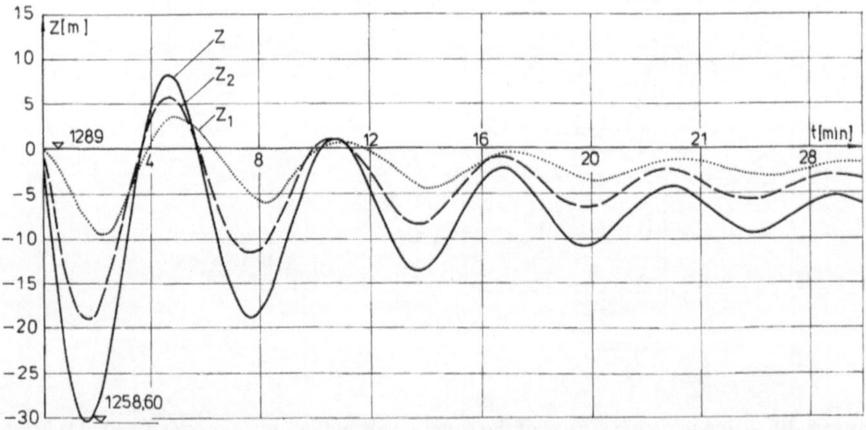

Figure 8.32. Damping oscillations of water level in the system for power manoeuvre at the central with $N_1 = 170$ MW, $N_2 = 340$ MW.

8.4.3. The SPR-2 pumping station of the Rasova-Vederoasa irrigation system

We shall present the results of a hydraulic study performed by numerical methods, for the SPR-2 pumping station of the Rasova-Vederoasa irrigation system, conceived as a hydraulic scheme of the type II [7, 27]. The study aimed at a correct selection and hydraulic dimensioning of the protection devices for the suction and discharge pipes,

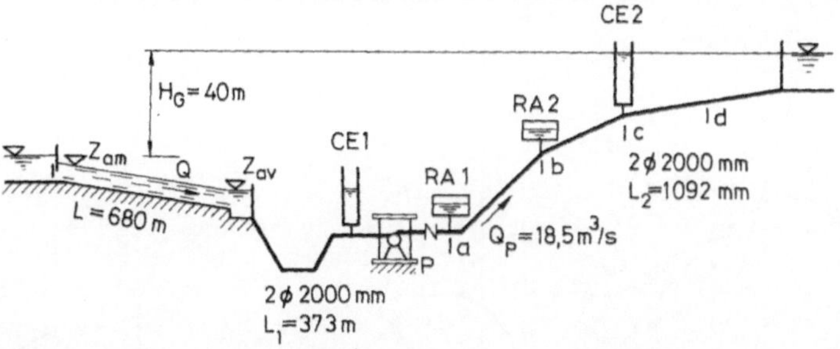

Figure 8.33. Scheme of the pumping station SRP 2 Rasova-Vederoasa.

in the conditions of a secure and efficient hydraulic operation of the station at transient regimes. The main characteristics of the station are presented in Fig. 8.33.

The water conveyance canal is of trapezoidal cross-section ($b = 3.50$ m; $m = 1.5$) and has a longitudinal slope $i = 0.0001$; the station is equipped with 12 electropumps of type 24 NDS/860 mm ($Q_o = 1.5\,m^3/s$, $H_o = 49.5$ m) and endowed with two independent pumping pipe runs. Computations have been performed for a single pipe run. If check valves exist at the pumps on the discharge pipes, the oscillatory motion in the system channel suction pipe may be considered independent of the water motion in the discharge pipe, after closing the check valves.

The canal suction pipe–pump has been studied on a scheme similar to the ones occurred in hydroelectric power plants. As a protective solution for the suction pipes one has chosen the solution with a single surge tank on both pipes. In the computations, a single total discharge manoeuvre at the station was considered namely the shutdown of the station (the start-up of the station is made by gradual operation of the pumps).

The surge tank must accordingly satisfy only the hydraulic conditions concerning the maximum jump. The following surge tanks types have been analysed:

A. cylindrical surge tank;
B. surge tank with cylindrical upper chamber;
C. surge tank with tronconic upper chamber;
D. surge tank with spherical upper chamber

with the remark that all the surge tanks have been provided with a hydraulic chamber at their bases. In order to obtain an efficient response of the system to transient regimes by varying the water level in the surge tank and the water pressure in the pipe within an optimum domain of values, several variants have been analysed for each surge tank type.

For instance, for the surge tank of type A, several variants for the tank diameter and the connecting pipes diameter values have been analysed, namely: $D = 3.00$; 3.50; 4.00 m and $d = 500$; 600; 700; 800 mm. From the analysed constructive variants, the following hydroelectric solution have been considered optimum:

- surge tank of the A type

$H = 15$ m, $\quad D = 3.00$ m, $\quad d = 700$ mm

- surge tank of the B type

$H = 12$ m, $\quad D_1 = 2.50$ m, $\quad D_2 = 5.00$ m, $\quad h = 5.00$ m, $\quad d = 800$ mm

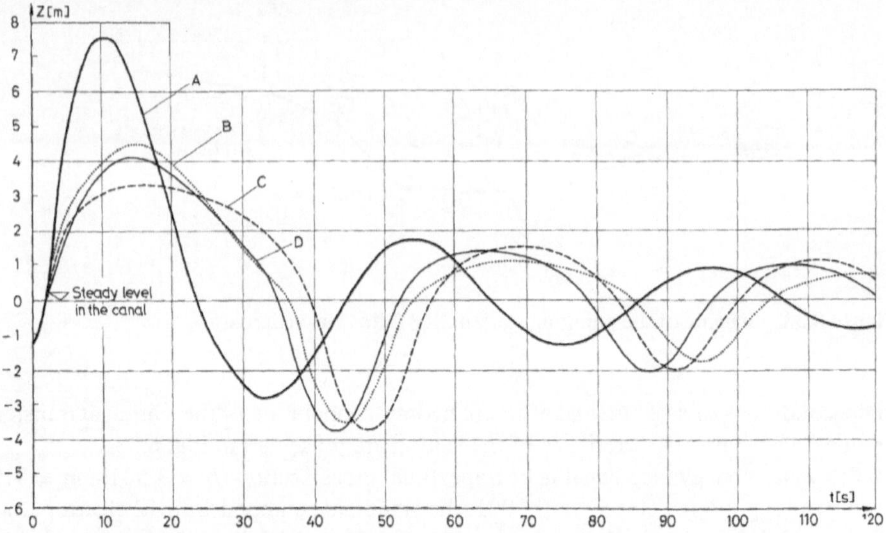

Figure 8.34. Time variation curves of water level in the surge tank.

- surge tank of the C type

  $H = 12\,\text{m}, \quad D_1 = 2.50\,\text{m}, \quad D_2 = 10.5\,\text{m}, \quad h = 4.00\,\text{m}, \quad d = 800\,\text{mm}$

- surge tank of the D type

  $H = 12\,\text{m}, \quad D_1 = 2.50\,\text{m}, \quad D_2 = 5.5\,\text{m}, \quad d = 800\,\text{mm}$

The hydraulic operation of the surge tanks is illustrated in Fig. 8.34 by means of the time variation curves of the water levels in the tanks. The surge tank of type B is generally applied in the underground constructive variant while types C and D give rise to more difficult constructive problems. For the design, the variant with surge tank of type A was recommended.

Two types of protective solutions for the discharge pipes have been analysed, namely:

R: with air chambers,
RC: with air chambers and a single surge tank

From the analysed protective solutions, we have retained those that ensure an efficient hydraulic operation at transient regimes.

Computations results are graphically presented in Fig. 8.35 for the protective solution of type R and in Fig. 8.36 for the protective solution of type RC. The following hydraulic parameters of the water motion have been graphically represented:

- water pressure (p/γ) in the discharge pipes;
- air cushion pressure (p_{RA}/γ) and transfer flow rate chamber pipe (Q_R);
- water level in the tank (Z) in absolute height (we quote the water level in the discharge basin: 143.60 m).

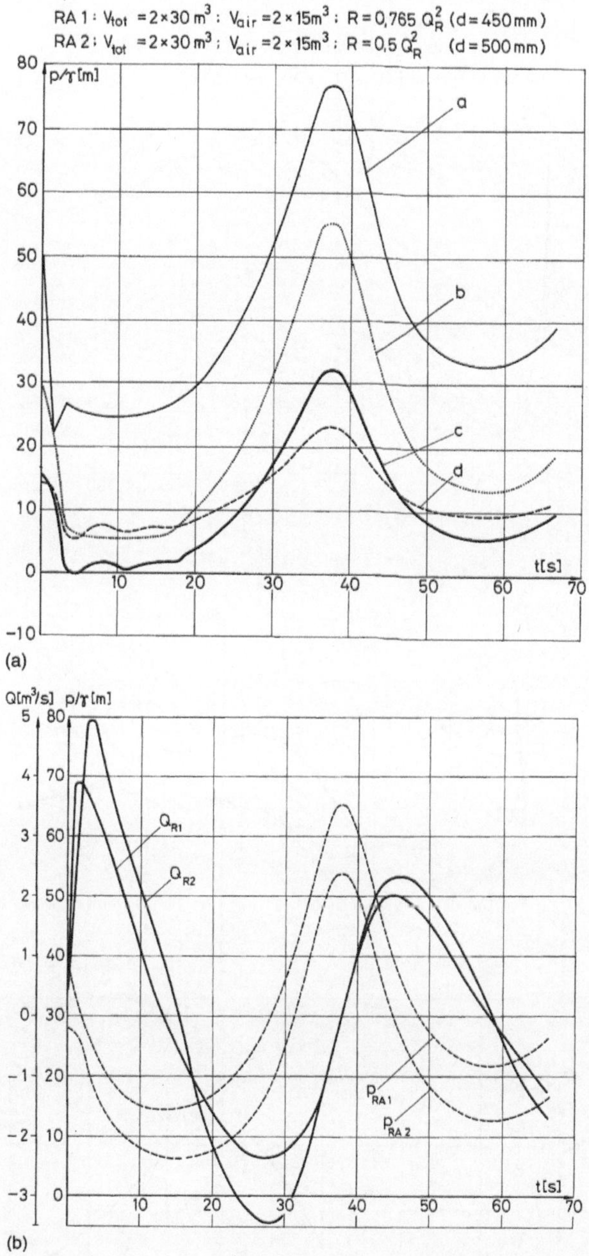

Figure 8.35. Protection solution with air chambers. Time variation curves of some hydraulic parameters of the motion.

The geometric and hydraulic characteristics of the protective installations illustrated in Figs. 8.35 and 8.36 refer to a single pipeline. Modulated air chambers have been used, with a volume of 30 m^3. The two above-mentioned solutions are almost similar from the economic point of view; therefore, in the design, the solution of type RC, with air chamber and a single surge tank was recommended, as being safer in operation.

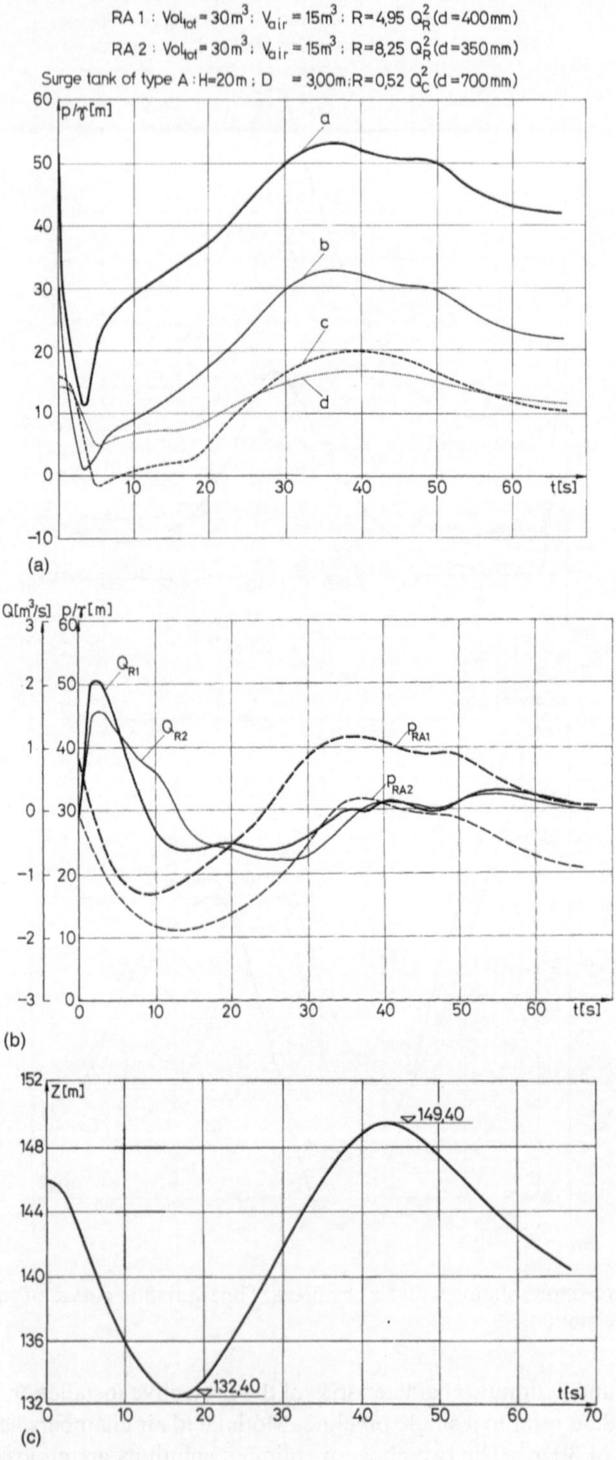

Figure 8.36. Protection solution with air chambers and a single surge tank. Time variation curves of some hydraulic parameters of the motion.

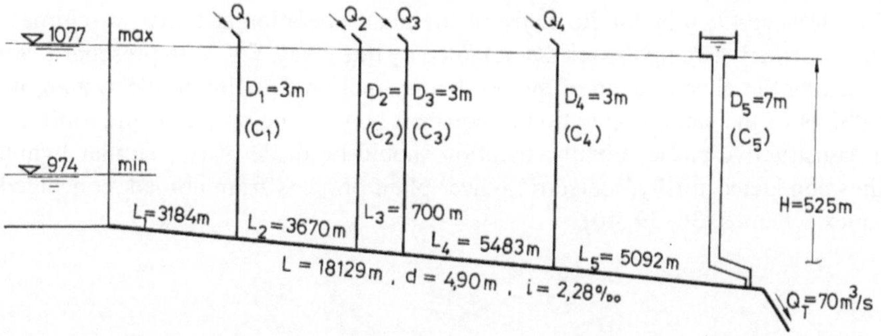

Figure 8.37. Scheme of the Riul Mare-Retezat hydroelectric power plant (variant).

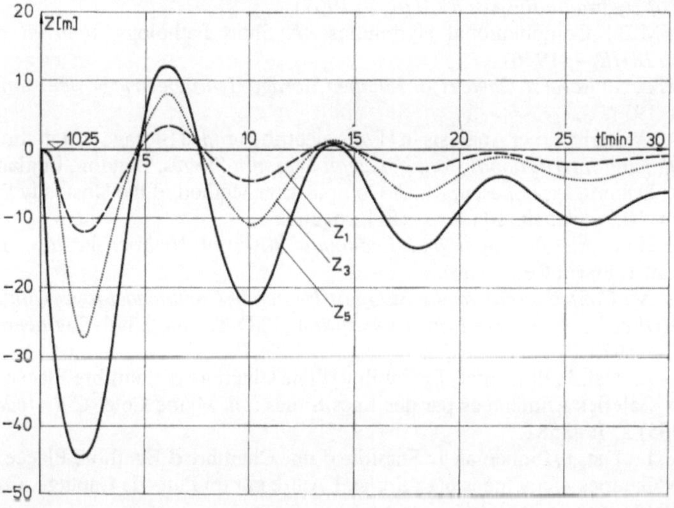

Figure 8.38. Damping of water levels oscillations in surge tanks at a power manoeuvre in the central station with $N_1 = 0$ and $N_2 = 168$ MW.

8.4.4. The Riul Mare-Retezat hydroelectric power plant

With a power of 320 MW, the Riul Mare-Retezat is the second largest interior rivers plant in Romania built on, after the Lotru HPP. Among the main technical objectives, of this plant we mention the following together with some technical characteristics (variant): dam built of local materials, with claycore $H = 170$ m; main pressure tunnel with $L = 18.1$ km and $d = 4.9$ m; surge tank with upper and lower chambers and hydraulic resistance at the base; penstock with vertical shaft; underground station equipped with two groups of Francis type turbines, of 160 MW each; tunnel and tailrace of over 2 km length; 19 secondary intakes and about 26 km of tunnels contribute to supplementing the discharge of the Riul Mare River from 8.5 m^3/s to 14 m^3/s by the intake and deviation of the tributaries from its own watershed or from the watershed of the Strei river.

The hydraulic scheme of the plant represents a modern approach, the intake pipes from the small rivers Netis, Bodu, Valea Mare and Riusor, being directly connected to the main pressure tunnel similarly to the Lotru hydroelectric power plant, it belongs to the group of hydraulic schemes of type C (Fig. 8.37).

In reference [7] a hydraulic study of the plant operation at transient regimes has been performed. Among the results obtained by this study, Fig. 8.38 presents a variant of checking the phenomenon of the oscillatory motion damping in the system, in the hypothesis of the station operation at constant power. The project is presently in the final constructive phase. Finally, mention should be made of two similar hydraulic studies conducted for hydroelectric power plant projects from abroad, conceived on complex schemes [36, 39, 40].

References

1. Abbott, M.B., Ionescu, F.: On the Numerical Computation of Nearly Horizontal Flows. *Journal of Hydraulic Research IAHR*, 5 (1967).
2. Abbott, M.B.: Computational Hydraulics—A Short Pathology. *Journal of Hydraulic Research IAHR*, 4 (1976).
3. Arsenevski, N.: *Reverse Devices for Pumped Storage Hydroelectric Power Plants.* Energhia, Moskva, 1977.
4. Boldy, A.: Waterhammer Analysis in Hydroelectric Pumped Storage Installations. *Proceedings of the 2nd International Conference on Pressure Surges*, London, England, 1976.
5. Cioc, D., Dogaru, L., Popescu, M.: A Computation Method of the Unsteady Water Motion in Rivers. *Hidrotehnica*, 10 (1967) (in Romanian).
6. Cogalniceanu, A.: *Technical and Economic Bases of Hydroenergetics.* Ed. Tehnica, Bucharest, 1986 (in Romanian).
7. Cristidis, V., *Contributions to the Study of Unsteady Combined Flows with Applications to the Hydroelectric Power Plants Operation.* PhD Thesis, Civil Engineering Institute, Bucharest, 1978.
8. Escande, L., Dat, J., Piquemal, J.: Stabilité d'une Chambre d'Équilibre Placée à la Jonction de Deux Galeries Alimentées par des Lacs Situés à la Même Cote. *C.R. Acad. Sci. Paris*, 261 (1965) 2579–2581.
9. Escande, L., Dat, J., Piquemal, J.: Stabilité d'une Chambre d'Équilibre Placée à la Jonction de Deux Galeries Alimentées par un Lac, l'Autre par un Puits de Captage. *C.R. Acad. Sci. Paris*, 262 (1965) 2579–2581.
10. Gerber, S.: *Etude Théorique et Expérimentale de la Stabilité des Chambres d'Équilibre Situées en Aval d'une Galerie en Charge Alimentée par un Canal à Écoulement Libre.* Thèse de doctorat, Université de Toulouse, 1954.
11. Halanay, A.: *Differential Equations. Stability, Oscillations, Time Lags.* Academic Press, New York, London, 1966.
12. Hancu, S., Popescu, M., Zaharescu, E.: *Applied Hydraulics.* Ed. Tehnica, Bucharest, 1985.
13. Henry, P., Graeser, J.: Energy Storage: Developments in Pumped Storage. *Water Power and Dam Construction*, 6 (1985).
14. Ionescu, F., Popescu, M.: On the Use of Numerical Models for Unsteady Regimes in River Channels. *International Conference on Fluvial Hydraulics IAHR*, Budapest, Hungary, 30 May–3 June, 1988.
15. Jaeger, C.: *Fluid Transients in Hydroelectric Engineering Practice.*. Blackie, London, 1977.
16. Klemm, D., Schilling, R.: Numerical Refinement of Palmiet's Pump–Turbine Design. *Water Power and Dam Construction*, 5 (1988).
17. Mahmoud, K., Yevjevich, J.: *Unsteady Flow in Open Channels.* Water Resources Publications, Fort Collins, USA, 1975.
18. Mateescu, C., Pavel, D.: Representative Hydropower Projects in Romania, Built between 1947–1972. *Hidrotehnica*, 12 (1972) (in Romanian).
19. McIunis, D, Karney, B.: Transients in Distribution Networks: Field Tests and Demands Models. *Journal of Hydraulic Engineering*, 3 (1995).
20. Nistreanu, V.: *Hydroelectric Power Plant and Pumping Stations.* Polytechnic University, Bucharest, 1988 (in Romanian).

21. Pavel, D.: Perspectives in the Creation of Hydroelectrical and Complex Transformers. *Hidrotehnica*, 6 (1973) (in Romanian).
22. Popescu, M.: Modèle Mathématique pour l'Étude de la Stabilité d'un Système de Deux Chambres d'Équilibre. *C.R. Acad. Sci. Paris*, 273 (1971) 129–132.
23. Popescu, M.: Hydraulic Computation of Transient Regimes in Hydroelectric Power Plants and Pumping Stations with Mixed Intakes Provided with Automatic Regulators. *Studii si Cercetari de Mecanica Aplicata*, 36 (1977) 502–512 (in Romanian).
24. Popescu, M.: Computation of Hydraulic Transients in Complex Hydro-Schemes. *Water Power and Dam Construction*, 9 (1986).
25. Popescu, M.: Etude de la Stabilité Hydraulique de Fonctionnement aux Régimes Transitoires des Usines Hydroélectriques Conçues par des Schémas Complexes. *Proceedings of the IAHR Symposium on Fluid Motion Stability in Hydraulic Systems with Automatic Control*, Bucharest, 1976, pp. 229–247.
26. Popescu, M., Arsenie, D.: Hydraulic Computation of Transient Regimes in Pumped-Storage Hydroelectric Power Plants. *Hidrotehnica*, 25 (1980) 245–248 (in Romanian).
27. Popescu, M., Cristidis, V.: Stabilité du Fonctionnement des Usines Hydroélectriques et des Stations de Pompage à Adductions Mixtes Munies de Régulateurs Automatiques. *Proceedings of IAHR Symposium on Fluid Motion Stability in Hydraulic Systems with Automatic Control*, Bucharest, 1976, pp. 257–266.
28. Popescu, M., Halanay, A.: Analyse de la Stabilité Hydraulique Générale de Fonctionnement aux Regimes Transitoires des Usines Hydroélectriques Munies d'Un où de Deux Chambres d'Équilibre. *Proceedings of the XVIII Congres AIRH*, Cagliari, Italia, 1979, pp. 85–97.
29. Popescu, M., Halanay, A.: Hydraulic Stability of Hydropower Stations Working in Complex Schemes. *Proceedings of the Second IAHR Symposium on Fluid Motion Stability in Hydraulic Systems with Automatic Control*, Bucharest, 1984, pp. 185–200.
30. Popescu, M., Halanay, A.: Hydraulic Stability of Hydroelectric Power Plants Working in Complex Schemes. *Proceedings of the IAHR Work Group on the Behaviour of Hydraulic Machinery Under Steady Oscillatory Conditions*, Fort Collins, USA, August 16–18, 1989.
31. Popescu, M., Jelev, I.: Hydraulic Computation of Transient Regimes in Complex Hydroelectric Power Plants. *Studii si Cercetari de Mecanica Aplicata*, 36 (1977) 819–836.
32. Popescu, M., Ionescu, F.: On the Use of Numerical Models for Unsteady Regimes in Open Channels. *3rd International Symposium on Ecohydraulics*, Salt Lake City, Utah, USA, July 13–16, 1999.
33. Popescu, M., Stanuca, A., Stabilité Hydraulique de Fonctionnement aux Regimes Transitoires Hydroélectriques Lotru et Tismana. *Proceedings of the IAHR Symposium on Fluid Motion Stability in Hydraulic Systems with Automatic Control*, Bucharest, 1976, pp. 249–256.
34. Popescu, M., Stanuca, A.: The High-Head Power Plant Lotru in Romania. Hydraulic transients. *L'Energia Elettrica*, 7–8 (1988).
35. Priscu, R.: *Hydraulic Structures*, Vol. II. Ed. Didactica si Pedagogica, Bucharest, 1974 (in Romanian).
36. Sabljak, R., Doran, W.: Hydraulic Analysis of Double Surge Shaft System by Digital Computer. *IAHR Congress*, Dubrovnik, Yugoslavia, 1961.
37. Spiridon, D.: *Types of Pumped Storage Hydroelectric Power Plant Projects and their Possibility of Application in Romania.* Work Session, ISPH, 1979 (in Romanian).
38. Popescu, M., Stere, C.: *Analysis of Transient Regimes within Complex Surge Tank Hydraulic Schemes with Application to Two High Head Hydroelectric Power Plants*, Conference Waterpower-XIII, Buffalo, USA, 29–31 July, 2003.
39. Tekle, T., Vigander, K.: Siso Water Power Plant Mass Oscillation in the Head Race System. *Bulletin of the River and Harbour Laborator Technical University of Norway*, 13 (1972).
40. Torell, P.: Surge in Multiple Tanks. *Water Power and Dam Construction*, 12 (1966).
41. Vladimirescu, I.: *Hydraulic Devices and Pumping Stations.* Ed. Didactica si Pedagogica, Bucharest, 1974 (in Romanian).
42. Popescu, M., Arsenie, D., Problemes Dynamiques dans le Projet et le Fonctionement de Usines Hydroelectriques et des Stations de Pompage. *Proceedings of the XXVI IAHR Congress*, London, Vol. 1, 1995, pp. 613–615.

43. Breaban, V.: *Amenajari Hidroenergetice—Aspecte Energetice si Economice* (Hydropower Works—Energetical and Economical Aspects). Ed. Ovidius University of Constantza, 1997.
44. Popovici, A.: *Baraje pentru Acumulari de Apa* (Dams for Water Storage). Ed. Tehnica, Bucharest, 2002.
45. Luca, O., Luca, B.: *Hydraulics of Constructions*. University Horizonts, Timisoara, 2002 (in Romanian).

Chapter 9

Computation of Unsteady Motions in the Intermediate Domain between Rapidly and Slowly Varying Water Motions

In Chapters 1 and 5, when dealing with the rapidly and slowly varying types of motions, we have mentioned their main characteristics as well as the conditions under which they can appear in pressure hydraulic systems. In order to determine the two types of unsteady motion, we selected the amplitudes and frequencies of underpressure oscillations of water in the system as characteristic quantities and retained as main factors influencing the motion type the *fluid and pipe wall elasticity* as well as the *mass transport of water* in the system.

Thus, we have defined *rapidly varying motions* (FV) and *slowly varying motions* (SV), depending on which of the above-mentioned factors influence decisively the phenomenon of unsteady water motion in the system. As far as the hydraulic computation of the two types of motion is concerned, we have adopted the compressible or incompressible fluid model conducting special studies to access the validity of one of these models. There is no strict delimitation in the specialised literature between the rapidly and slowly varying motions in hydraulic systems under pressure [1, 2]. In the present paper, we shall introduce the concept of *unsteady motion in the intermediate domain between rapidly and slowly varying motions* (ID), for hydraulic systems under pressure.

This new type of unsteady motion is influenced to some extent by both factors that generally define the character of unsteady water motion in underpressure systems, hence by the *elasticity of the fluid and of the pipe walls* and by the *mass transport of water* in the system as well and may be generated in a hydraulic system with surge tank by variation of the local hydraulic resistance at the tank base, and in a hydraulic system with air chamber by variation of the air cushion volume.

In our study of the water motions in the intermediate domain, we have used the *experimental laboratory installation* ICH, which allowed us to document the peculiarities of these motions [3, 5]. For the theoretical analysis, the problem was raised of validating a mathematical model capable of describing the considered class of motions and to allow the analysis of the evolution these motion types according to the adopted parameters of local hydraulic resistance for the surge tank and air cushion volume for the air chamber, respectively [4, 6]. The adopted mathematical model describes the unsteady water motion in the pipe by linear partial differential equations, while the surge tank and the air chamber involve the boundary conditions where the above-mentioned parameters appear as well.

Taking into account the objective set forth to reveal the peculiarities of the motion function of these parameters and the relations between the slow motions in the tank or the air chamber and the rapidly varying motions in the pipe, we have considered an analytic approach preferable to a direct numerical procedure. Therefore, we have adopted a method relying on *asymptotic expansions*, which proved to be able to yield conclusions in good agreement with experimental results [6–8].

We shall analyse experimentally as well as computationally the unsteady motion in the intermediate domain (ID), which can arise in a hydraulic system consisting of reservoir, pressure tunnel, and surge tank with diaphragm at the base and valve (Fig. 2.4, IA).

9.1. Experimental results

Experiments have been performed on the *laboratory experimental installation* ICH. We denote by d and D the diameter of the diaphragm and of the connecting pipe between tank and pressure tunnel, respectively. For extreme values of the diaphragm, diameter $d = 0$ and $d = D$, we obviously find again the types of rapidly or slowly varying motions previously studied and analysed for the hydraulic schemes I-O and I-A (Fig. 2.4) and illustrated by measurements in the corresponding figures of Chapters 2 and 7.

The experiments performed to study unsteady flows in the intermediate domain (ID) covered a large scale of values at the surge tank base, using diaphragm of diameter $d = 1, 2, 3, \ldots 8$ cm. Thus, we have obtained local hydraulic resistances $\zeta = 8$–1800.

At a sudden valve-closing manoeuvre in hydraulic scheme I-A, the resulting water level oscillation, function of the value of the hydraulic resistance at the tank base, was of the following types: damped harmonic oscillation, periodic oscillation and highly damped oscillation. Under these conditions, the water pressure variation in the pipe has been strongly influenced by the magnitude of the hydraulic resistance at the tank base.

The investigated intermediate domain (ID) of unsteady motions corresponds to a hydraulic resistance domain of values at the tank base, associated with non-periodic and highly damped oscillations of the water in the tank, values which are inferior to those for which classical effects of waterhammer appear. Measurements performed with diaphragms of diameters $d = 6$, 7, and 8 cm did not differ qualitatively from experimental results obtained in the case of a diaphragm of $d = 5$ cm.

The results of laboratory measurements performed at two regime flow rates $q_o = 13.5$ l/s and $q_o = 19.2$ l/s are shown in Figs. 9.1–9.4 and point out some very interesting qualitative aspects, specific to this type of unsteady motions (ID). As one can notice from the above, diagrams for the pressure variation in the pipe have a single pressure peak, followed by a slow decrease towards the regime pressure. The occurrence of these types of motion similar to the rapidly varying ones is signalled by the appearance of several pressure peaks, as for instance in the case of the diaphragm with $d = 1$ cm.

If we stick to this domain, we must mention an experimental hydraulic study conducted on a similar laboratory installation which analysed the phenomenon of waterhammer transmission along the pressure tunnel of a hydroelectric power plant, as depending on the magnitude of the hydraulic resistance at the base of the surge tank [1].

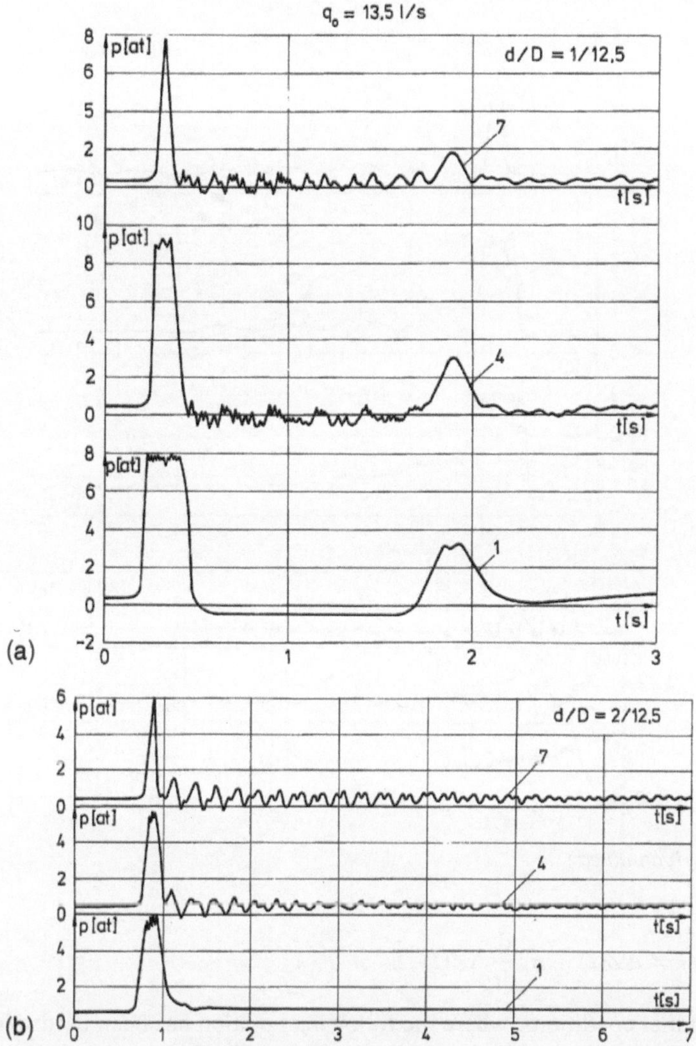

Figure 9.1. Scheme I-A, with diaphragm-equipped surge tank: pressure measurements at three points of the pipe, for regime flow rate $q_o = 13.5$ l/s and various hydraulic resistances at the tank basis (sudden valve closing).

9.2. Computation method

The mathematical model of the unsteady water motion in a hydraulic system consisting of reservoir, pressure tunnel, and surge tank with diaphragm and valve (Fig. 2.4, I-A) as a response to a step discharge has been considered in the following form, once the equations system has been linearised:

$$\frac{\partial}{\partial t}\Delta H + \frac{c^2}{gf}\frac{\partial}{\partial x}\Delta Q = 0, \qquad \frac{\partial}{\partial x}\Delta H + \frac{1}{gf}\frac{\partial}{\partial t}\Delta Q = 0$$

$$\Delta H(0,t) = 0, \quad \Delta Q(\ell,t) = \hat{q} + F\frac{\mathrm{d}}{\mathrm{d}t}\Delta Z(t) \tag{9.1}$$

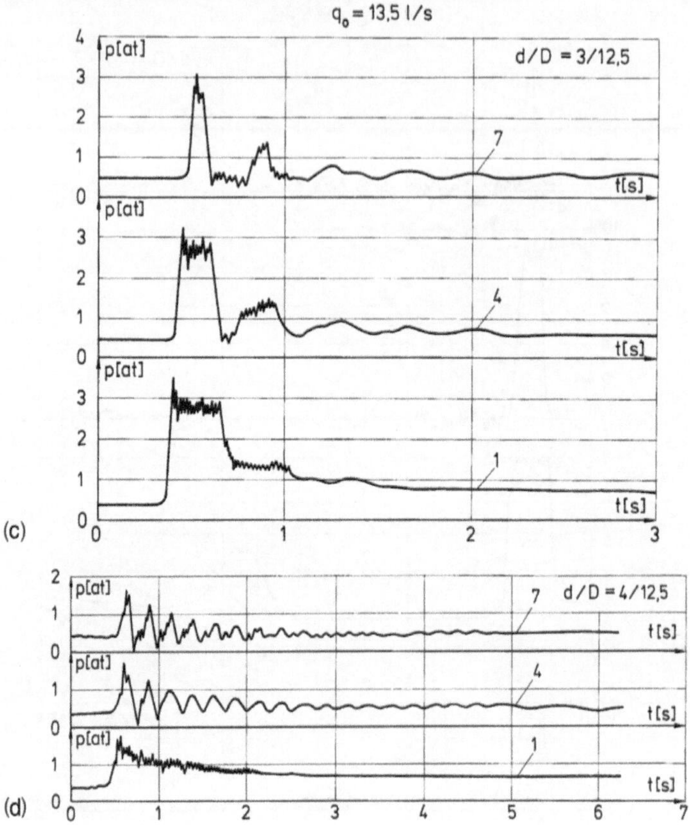

Figure 9.1. (*continued*).

$$\Delta H(\ell, t) = \Delta Z(t) + R_c' \frac{\mathrm{d}}{\mathrm{d}t} \Delta Z(t)$$

with null initial conditions, where the following notation has been used: $H(x, t)$ is the instantaneous piezometric head in the pressure conduit with respect to a horizontal reference plane at the valve level downstream the tank, $Q(x,t)$ the instantaneous discharge in the pressure conduit, $Z(t)$ the water level in the surge tank, $\Delta H(x,t)$, $\Delta Q(x,t)$, $\Delta Z(t)$ are deviations from the steady regime of the pressure, the flow, and the level in the tank, ℓ is the length of the pressure conduit, f the cross-section area of the pressure conduit, F the horizontal cross-section area of the surge tank, R_c' the local head loss coefficient for the surge tank strangulation, C the celerity, g the gravitational acceleration, x the distance measured in the flow direction, t the time, and q the step discharge. The model neglected the head losses in the pressure conduit.

Then, a non-dimensional model, is approached, assuming

$$y(\tau) = \frac{1}{\ell} \Delta Z\left(\frac{\tau \ell f}{|\hat{q}|}\right), \quad h(\xi, \tau) = \frac{1}{\ell} \Delta H\left(\xi_1 \frac{\tau \ell f}{|\hat{q}|}\right),$$

$$q(\xi, \tau) = \frac{1}{|\hat{q}|} \Delta Q\left(\ell \xi, \frac{\tau \ell f}{|\hat{q}|}\right) \tag{9.2}$$

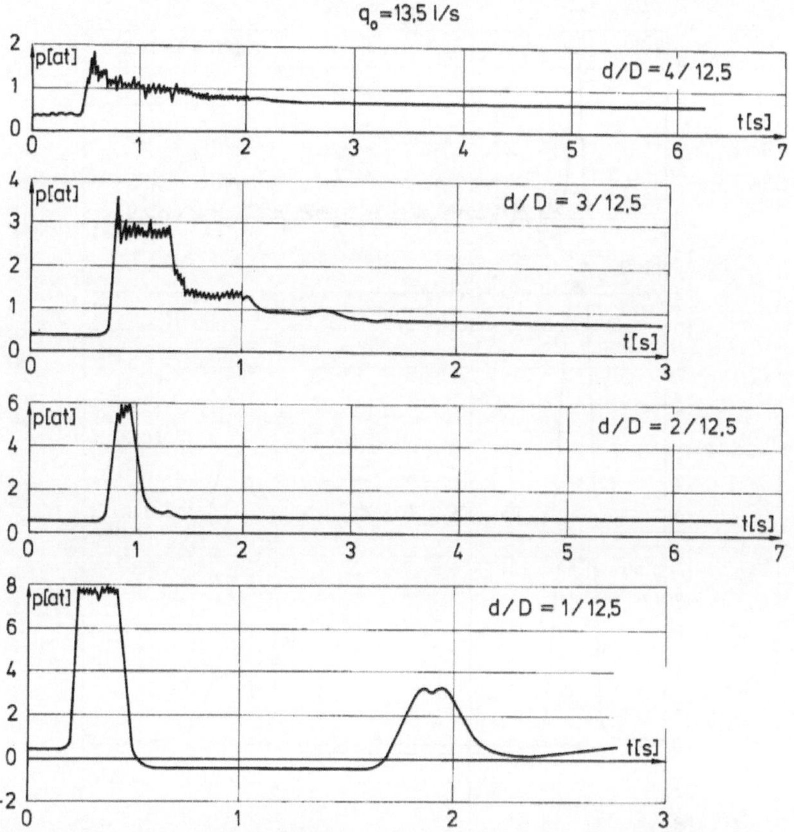

Figure 9.2. Scheme I-A, with diaphragm-equipped surge tank: pressure measurements in section 1 of the pipe, for regime flow rate $q_o = 13.5$ l/s and various hydraulic resistances at the tank basis (sudden valve closing).

The following equations are obtained

$$\frac{\partial h}{\partial \tau} + \frac{c^2}{g\ell}\frac{\partial q}{\partial \xi} = 0, \qquad \frac{\partial h}{\partial \xi} + \frac{|\hat{q}|^2}{g\ell f^2}\frac{\partial q}{\partial \tau} = 0 \tag{9.3}$$

with the boundary conditions

$$\begin{aligned} &h(0,\tau) = 0, \quad q(1,\tau) = \frac{\hat{q}}{|\hat{q}|} + \frac{F}{f}y'(\tau) \\ &h(1,\tau) = y(\tau) + \frac{R_c'|\hat{q}|}{\ell f}y'(\tau) \end{aligned} \tag{9.4}$$

where ξ, τ, $h(\xi,\tau)$, $q(\xi,\tau)$, and $y(\tau)$ denote the length, the time, the pressure, the discharge and the water level surge tank, respectively as dimensionless quantities.

Riemann invariants enter through

$$q = u - v, \quad h = \frac{c|\hat{q}|}{g\ell f}(u + v) \tag{9.5}$$

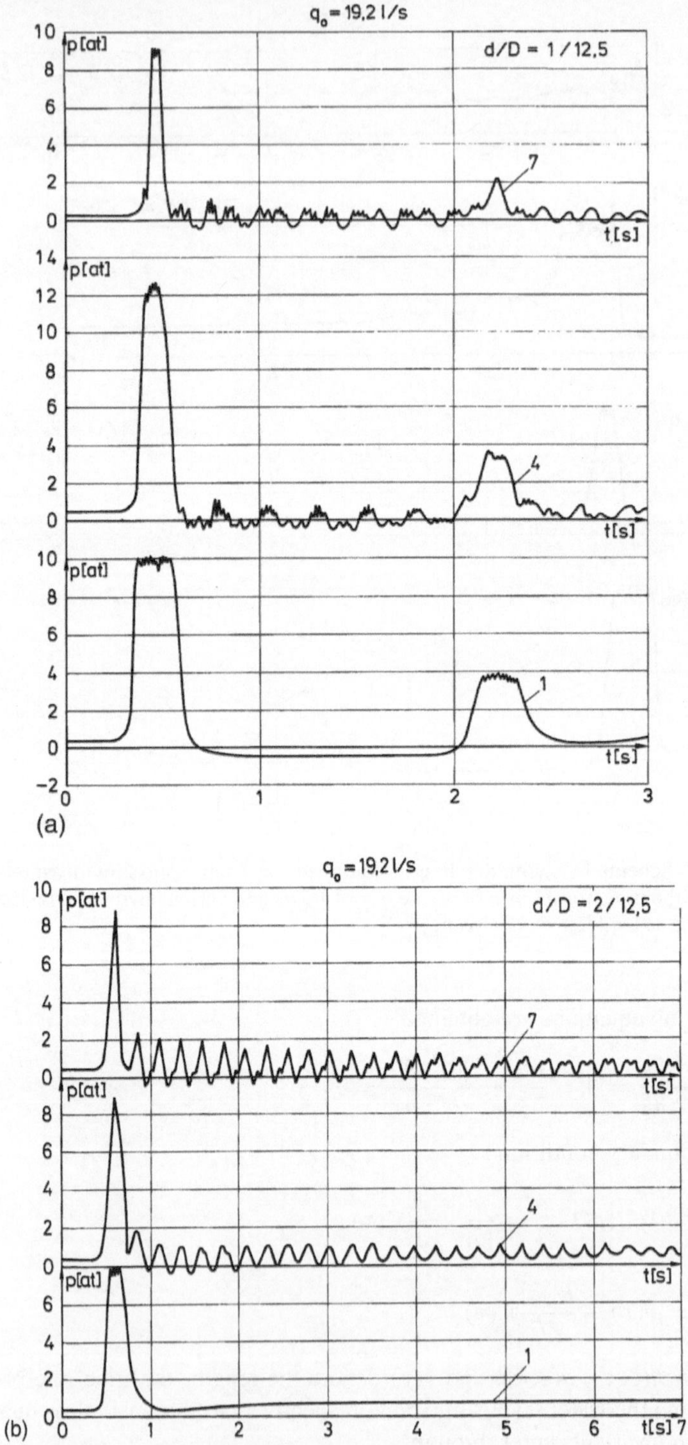

Figure 9.3. Scheme I-A, with diaphragm-equipped surge tank: pressure measurements at three points of the pipe, for regime flow rate $q_o = 19.2$ l/s and various hydraulic resistances at the tank basis (sudden valve closing).

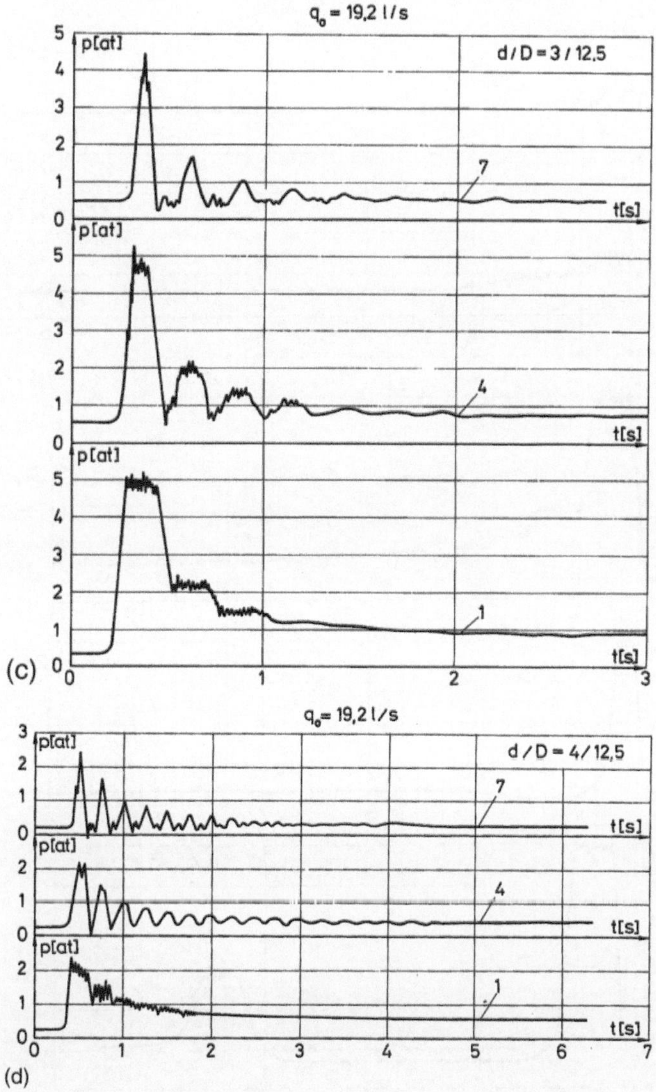

Figure 9.3. *(continued).*

Equations (9.3) become

$$\frac{\partial u}{\partial \xi} + \frac{|\hat{q}|}{ef}\frac{\partial u}{\partial \tau} = 0, \qquad \frac{\partial v}{\partial \xi} - \frac{|\hat{q}|}{cf}\frac{\partial v}{\partial \tau} = 0 \tag{9.6}$$

and the boundary conditions yield

$$\begin{gathered} u(0,\tau) + v(0,\tau) = 0, \quad u(1,\tau) - v(1,\tau) = \frac{\hat{q}}{|\hat{q}|} + \frac{F}{f}y'(\tau) \\ \frac{c|\hat{q}|}{g\ell f}[u(1,\tau) + v(1,\tau)] = y(\tau) + \frac{R_c'|\hat{q}|}{\ell f}y'(\tau) \end{gathered} \tag{9.7}$$

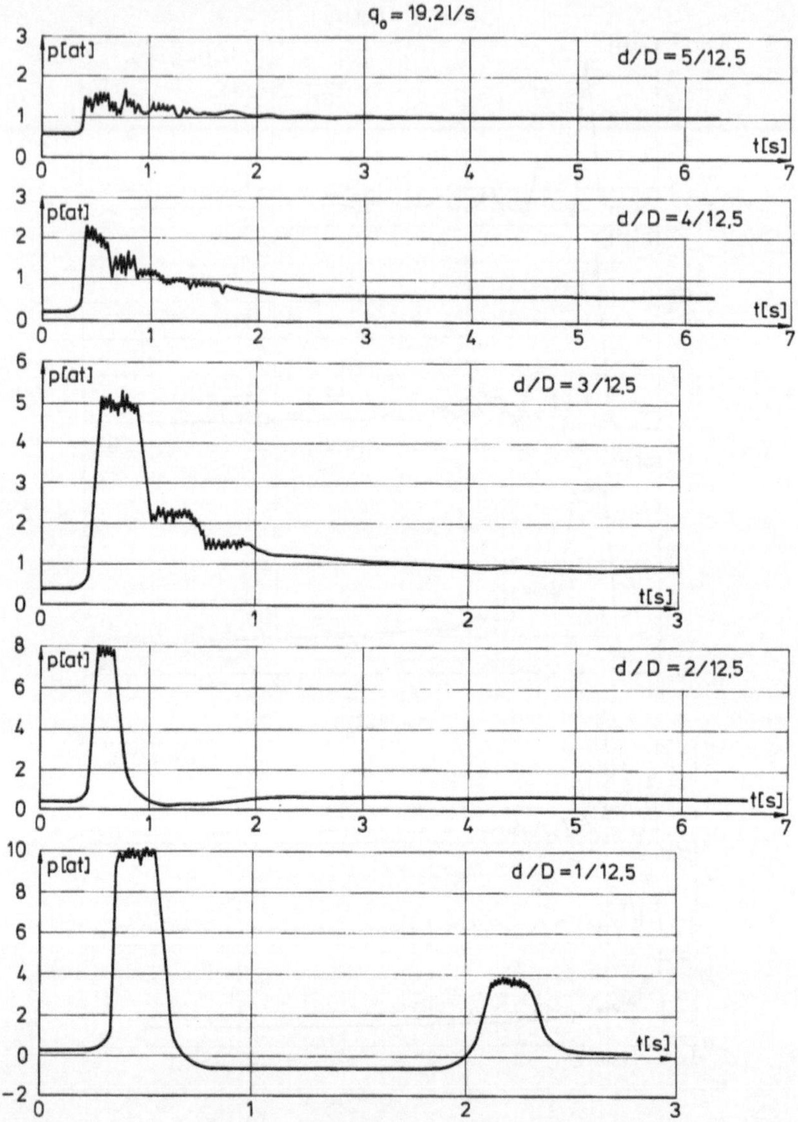

Figure 9.4. Scheme I-A, with diaphragm-equipped surge tank: pressure measurements in cross-section 1 of the pipe, for regime flow rate $q_o = 19.2$ l/s and various hydraulic resistances at the tank basis (sudden valve closing).

The method of characteristics hence leads to

$$\frac{F}{f}y'(\tau) + w(\tau) + w\left(\tau - \frac{2|\hat{q}|}{cf}\right) + \frac{\hat{q}}{|\hat{q}|} = 0$$

$$\frac{R_c'|\hat{q}|}{\ell f}y'(\tau) + y(\tau) = \frac{c|\hat{q}|}{g\ell f}\left[w(\tau) - w\left(\tau - \frac{2|\hat{q}|}{ef}\right)\right] \tag{9.8}$$

$$y(0) = 0, \quad w(\tau) \equiv 0 \ \text{for} \ \tau < 0$$

where we denoted $w(\tau) = v(1, \tau)$. Note that

$$\begin{aligned} u(\xi,\tau) &= -w\left(\tau - \frac{|\hat{q}|}{cf}(1+\xi)\right) \\ v(\xi,\tau) &= w\left(\tau - \frac{|\hat{q}|}{cf}(1-\xi)\right) \end{aligned} \tag{9.9}$$

hence

$$h(\xi,\tau) = \frac{c|\hat{q}|}{gf\ell}\left[w\left(\tau - \frac{|\hat{q}|}{cf}(1-\xi)\right) - w\left(\tau - \frac{|\hat{q}|}{cf}(1+\xi)\right)\right] \tag{9.10}$$

Thus, eqns. (9.8) for y and w allow computing the water level in the tank and the pressure in the pressure conduit.

The fundamental parameters entered

$$\varepsilon = \frac{2|\hat{q}|}{cf}, \quad k = \frac{c|\hat{q}|}{g\ell f}, \quad \zeta = \frac{2gf^2 R'_c}{|\hat{q}|F} \tag{9.11}$$

By means of these parameters, the model can be rewritten as

$$\begin{aligned} y'(\tau) &= -\frac{\dfrac{f}{kF}}{1+\dfrac{\zeta\varepsilon}{4}}y(\tau) - \frac{\dfrac{2f}{F}}{1+\dfrac{\zeta\varepsilon}{4}}w(\tau-\varepsilon) - \frac{\dfrac{f}{F}}{1+\dfrac{\zeta\varepsilon}{4}}\frac{\hat{q}}{|\hat{q}|} \\ w(\tau) &= \frac{1-\dfrac{\zeta\varepsilon}{4}}{1+\dfrac{\zeta\varepsilon}{4}}w(\tau-\varepsilon) + \frac{1}{k}\frac{1}{1+\dfrac{\zeta\varepsilon}{4}}y(\tau) - \frac{\dfrac{\zeta\varepsilon}{4}}{1+\dfrac{\zeta\varepsilon}{4}}\frac{\hat{q}}{|\hat{q}|} \end{aligned} \tag{9.12}$$

Note that parameter k represents the non-dimensional pressure value given by water-hammer in the pressure conduit in the absence of the surge tank, and

$$\frac{\zeta\varepsilon}{4} = \frac{gfR'_c}{cF} \tag{9.13}$$

The suggested computation method consists in the asymptotic expansion of the solution of system (9.12) with respect to the small parameter ε (in the case of the laboratory hydraulic scheme with $q_o = 13.5$ l/s, we get $\varepsilon \approx 1/625$).

Asymptotic expansions have been rigorously studied for general problems of the form

$$\begin{aligned} x_1'(\tau) &= \boldsymbol{A}x_1(\tau) + \boldsymbol{B}x_2(\tau-\varepsilon) + f_1(\tau), \quad x_1(0) = x_1^o \\ x_2(\tau) &= \boldsymbol{G}x_1(\tau) + \boldsymbol{H}x_2(\tau-\varepsilon) + f_2(\tau), \quad x_2(\tau) = \phi(\tau) \\ & -\varepsilon \le \tau < 0 \end{aligned} \tag{9.14}$$

where $\boldsymbol{A}, \boldsymbol{B}, \boldsymbol{G}, \boldsymbol{H}$ are matrices; the fundamental hypotheses consists in the fact that the eigenvalues of matrix $\boldsymbol{H}$ fall inside the unit circle (in the case of the hydraulic scheme

$$\boldsymbol{H} = \frac{1-\dfrac{\zeta\varepsilon}{4}}{1+\dfrac{\zeta\varepsilon}{4}}$$

and we find that $|\boldsymbol{H}| < 1$; the results are valid for sufficiently large head losses in the surge tank, so that $\boldsymbol{H}$ is not close to 1).

With the given system of equations we associate the reduced model

$$\begin{aligned} x_1' &= \tilde{\boldsymbol{A}}x_1 + \tilde{f}_1, \quad \tilde{\boldsymbol{A}} = \boldsymbol{A} + \boldsymbol{B}(I - \boldsymbol{H})^{-1}\boldsymbol{G} \\ \tilde{f} &= f_1 + \boldsymbol{B}(I - \boldsymbol{H})^{-1}f_2, \quad x_1(0) = x_1^{\mathrm{o}} \end{aligned} \tag{9.15}$$

We shall denote the corresponding solution by $\tilde{x}_1$ and compute x_2 from the difference equation

$$\begin{aligned} x_2(\tau) &= \boldsymbol{G}\tilde{x}_1(\tau) + \boldsymbol{H}x_2(\tau - \varepsilon) + f_2(\tau) \\ x_2(\tau) &= \phi(\tau), \quad -\varepsilon \leq \tau < 0 \end{aligned} \tag{9.16}$$

Thus, the problem is separately treated: a system of ordinary differential equations and a system of difference equations are solved individually.

In order to solve the problem, the asymptotic expansion is written under the form

$$\begin{aligned} x_1(\tau) &= \tilde{x}_1(\tau) + \varepsilon\xi_1(\tau) \\ x_2(\tau) &= \tilde{x}_2(\tau) + \varepsilon\xi_2(\tau) \end{aligned} \tag{9.17}$$

and we can prove that ξ_1 and ξ_2 are bounded, which shows that $\tilde{x}_1$ and $\tilde{x}_2$ approximate the exact solutions x_1 andx_2 with an error of the order ε. If the matrix $\tilde{\boldsymbol{A}}$ is of Hurwitz type (its eigenvalues have negative real elements), the functions ξ_1 and ξ_2 result in being uniformly bounded on the semiaxis $\tau \geq 0$, demonstrating that the approximation holds on large time intervals.

The analysis of functions ξ_1 and ξ_2 behaviour is performed in several steps. The solution representation formula of the difference equation yields

$$\left|\tilde{x}_2(\tau) - (1 - \boldsymbol{H})^{-1}[\boldsymbol{G}\tilde{x}_1(\tau) + f_2(\tau)]\right| \leq \mathrm{e}\left(\varepsilon + \mathrm{e}^{-\frac{\alpha\tau}{\varepsilon}}\right) \tag{9.18}$$

Then, the system of equations satisfied by functions ξ_1 and ξ_2 is written. This system gives the relationship

$$\xi_1(\tau) = \frac{1}{\varepsilon}\int_{\mathrm{o}}^{\tau} X^*(\tau - s)\boldsymbol{B}[\tilde{x}_2(s - \varepsilon) - \hat{x}_2(s)]\,\mathrm{d}s \tag{9.19}$$

where

$$\hat{x}_2(s) = (I - \boldsymbol{H})^{-1}[\boldsymbol{G}\tilde{x}_1(s) + f_2(s)] \tag{9.20}$$

and X is defined from the problem *

$$\begin{aligned} \frac{\mathrm{d}}{\mathrm{d}t}X(t) &= \boldsymbol{A}^*X(t) + \boldsymbol{G}^*Y(t) \\ Y(t) &= \boldsymbol{B}^*X(t - \varepsilon) + \boldsymbol{H}^*Y(t - \varepsilon) \\ X(t) &\equiv 0, \quad t < 0, \quad X(0) \equiv I, \quad Y(t) \equiv 0, \quad t < 0. \end{aligned} \tag{9.21}$$

By $\boldsymbol{B}^*$ we denote the transpose of matrix $\boldsymbol{M}$.

For the X kernel an estimation of the form

$$|X(\tau)| \leq ce^{-\tilde{\alpha}\tau} \tag{9.22}$$

is obtained. This estimation, together with the estimation for

$$|\tilde{x}_2(\tau) - \hat{x}_2(\tau)|$$

shows that ξ_1 is bounded.

The equation satisfied by ξ_2 then allows us to conclude that ξ_2 is bounded too. The hardest part of the demonstration consists in obtaining the estimation for the kernel X. In our case, we have

$$\boldsymbol{A} = -\frac{\dfrac{f}{kF}}{1+\dfrac{\zeta\varepsilon}{4}}, \quad \boldsymbol{B} = -\frac{\dfrac{2f}{F}}{1+\dfrac{\zeta\varepsilon}{4}}, \quad \boldsymbol{G} = \frac{\dfrac{1}{k}}{1+\dfrac{\zeta\varepsilon}{4}}, \quad \boldsymbol{H} = \frac{1-\dfrac{\zeta\varepsilon}{4}}{1+\dfrac{\zeta\varepsilon}{4}}$$

$$f_1 = -\frac{\dfrac{f}{F}}{1+\dfrac{\zeta\varepsilon}{4}}\frac{\hat{q}}{|\hat{q}|}, \quad f_2 = -\frac{\dfrac{\zeta\varepsilon}{4}}{1+\dfrac{\zeta\varepsilon}{4}}\frac{\hat{q}}{|\hat{q}|} \tag{9.23}$$

By $\boldsymbol{A}^*$ we denote the transposed of the matrix $\boldsymbol{A}$.

A direct computation yields

$$\tilde{\boldsymbol{A}} = -\frac{4f}{kF\zeta\varepsilon}, \quad \tilde{f} = 0. \tag{9.24}$$

Hence, the reduced model is

$$\tilde{y}(\iota) = -\frac{4f}{kF\zeta\varepsilon}\tilde{y}(\iota), \quad \tilde{y}(0) = 0 \tag{9.25}$$

yielding $\tilde{y}(\tau) \equiv 0$, which shows that the water level in the tank is stationary in the first approximation (the oscillations in the tank are strongly damped).

Hence, $\tilde{w}$ is computed from the equation

$$\tilde{w}(\tau) = \frac{1-\dfrac{\zeta\varepsilon}{4}}{1+\dfrac{\zeta\varepsilon}{4}}\tilde{w}(\tau-\varepsilon) - \frac{\dfrac{\zeta\varepsilon}{4}}{1+\dfrac{\zeta\varepsilon}{4}}\frac{\hat{q}}{|\hat{q}|} \tag{9.26}$$

$$\tilde{w}(\tau) = 0 \text{ for } \tau < 0.$$

The solution of such a difference equation can be explicitly written

$$\tilde{w}(\tau) = -\frac{\hat{q}}{2|\hat{q}|}\left[1 - \left(\frac{1-\dfrac{\zeta\varepsilon}{4}}{1+\dfrac{\zeta\varepsilon}{4}}\right)^{[\frac{\tau}{\varepsilon}]+1}\right] \quad \text{for } \tau \geq 0 \tag{9.27}$$

Here $[\tau/\varepsilon]$ denote the largest integer number below τ/ε.

The explicit formula for $\tilde{w}$ allows us to write the main part of the pressure value in the form

$$\bar{h}(\xi,\tau) = \frac{c\hat{q}}{2g\ell f}\left(\frac{1-\dfrac{\zeta\varepsilon}{4}}{1+\dfrac{\zeta\varepsilon}{4}}\right)^{\left[\frac{\tau}{\varepsilon}-\frac{1+\varepsilon}{2}\right]}\left\{1-\left(\frac{1-\frac{\zeta\varepsilon}{4}}{1+\frac{\zeta\varepsilon}{4}}\right)^{\left[\frac{\tau}{\varepsilon}-\frac{1-\xi}{2}\right]-\left[\frac{\tau}{\varepsilon}-\frac{1+\xi}{4}\right]}\right\} \tag{9.28}$$

Reverting to the initial non-dimensional form, it follows that, in this case, the pressure increase ΔH is approximately given by the formula

$$\Delta\tilde{H}(x,t) = \frac{c\hat{q}}{2gf}\left(\frac{cF-gfR_c'}{cF+gfR_c'}\right)^{\left[\frac{ct-x}{2\ell}-\frac{1}{2}\right]}\left\{1-\left(\frac{cF-gfR_c'}{cF+gfR_c'}\right)^{\left[\frac{ct+x}{2\ell}-\frac{1}{2}\right]\left[\frac{ct-x}{2\ell}-\frac{1}{2}\right]}\right\} \tag{9.29}$$

In particular, for $x = \ell$(at the valve):

$$\Delta\tilde{H}(x,t) = \frac{c\hat{q}}{2gf}\left(\frac{cF-gfR_c'}{cF+gfR_c'}\right)^{\left[\frac{ct-x}{2\ell}-1\right]}\left[1-\frac{cF-gfR_c'}{cF+gfR_c'}\right] \tag{9.30}$$

It is evident form this formula that for $R_c' < cF/gf$ (i.e. for $\zeta\varepsilon/4 < 1$, hence in the case of laboratory experiments $\zeta < 2500$) the pressure deviation decreases monotonously with time, tending to zero in first approximation.

If the head losses in the tank diaphragm are larger, hence for $\zeta > 4/\varepsilon$, the pressure deviations become oscillatory and, for very large ζ, the pressure deviations are close to the waterhammer level. It is worthwhile to analyse this computation model in the case of smaller head losses.

If ζ is not much higher than $1/\sqrt{\varepsilon}$ (in the case of laboratory experiments, it can be assumed up to 1/100), it is convenient to consider

$$y(\tau) = \sqrt{\varepsilon}\hat{y}\left(\frac{\tau}{\sqrt{\varepsilon}}\right), \quad w(\tau) = \hat{w}\left(\frac{\tau}{\sqrt{\varepsilon}}\right). \tag{9.31}$$

For the new unknown quantities, the following equations are obtained:

$$\begin{aligned}
\hat{y}'(\sigma) &= -\sqrt{\varepsilon}\frac{\dfrac{f}{kF}}{1+\dfrac{\zeta\varepsilon}{4}}\hat{y}(\sigma) - \frac{\dfrac{2f}{F}}{1+\dfrac{\zeta\varepsilon}{4}}\hat{w}(\sigma-\sqrt{\varepsilon}) - \frac{\dfrac{f}{F}}{1+\dfrac{\zeta\varepsilon}{4}}\frac{\hat{q}}{|\hat{q}|}\\
\hat{w}(\sigma) &= \frac{1-\dfrac{\zeta\varepsilon}{4}}{1+\dfrac{\zeta\varepsilon}{4}}\hat{w}(\sigma-\sqrt{\varepsilon}) + \frac{\sqrt{\varepsilon}}{k\left(1+\dfrac{\zeta\varepsilon}{4}\right)}\hat{y}(\sigma) - \frac{\dfrac{\zeta\varepsilon}{4}}{1+\dfrac{\zeta\varepsilon}{4}}\frac{\hat{q}}{|\hat{q}|}
\end{aligned} \tag{9.32}$$

These equations belong to the general scheme

$$\begin{aligned}
\hat{y}'(\sigma) &= \sqrt{\varepsilon}\boldsymbol{A}\hat{y}(\sigma) + \boldsymbol{B}\hat{w}(\sigma-\sqrt{\varepsilon}) + f_1\\
\hat{w}(\sigma) &= \sqrt{\varepsilon}\boldsymbol{G}\hat{y}(\sigma) + (1+\sqrt{\varepsilon}\boldsymbol{H})\hat{w}(\sigma-\sqrt{\varepsilon}) + \sqrt{\varepsilon}f_2
\end{aligned} \tag{9.33}$$

In this case, the reduced model is

$$\begin{aligned} &y'(\sigma) = \boldsymbol{B}y(\sigma) + f_1(\sigma), \quad w'(\sigma) = \boldsymbol{G}y(\sigma) + \boldsymbol{H}w(\sigma) + f_2(\sigma) \\ &y(0) = 0, \quad w(0) = 0 \end{aligned} \tag{9.34}$$

This reduced model yields for y an approximation with an error of the order ε and for w, an approximation of the order $\sqrt{\varepsilon}$. This means that the approximation for the level deviation in the tank is of the order ε, and that the one for the pressure, is only of the order $\sqrt{\varepsilon}$.

For the analysed hydraulic scheme, the reduced model is

$$\begin{aligned} y'(\sigma) &= -\frac{2f}{F}w(\sigma) - \frac{f}{F}\frac{\hat{q}}{|\hat{q}|} \\ w'(\sigma) &= \frac{1}{k}y(\sigma) - \frac{\zeta\sqrt{\varepsilon}}{2}w(\sigma) - \frac{\zeta\sqrt{\varepsilon}}{4}\frac{\hat{q}}{|\hat{q}|} \end{aligned} \tag{9.35}$$

where the terms of the order $\sqrt{\varepsilon}$ have been neglected.

For $\tilde{y}(\tau)$ this yields the second-order equation below:

$$\frac{F\ell}{fg}\frac{\mathrm{d}^2 Z}{\mathrm{d}t} + R'_{\mathrm{c}}\frac{\mathrm{d}Z}{\mathrm{d}t} + Z = 0 \tag{9.36}$$

thus, the mass-oscillation equation in the pressure conduit tank system has been recovered. This approximation holds even in the absence of head losses. With an error of the order $\sqrt{\varepsilon}$, the pressure at the valve can be approximated by $h(1,\tau) \approx \tilde{y}(\tau)$. This type of approximation operates when ζ assumes values of the order $1/\sqrt{\varepsilon}$ (which under the conditions of the performed laboratory experiments corresponds to $\zeta \leq 100$).

It is worthwhile emphasising that, for larger values of ζ, the level deviations in the tank are insignificant in the first approximation (non-periodic oscillations in the tank), and the two computation models considered above (for small or large head losses) lead, in the first approximation, to the same results for the pressures.

To conclude, in this last chapter we have demonstrated a new domain of unsteady motion which we call the intermediate domain (ID) between the rapidly and slowly varying motions, that we have studied theoretically and experimentally. Taking into account that the obtained theoretical results are in good agreement with the experimental ones, the deduced formulae can be directly used for design computations of hydroelectric power plants and pumping stations with surge tanks equipped with large hydraulic resistances at their bases.

References

1. Mosonyi, E., Seth, H.: The Surge Tank—a Device for Controlling Waterhammer. *Water Power and Dam Construction*, 2–3 (1975).
2. Pelis, H., Shand, M.: Combined Analysis of Surge and Waterhammer Pressures. *Proceedings of the 3rd International Conference on Pressure Surges*, Canterbury, England, 1980.
3. Popescu, M.: Experimental Results in the Study of Unsteady Water Motion in Hydraulic Pressure Systems. *Studii si Cercetari de Mecanica Aplicata* 41 (1982) 443–458, 595–609.
4. Popescu, M.: Several Results on the Unsteady Motions in Hydraulic Pressure Systems. *Hidrotehnica*, 11–12 (1985) (in Romanian).

5. Popescu, M.: Résultats Expérimentaux dans l'Études des Mouvements Non-Permanents des Systèmes Hydrauliques sous Pression avec des Applications aux Usines Hydroélectriques et aux Stations de Pompage. *XX Convegno di Idraulica e Construzioni Idrauliche*, Padova, Italia, September 8–10, 1986.
6. Popescu, M., Halanay, A.: Unsteady motions in the Intermediary Domain between Slow and Fast Motions in Hydraulic Surge Systems. *International Conference on Hydraulic Engineering Software*, Portoroz, Yugoslavia, September 10–14, 1984.
7. Popescu, M., Halanay, A., Dragan, V.: An Intermediary Domain between Fast and Slow Motions in Hydraulic Surge Systems. *Revue Roumaine des Sciences Techniques—Mécanique Appliquée*, 30 (1985) 101–115.
8. Popescu, M., Ionescu, F., Vlase, P.: Un Cas d'Integration Modelisation Phisique—Modelisation Numerique au Support Informatique. *Proceedings of the XXVI IAHR Congress*, London, England 1995, pp. 308–313.

Appendix 1

Validity of the Model of Incompressible Fluid in the Study of Hydraulic Stability of Surge Tanks

A1.1. Introduction

The study of stability of surge systems provided with surge tanks automatically governed for constant power output of the turbines, initiated by D. Thoma and considered further by L. Escande and Ch. Jaeger, has been performed by using the model of incompressible fluid [1, 4]. In the following it is shown that the Thoma's least area of cross section of a surge tank gives a condition for stable behaviour against small oscillations even if we use the model of compressible fluid; in this way a new argument for the validity of the model of the incompressible fluid [3] is yielded.

The mathematical treatment is based on deriving a small parameter waterhammer equation; this suggests that the space dependence of the discharge in the tunnel may be neglected. An analogous approach, for a different problem has been adopted by C. Thirriot [7]. The compressible–incompressible fluid model relationship was studied by A. Halanay and M. Popescu in some other framework. These authors pointed out a special type of unsteady flows in surge hydraulic systems [6].

A1.2. The mathematical model for the compressible fluid

We consider the motion in a basic hydraulic surge system consisting of a reservoir, pressure tunnel, surge tank and turbine (Fig. A1.1). We start with the mathematical model of unsteady motion of the water in the system

$$
\begin{aligned}
&\frac{\partial H(x,t)}{\partial x} + \frac{1}{gf}\frac{\partial Q(x,t)}{\partial t} + PQ^2(x,t) = 0 \\
&\frac{\partial H(x,t)}{\partial t} + \frac{c^2}{gf}\frac{\partial Q(x,t)}{\partial x} = 0 \\
&F\frac{\mathrm{d}Z(t)}{\mathrm{d}t} + Q_{\mathrm{T}}(t) = Q(\ell,t) \\
&H(0,t) = H, \quad H(\ell,t) = H + Z(t)
\end{aligned}
\tag{A1.1}
$$

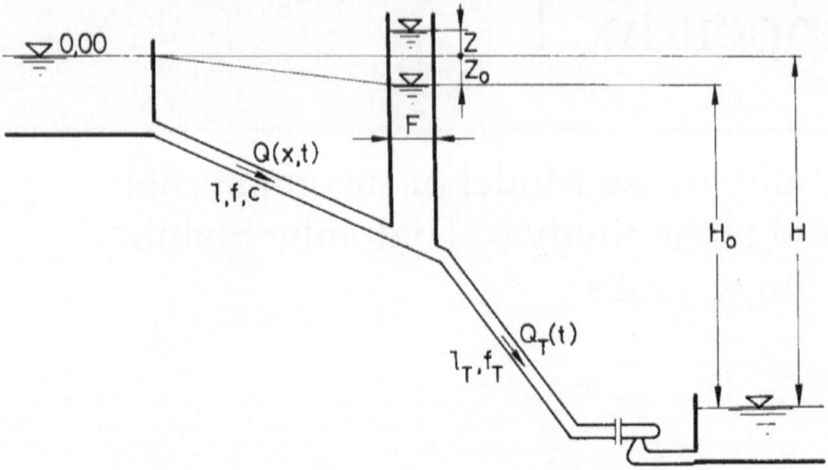

Figure A1.1. Scheme of a hydroelectric power plant provided with a surge tank.

We shall assume the system automatically governed for constant output power of the turbine; this situation is described mathematically by the condition

$$\eta g Q_{\mathrm{T}} H(\ell, t) = N \tag{A1.2}$$

The corresponding steady state is given by the equations:

$$H(x) = H - PQ_{\mathrm{o}}^2 x, \quad Z_{\mathrm{o}} = -PQ_{\mathrm{o}}^2 \ell, \quad \eta g Q_{\mathrm{o}} H_{\mathrm{o}} = N, \quad H_{\mathrm{o}} = H + Z_{\mathrm{o}} \tag{A1.3}$$

hence

$$\frac{\eta^2 g^2}{P\ell} Z_{\mathrm{o}}(H + Z_{\mathrm{o}})^2 + N^2 = 0 \tag{A1.4}$$

and we shall study the stability of the steady state for which

$$Z_{\mathrm{o}} \in \left(-\frac{H}{3}, 0\right) \tag{A1.5}$$

If equations are linearised around the steady state and keeping the same notation as in (A1.1) for the difference motion between the perturbed and the steady motion, the following mathematical model follows

$$\begin{aligned}
&\frac{\partial H(x,t)}{\partial x} + \frac{1}{gf}\frac{\partial Q(x,t)}{\partial t} + 2PQ_{\mathrm{o}}Q(x,t) = 0 \\
&\frac{\partial H(x,t)}{\partial t} + \frac{c^2}{gf}\frac{\partial Q(x,t)}{\partial x} = 0 \\
&F\frac{\mathrm{d}Z(t)}{\mathrm{d}t} = \frac{Q_{\mathrm{o}}}{H_{\mathrm{o}}}Z(t) + Q(\ell, t) \\
&H(0,t) = 0, \quad H(\ell, t) = Z(t)
\end{aligned} \tag{A1.6}$$

A1.3. The non-dimensional form of the mathematical model

We write the mathematical model in non-dimensional form by choosing the following forms for the main parameters of motion:

$$
\begin{aligned}
y(\tau) &= \frac{1}{Z_o} Z\left(\frac{\tau f \ell}{Z_o}\right) \\
h(\xi, \tau) &= \frac{1}{Z_o} H\left(\ell \xi, \frac{\tau f \ell}{Q_o}\right) \\
q(\xi, \tau) &= \frac{1}{Q_o} Q\left(\ell \xi, \frac{\tau f \ell}{Q_o}\right)
\end{aligned} \tag{A1.7}
$$

After some computations the non-dimensional equations of motion and the boundary conditions (A1.6) take the form

$$
\begin{aligned}
&\frac{\partial h}{\partial \xi} + \frac{Q_o^2}{g f^2 Z_o} \frac{\partial q}{\partial \tau} + \frac{2\ell}{Z_o} P Q_o q = 0 \\
&\frac{\partial h}{\partial \tau} + \frac{c^2}{g Z_o} \frac{\partial q}{\partial \xi} = 0 \\
&F y'(\tau) = \frac{f \ell}{H_o} y(\tau) + \frac{f \ell}{Z_o} q(\ell, \tau) \\
&h(0, \tau) = 0, \quad h(\ell, \tau) = y(\tau)
\end{aligned} \tag{A1.8}
$$

A1.4. Approximation by the model of the incompressible fluid

We rewrite these basic equations by introducing the following fundamental constants of the hydraulic scheme under consideration [2, 5, 6]:

$$
\varepsilon = \frac{2 Q_o}{c f}, \quad k = \frac{c Q_o}{g \ell f} \tag{A1.9}
$$

and obtain

$$
\begin{aligned}
&\frac{\partial h}{\partial \xi} + \frac{k \varepsilon}{2} \frac{\ell}{Z_o} \frac{\partial q}{\partial \tau} + \frac{2\ell}{Z_o} P Q_o^2 q = 0 \\
&\frac{\varepsilon Z_o}{2 k \ell} \frac{\partial h}{\partial \tau} + \frac{\partial q}{\partial \xi} = 0
\end{aligned} \tag{A1.10}
$$

We remark that for hydraulic systems which are met in the engineering practice the number $\varepsilon Z_o / 2 k \ell$ is small, namely of order 10^{-3}–10^{-4} (for example, in the case of the Bicaz and Somes-Mariselu hydroelectric power plants this number is 10^{-4} and 8×10^{-4}, respectively [5]), suggesting that we should change the space scale.

Denote

$$
\frac{\varepsilon Z_o}{2 k \ell} = \hat{\varepsilon} \tag{A1.11}
$$

and write

$$h(\xi,\tau)=\hat{h}(\hat{\varepsilon}\xi,\tau),\quad q(\xi,\tau)=\hat{q}(\hat{\varepsilon}\xi,\tau) \tag{A1.12}$$

We obtain the equations

$$\begin{aligned}&\frac{\partial\hat{h}}{\partial\tau}(\hat{\xi},\tau)+\frac{\partial\hat{q}}{\partial\hat{\xi}}(\hat{\xi},\tau)=0\\&\frac{\partial\hat{h}}{\partial\hat{\xi}}(\hat{\xi},\tau)+\left(\frac{k\ell}{Z_o}\right)^2\frac{\partial\hat{q}}{\partial\tau}(\hat{\xi},\tau)+\frac{2\ell}{Z_o\hat{\varepsilon}}\ell Q_o^2\hat{q}(\hat{\xi},\tau)=0\end{aligned} \tag{A1.13}$$

By integrating the last equation from 0 to ε, we deduce

$$\hat{h}(\hat{\varepsilon},\tau)-\hat{h}(0,\tau)+\left(\frac{k\ell}{Z_o}\right)^2\int_o^{\hat{\varepsilon}}\frac{\partial\hat{q}}{\partial\hat{\tau}}(\hat{\xi},\tau)\mathrm{d}\hat{\xi}+\frac{2\ell}{Z_o}PQ_o^2\frac{1}{\varepsilon}\int_o^{\hat{\varepsilon}}\hat{q}(\hat{\xi},\tau)\mathrm{d}\hat{\xi}=0 \tag{A1.14}$$

Denote:

$$q(\tau)=\frac{1}{\hat{\varepsilon}}\int_o^{\hat{\varepsilon}}\hat{q}(\hat{\xi},\tau)\mathrm{d}\hat{\xi} \tag{A1.15}$$

Assuming that q has a continuous first-order partial differential, eqn. (A1.14) may be written in the form

$$\frac{Q_o^2}{gZ_of^2}q(\tau)+\frac{2\ell}{Z_o}PQ_o^2q(\tau)+y(\tau)=0 \tag{A1.16}$$

Next, we consider the equation

$$Fy'(\tau)=\frac{f\ell}{H+Z_o}y(\tau)+\frac{f\ell}{Z_o}q(\ell,\tau) \tag{A1.17}$$

which may be written in the form

$$Fy'(\tau)=\frac{f\ell}{H+Z_o}y(\tau)+\frac{f\ell}{Z_o}q(\tau)+\frac{f\ell}{Z_o}[q(\ell,\tau)-q(\tau)] \tag{A1.18}$$

By using eqns. (A1.12) and (A1.15) we have

$$q(\ell,\tau)-q(\tau)=\frac{1}{\hat{\varepsilon}}\int_o^{\hat{\varepsilon}}[\hat{q}(\hat{\varepsilon},\tau)-\hat{q}(\hat{\xi},\tau)]\mathrm{d}\hat{\xi} \tag{A1.19}$$

It follows that

$$|q(\ell,\tau)-q(\tau)|\le\frac{L}{\hat{\varepsilon}}\int_o^{\hat{\varepsilon}}(\hat{\varepsilon}-\hat{\xi})\mathrm{d}\hat{\xi}=\frac{L\hat{\varepsilon}}{2} \tag{A1.20}$$

if the component q of the solution of problem (A1.8) is Lipschitz continuous with respect to the space variable.

In this way, we may consider the given problem as a perturbation of the system

$$\begin{aligned}&q'=-2\ell Pgf^2q-\frac{gZ_o^2f^2}{Q_o^2}y\\&y'=\frac{f\ell}{F(H+Z_o)}y+\frac{f\ell}{fZ_o}q\end{aligned} \tag{A1.21}$$

Let us show the stability criteria for this system. To this aim, for this system we write the characteristic equation and stability is obtained if

$$2\ell Pgf^2 - \frac{f\ell}{F(H+Z_o)} > 0, \quad \text{hence} \quad F > \frac{1}{2g(H+Z_o)Pf} = F_{\mathrm{Th}} \tag{A1.22}$$

and

$$\frac{gf^3\ell}{FQ_o^2} - \frac{2\ell^2 pgf^3}{F(H+Z_o)} > 0, \quad \text{hence} \quad \frac{H+3Z_o}{H+Z_o} > 0 \tag{A1.23}$$

We note that eqn. (A1.22) is just Thoma's classical formula for stability against small oscillations of the level of water in the surge tank, the coefficient p having here the meaning resulting from the basic mathematical model (A1.1). The behaviour of the pressure in the considered hydraulic system follows from the model consisting of eqns. (A1.8). The boundary condition $h(\ell,\tau) = y(\tau)$ is given with a high accuracy by the reduced model (A1.21).

The above considerations correspond to a method of asymptotic expansions, the leading term being obtained from

$$\frac{\partial \hat{h}}{\partial \tau}(\hat{\xi},\tau) + \frac{\partial \hat{q}}{\partial \hat{\xi}}(\hat{\xi},\tau) = 0$$

$$\frac{\partial \hat{h}}{\partial \tau}(\hat{\xi},\tau) + \left(\frac{k\ell}{Z_o}\right)^2 \frac{\partial \hat{q}}{\partial \tau}(\hat{\xi},\tau) + \frac{2\ell}{Z_o\hat{\varepsilon}} PQ_o^2 \hat{q}(\hat{\xi},\tau) = 0 \tag{A1.24}$$

$$\hat{h}(\hat{\varepsilon},\tau) = y(\tau), \quad \hat{h}(0,\tau) = 0$$

with $y(\tau)$ given by the model (A1.21), which is based on the incompressibility assumption. If F is greater that the Thoma's cross section, then the basic motion will be exponentially stable.

References

1. Escande, L.: *Méthodes Nouvelles pour le Calcul des Chambres d'Équilibre.* Dunod, Paris, 1950.
2. Halanay, A.: *Differential Equations. Stability, Oscillations, Time Lags.* Academic Press, New York, 1966, Chapter 1, Section 1.8.
3. Halanay, A., Popescu, M.: Validity of the Model of the Incompressible Fluid in the Study of the Hydraulic Stability of Surge Tanks. *Proceedings of the Second IAHR Symposium on Fluid Motion Stability in Hydraulic Systems with Automatic Control*, Bucharest, 1–3 October, 1984.
4. Jaeger, Ch.: *Hydraulique Technique.* Dunod, Paris, 1954.
5. Popescu, M., Arsenie, D.: *Methods of Hydraulic Computation for Hydroelectric Power Plants and Pumping Stations.* Ed. Tehnica, Bucharest, 1987 (in Romanian).
6. Popescu, M., Halanay, A.: Unsteady Motions in the Intermediary Domain between Slow and Fast Motions in Hydraulic Surge Systems. *International Conference on Hydraulic Engineering Software*, Portoroz, Yugoslavia, 1984.
7. Thirriot, C.: Considérations sur la Calcul des Surpressions dans les Conduites Provoquées par des Manoeuvres Graduelles, *C.R. Acad. Sci. Paris,* 260 (1965) 5699–5702.

Appendix 2

The Method of Liapunov's Function in the Stability Theory

The method of Liapunov's function was introduced almost a century ago. It proves to be one of the most frequently used tool in various technical domains. Although the construction of a Liapunov function that provides good stability conditions is a kind of art, there exist many cases when the physical intuition yields natural Liapunov functions.

A2.1. Definitions

Consider the following system of differential equations

$$x_i' = f_i(x_1, x_2, \ldots, x_n), \quad i = 1, 2, \ldots, n \tag{A2.1}$$

An equilibrium point (or a steady solution or a constant solution) for (A2.1) may be obtained by solving the system

$$f_i(x_1, x_2, \ldots, x_n) = 0, \quad i = 1, 2, \ldots, n \tag{A2.2}$$

The equilibrium point $(\hat{x}_1, \ldots, \hat{x}_n)$ is *stable* if for any $\varepsilon > 0$ there exists $\delta > 0$ so that for $|x_1^o - \hat{x}_1| < \delta, i = 1, 2, \ldots, n$ the solution of the system (A2.1) which at $t = 0$ is $(x_1^o, x_2^o, \ldots, x_n^o)$ differs from the constant solution $(\hat{x}_1, \hat{x}_2, \ldots, \hat{x}_n)$ by less that ε for any $t \geq 0$, that is

$$|x_i(t; (x_1^o, \ldots, x_n^o)) - \hat{x}_i| < \varepsilon \quad \text{for all } t \geq 0 \tag{A2.3}$$

This means that the initial perturbations, small with respect to the equilibrium, do not move essentially further away from the equilibrium position. The equilibrium point is called *asymptotically stable* if it is stable and, in addition, there exists $\delta_o > 0$ so that $|x_1^o - \hat{x}_1| < \delta_o$ implies $\lim_{t\to\infty} x_i(t; (x_1^o, \ldots, x_n^o)) = \hat{x}_i$; this means that initial perturbations below some threshold are damped out.

Usually, the study of the *stability* or *asymptotic stability* for an equilibrium point reduces to the study of the corresponding properties of the null solution. This is obtained by passing to the system of equations for the difference motion (i.e. deviations). If $x_i(t; (x_1^o, \ldots, x_n^o))$ is the solution of the given system then denoting

$$y_i(t; (x_1^o, \ldots, x_n^o)) = x_i(t; (x_1^o, \ldots, x_n^o)) - \hat{x}_i \tag{A2.4}$$

we deduce

$$\begin{aligned} y_i'(t;(x_1^o,\dots,x_n^o)) &= x_1'(t;(x_1^o,\dots,x_n^o)) \\ &= f_i(x_1(t;(x_1^o,\dots,x_n^o)),\dots,x_n(t;(x_1^o,\dots,x_n^o))) \\ &= f_i(\hat{x}_1 + y_1(t;(x_1^o,\dots,x_n^o)),\dots,\hat{x}_n + y_n(t;(x_1^o,\dots,x_n^o))) \\ &= g_i(y_1(t;(x_1^o,\dots,x_n^o)),\dots,y_n(t;(x_1^o,\dots,x_n^o))) \end{aligned} \tag{A2.5}$$

therefore $(y_1,\dots,y_n)$ represents a solution of the system in deviations:

$$\begin{aligned} y_i' &= g_i(y_i,\dots,y_n), \quad i = 1,2,\dots,n \\ g_i(y_i,\dots,y_n) &= f_i(\hat{x}_1 + y_1,\dots,\hat{x}_n + y_n) \end{aligned} \tag{A2.6}$$

therefore $g_i(0,\dots,0) = 0$.

A2.2. The stability theorem by the first approximation

Usually, the system in deviations is written by means of the first (linear) approximation of

$$y_i' = \sum_{j=1}^{n} a_{ij}y_j + h_i(y_1,\dots,y_n), \quad i = 1,2,\dots,n \tag{A2.7}$$

The terms denoted by h_i are of higher order and the fundamental theorem due to Liapunov shows that their influence upon the asymptotic stability is null for the given (non-linear) system if the first-order linear system has the property of *asymptotic stability.*

For the linear system, this stability property holds if all the eigenvalues of the matrix with entries a_{ij} have strictly negative real parts. These eigenvalues are given by the equation

$$\begin{vmatrix} \lambda - a_{11} & a_{12} & a_{1n} \\ a_{21} & \lambda - a_{22} & a_{2n} \\ a_{n1} & a_{n2} & \lambda - a_{nn} \end{vmatrix} = 0 \tag{A2.8}$$

which may be written in the form

$$\lambda^n + C_1\lambda^{n-1} + \dots + C_{n-1}\lambda + C_n = 0 \tag{A2.9}$$

In this way, the question occurs to recognize, without solving the equation, if the roots have strictly negative real parts. If $n = 2$, the necessary and sufficient conditions read $C_1 > 0$, $C_2 > 0$. If $n = 3$, there appears an extra condition and, the more the degree increases, the more the conditions become complicated and are called Hurwitz conditions. Liapunov proved the following results:

$$\sum_{i,j} A_{ij}y_iy_j, \; A_{ij} = A_{ji}$$

which is positively defined (i.e. takes strictly positive values when the vector $(y_1, \ldots, y_n)$ is other than zero) so that "its derivative by virtue of the system"

$$\sum_{i=1}^{n}\sum_{j=1}^{n}\sum_{k=1}^{n}[A_{ij}a_{jk}y_iy_k + A_{ij}a_{ik}y_ky_j] = \sum_{i=1}^{n}\sum_{k=1}^{n}\left(\sum_{j=1}^{n}A_{ij}a_{jk} + A_{ij}a_{jk}\right)y_jy_k$$

$$= 2\sum_{i=1}^{n}\sum_{k=1}^{n}\left(\sum_{j=1}^{n}A_{ij}a_{jk}\right)y_iy_k \qquad \text{(A2.10)}$$

is equal to $-\sum_{k=1}^{n}y_k^2$, then the null solution of the linear system is *asymptotically stable* and conversely; thus instead of the Hurwitz conditions, there occurs a condition of the existence of a quadratic form with the above-mentioned properties. From the very assertion, it is obvious how such conditions are obtained.

The relations

$$\sum_{i=1}^{n}\sum_{k=1}^{n}\left(\sum_{j=1}^{n}A_{ij}a_{jk}\right)y_jy_k = -\frac{1}{2}\sum_{k=1}^{n}y_k^2 \qquad \text{(A2.11)}$$

are equivalent to a system of equations which determine the coefficients A_{ij} in terms of the given coefficients a_{ij}; then the condition that the quadratic form be positively defined, is written by using the main diagonal minors (Sylvester's criterion).

This procedure is not simpler than the direct writing of the Hurwitz conditions (which, generally, reduces also to the calculus of some determinants). However, if we give up the idea of obtaining necessary and sufficient conditions for stability, instead of the above quadratic form, we may use another Liapunov function starting from the specific properties of the problem derived from the physical nature. In this way, the construction of the Liapunov function becomes an art, mostly based on intuition.

A2.3. A general theorem

Definition

A function V is called a Liapunov function for the system

$$y_i' = f_i(y_1, \ldots, y_n) \quad \text{if} \sum_{i=1}^{n}\frac{\partial V}{\partial y_i}(y_1, \ldots, y_n)f_i(y_1, \ldots, y_n) \leq 0 \qquad \text{(A2.12)}$$

and $V(y_1, \ldots, y_n) > 0$ for $(y_1, \ldots, y_n) \neq (0, \ldots, 0)$

Theorem (Barbashin–Krasovskii–LaSalle)

Let V be a Liapunov function for the given system and G the set of the points $(y_1, \ldots, y_n)$ with the property that

$$\sum_{i=1}^{n}\frac{\partial V}{\partial y_i}(y_1, \ldots, y_n)f_i(y_1, \ldots, y_n) = 0 \qquad \text{(A2.13)}$$

If the set G does not contain integer half trajectories besides $(0, \ldots, 0)$ and $V(0, \ldots, 0) = 0$, then the point $(0, \ldots, 0)$ is *asymptotically stable.* An important question from the application viewpoint is the problem of the domain of attraction of an *asymptotically stable* equilibrium point which represents the set of the initial perturbations which are damped out with time.

The above theorem is completed by the remark that an arbitrary set of points in which the function V has the property in the definition and which is *positively invariant* (any solution emerging from it does not leave it) belongs to the domain of attraction of the equilibrium point.

References

1. Halanay, A.: *Stabilita*. Quaderni dell'Unione Matematica Italiana, 5, Pitagora Editrice, Bologna, 1978.
2. Hann, W.: *Stability of motion*. Springer, Berlin, 1967.
3. La Salle, S.P., Lefschitz, S.: *Stability by Liapunov's Direct Method with Applications*. Academic Press, New York, 1961.
4. Rouche, N., Habets, P., Laloy, M.: *Stability Theory by Liapunov Direct Method*. Springer, Berlin, 1977.

Author Index